Topics in Fixed Point Theory

Saleh Almezel • Qamrul Hasan Ansari
Mohamed Amine Khamsi
Editors

Topics in Fixed Point Theory

Springer

Editors
Saleh Almezel
Department of Mathematics
University of Tabuk
Tabuk, Saudi Arabia

Qamrul Hasan Ansari
Department of Mathematics
Aligarh Muslim University
Aligarh, India

Mohamed Amine Khamsi
Department of Mathematical Sciences
University of Texas at El Paso
El Paso, TX, USA

ISBN 978-3-319-01585-9 ISBN 978-3-319-01586-6 (eBook)
DOI 10.1007/978-3-319-01586-6
Springer Cham Heidelberg New York Dordrecht London

Library of Congress Control Number: 2013951234

Mathematics Subject Classification: 06F30, 46B20, 46E30, 47H04, 47H05, 47H09, 47H10, 47H20, 47J20, 47J25, 47J30, 47N10, 49J40, 49J53, 54E40, 54H25, 54M20, 58C30, 65J15, 65K10, 90C26

Printed on acid-free paper

Springer is part of Springer Science+Business Media (www.springer.com)

Preface

Fixed point theory is one of the most powerful and fruitful tools of modern mathematics and may be considered a core subject of nonlinear analysis. In the last 50 years, fixed point theory has been a flourishing area of research for many mathematicians. The origins of the theory, which date to the later part of the nineteenth century, rest in the use of successive approximations to establish the existence and uniqueness of solutions, particularly to differential equations. This method is associated with many celebrated mathematicians like Cauchy, Fredholm, Liouville, Lipschitz, Peano, and Picard. It is worth noting that the abstract formulation of Banach is credited as the starting point to metric fixed point theory. But the theory did not gain enough impetus till Felix Browder's major contribution to the development of the nonlinear functional analysis as an active and vital branch of mathematics. In recent years a number of excellent books, monographs, and surveys by distinguished authors about fixed point theory have appeared. This monograph is an attempt to explore some of its aspects after the theory has reached a level of maturity appropriate to an examination of its central themes. One of the objectives of this monograph is to offer the mathematical community an accessible self-contained document which can be used as an introduction to the subject and its development. This monograph is composed of eight chapters on different aspects of the fixed point theory, namely, metric fixed point theory, Ekeland's variation principle, hyperconvex metric spaces, modular function spaces, discrete fixed point theory, topological fixed point theory, and iterative methods in fixed point theory.

Each chapter is written by different authors who attempted to render the major results understandable to a wide audience, including nonspecialists, and at the same time to provide a source for examples, references, open questions, and sometimes new approaches for those currently working in this area of mathematics. This monograph should be of interest to graduate students seeking a field of interest, to mathematicians interested in learning about the subject, and to specialists.

We thank collectively our many friends and colleagues who, through their encouragement and help, influenced the development of this book. In particular, we are grateful to the Rector of the University of Tabuk Dr. Abdulaziz S. Al-Enazi and the Vice Rector for postgraduate studies and scientific research Prof.

F. Al-Solamy for their support in the organization of the International Mathematical Workshop on Fixed Point Theory and Applications, May 2012. Most of the authors who contributed in this monograph attended this workshop and agreed to be part of this project. Special thanks go to Ms. Kaitlin Leach (Associate Editor, Mathematics, Springer) for her interest to publish this monograph.

Tabuk, Saudi Arabia — Saleh Almezel
Aligarh, India — Qamrul Hasan Ansari
El Paso, TX, USA — Mohamed Amine Khamsi

Contents

1 Introduction to Metric Fixed Point Theory 1
M.A. Khamsi
1.1 Introduction 1
1.2 Metric Fixed Point Theory 2
1.3 Caristi–Ekeland Extension 3
1.4 The Converse Problem 4
1.5 Some Applications 5
1.5.1 ODE and Integral Equations 5
1.5.2 Cantor and Fractal Sets 6
1.6 Historical Note 6
1.7 Metric Fixed Point Theory in Banach Spaces 7
1.7.1 Classical Existence Results 8
1.7.2 The Normal Structure Property 10
1.7.3 More on Normal Structure Property 11
1.7.4 Normal Structure and Smoothness 13
1.7.5 Karlovitz–Goebel Lemma 14
1.8 Nonstandard Techniques 14
1.8.1 Extending Mappings to Ultrapowers 15
1.9 More on Metric Fixed Point Theory in Metric Spaces 17
1.9.1 Menger Convexity in Metric Spaces 17
1.9.2 Uniformly Convex Metric Spaces 19
1.9.3 Fixed Point Property in Uniformly Convex Metric Spaces 22
1.10 The Convexity Structures 24
1.10.1 Hyperconvex Metric Spaces 25
References 29

2 Banach Contraction Principle and Its Generalizations 33
Abdul Latif
2.1 Introduction 33
2.2 Contractions: Definition and Examples 33

2.3 The Banach Contraction Principle with Some Applications ... 35
2.4 Some Other Extensions of BCP for Single-Valued Mappings ... 40
2.5 Caristi's Fixed Point Theorem ... 47
2.6 Some Extensions of BCP Under Generalized Distances ... 48
2.7 Multivalued Versions of BCP ... 51
References ... 61

3 Ekeland's Variational Principle and Its Extensions with Applications ... 65
Qamrul Hasan Ansari
3.1 Introduction ... 65
3.2 Ekeland's Variational Principle in Complete Metric Spaces ... 66
3.3 Applications to Fixed Point Theorems ... 75
3.4 Applications to Optimization ... 80
3.5 Applications to Weak Sharp Minima ... 84
3.6 Equilibrium Problems and Extended Ekeland's Variational Principle ... 87
3.6.1 Equilibrium Problems ... 87
3.6.2 Extended Ekeland's Variational Principle ... 88
Appendix A ... 96
References ... 96

4 Fixed Point Theory in Hyperconvex Metric Spaces ... 101
Rafael Espínola and Aurora Fernández-León
4.1 Introduction and Basic Definitions ... 101
4.2 Some Basic Properties of Hyperconvex Metric Spaces ... 108
4.3 Hyperconvexity, Injectivity, and Retractions ... 114
4.4 Isbell's Hyperconvex Hull ... 120
4.5 Topological Fixed Point Property and Hyperconvexity ... 127
4.6 Metric Fixed Point Property and Hyperconvexity ... 134
4.7 Metric Fixed Point Property for Multivalued Mappings and Nonexpansive Selections ... 137
4.8 KKM Theory and Hyperconvex Spaces ... 140
4.9 Fixed Point Theory and $\mathbb{R}$-Trees ... 144
4.10 New Trends in Hyperconvexity ... 151
4.10.1 Two Long-Standing Open Problems ... 151
4.10.2 Ultrametrics and Hyperconvex Metric Spaces ... 152
4.10.3 Diversities and Hyperconvexity ... 152
4.10.4 Q-hyperconvexity ... 154
References ... 155

5 An Introduction to Fixed Point Theory in Modular Function Spaces ... 159
W.M. Kozlowski
5.1 Introduction ... 159

5.2 Modular Function Spaces and Modular Geometry 161
5.2.1 Introduction to Modular Function Spaces 162
5.2.2 Geometrical Properties of Modular Function Spaces 170
5.2.3 Nonlinear Mappings in Modular Function Spaces 183
5.3 Existence of Fixed Points 186
5.3.1 Generalized Contractions in Modular Function Spaces 186
5.3.2 Generalized Nonexpansive Mappings 193
5.4 Convergence of Fixed Point Iterative Algorithms 200
5.5 Semigroups of Mappings in Modular Function Spaces 214
References 219

6 Fixed Point Theory in Ordered Sets from the Metric Point of View 223
M.Z. Abu-Sbeih and M.A. Khamsi
6.1 Introduction 223
6.2 Generalized Metric Spaces 224
6.3 The Retraction Property 228
6.4 Fixed Point Property 230
6.5 Externally Complete Sets 233
References 235

7 Some Fundamental Topological Fixed Point Theorems for Set-Valued Maps 237
Hichem Ben-El-Mechaiekh
7.1 Introduction and Preliminaries 237
7.1.1 Elementary Proofs of the Brouwer and Schauder–Tychonoff Fixed Point Theorems 239
7.2 Ky Fan and Kakutani Maps 243
7.2.1 Ky Fan Maps 244
7.2.2 Kakutani Maps 245
7.3 Continuous Selections and Approximations 246
7.3.1 Continuous Selections for Ky Fan Maps 246
7.3.2 Continuous Approximations of Kakutani Maps 247
7.4 The Browder–Ky Fan and the Kakutani–Ky Fan Fixed Point Theorems and Their Consequences 249
7.4.1 The Browder–Ky Fan Fixed Point Theorem 249
7.4.2 The Kakutani–Ky Fan Fixed Point Theorem 250
7.4.3 Coincidence Theorems 251
7.5 Relaxing Compactness and Related Results 254
7.5.1 The Coercivity Condition (κ) 254
7.5.2 Intersection Theorems: KKM and Matching 256
7.5.3 Browder–Ky Fan Theorem on Star-Shaped Domains 258
7.5.4 A Leray–Schauder Alternative for Kakutani Maps 259
7.6 Systems of Nonlinear Inequalities and Applications 262
7.7 Concluding Remarks 269
References 270

8 Some Iterative Methods for Fixed Point Problems 273
Q.H. Ansari and D.R. Sahu
8.1 Introduction 273
8.2 Preliminaries 274
8.2.1 The LE Property and the AF Point Property for Nonlinear Operators 277
8.2.2 Nearly Lipschitzian Mappings 278
8.2.3 Asymptotically κ-strict Pseudocontractive Mappings in the Intermediate Sense 281
8.3 Picard Iterative Method 283
8.4 Mann Iterative Method 285
8.5 Ishikawa Iterative Method 288
8.6 Helpern Iterative Method 290
8.7 CQ Iteration Method 293
8.8 Browder Iterative Method 296
References 298

Index 301

Contributors

M.Z. Abu-Sbeih Department of Mathematics & Statistics, King Fahd University of Petroleum & Minerals, Dhahran, Saudi Arabia

Qamrul Hasan Ansari Department of Mathematics, Aligarh Muslim University, Aligarh, India

Hichem Ben-El-Mechaiekh Department of Mathematics, Brock University, St. Catharines, ON, Canada

Rafael Espínola Dpto. de Análisis Matemático-IMUS, Universidad de Sevilla, Sevilla, Spain

Aurora Fernández-León Dpto. de Análisis Matemático-IMUS, Universidad de Sevilla, Sevilla, Spain

Mohamed Amine Khamsi Department of Mathematical Sciences, University of Texas at El Paso, El Paso, TX, USA

W.M. Kozlowski School of Mathematics and Statistics, University of New South Wales, Sydney, NSW, Australia

Abdul Latif Department of Mathematics, King Abdulaziz University, Jeddah, Saudi Arabia

Daya Ram Sahu Department of Mathematics, Banaras Hindu University, Varanasi, India

Chapter 1
Introduction to Metric Fixed Point Theory

M.A. Khamsi

1.1 Introduction

This chapter is primarily intended to serve as an introduction to metric fixed point theory for those who might not have ready access to other sources, or perhaps as a group might have diverse mathematical backgrounds. In this sense the text is self-contained. Most of the chapter should be accessible to reasonably mature students who have had very little training in advanced mathematics. At the same time the chapter contains a large amount of material that might be of interest to more advanced students and even to serious scholars. At the same time most readers will find something new and they might find the inclusion of background material helpful as well. Despite the fact that the chapter is largely self-contained, extensive bibliographic references are included.

In terms of content this chapter overlaps in places with the following popular books on fixed point theory by Aksoy and Khamsi [1], Goebel and Kirk [42], Dugundji and Granas, [32], Khamsi and Kirk [55], and Zeidler [81].

Material on the general theory of Banach space geometry is drawn from many sources but the following books are worth special mention by Beauzamy [9] and Diestel [31].

The fixed point problem (at the basis of the Fixed Point Theory) may be stated as: *Let X be a set, A and B two nonempty subsets of X such that $A \cap B \neq \emptyset$, and $f : A \to B$ be a map. When does a point $x \in A$ such that $f(x) = x$, also called a fixed point of f, exist?*

A multivalued fixed point problem may be stated but in this chapter we will mainly focus on the single valued mappings.

M.A. Khamsi (✉)
Department of Mathematical Sciences, University of Texas at El Paso,
El Paso, TX 79968-0514, USA
e-mail: mohamed@utep.edu

S. Almezel et al. (eds.), *Topics in Fixed Point Theory*, DOI 10.1007/978-3-319-01586-6_1,

Fixed Point Theory is divided into three major areas:

- Topological Fixed Point Theory
- Metric Fixed Point Theory
- Discrete Fixed Point Theory

Historically the boundary lines between the three areas were defined by the discovery of three major theorems:

- Brouwer's Fixed Point Theorem [21]
- Banach's Fixed Point Theorem [8]
- Tarski's Fixed Point Theorem [78]

In this chapter, we will focus mainly on the second area though from time to time we may say a word on the other areas as well.

1.2 Metric Fixed Point Theory

In 1922 Banach [8] published his fixed point theorem, also known as *Banach Contraction Principle*, which uses the concept of Lipschitz mappings.

Definition 1.1. Let (M,d) be a metric space. The map $T : M \to M$ is said to be *Lipschitzian* if there exists a constant $k > 0$ (called *Lipschitz constant*) such that

$$d\Big(T(x),T(y)\Big) \leq k\, d(x,y), \quad \text{for all } x,y \in M.$$

A Lipschitzian mapping with a Lipschitz constant $k < 1$ is called *contraction*.

Theorem 1.1 (Banach Contraction Principle). *Let (M,d) be a complete metric space and let $T : M \to M$ be a contraction mapping, with Lipschitz constant $k < 1$. Then, T has a unique fixed point ω in M, and for each $x \in M$, we have*

$$\lim_{n\to\infty} T^n(x) = \omega$$

Moreover, for each $x \in M$, we have

$$d\left(T^n(x),\omega\right) \leq \frac{k^n}{1-k} d\left(T(x),x\right).$$

An easy implication of the Banach Contraction Principle is the following theorem.

Theorem 1.2. *Suppose (M,d) is a complete metric space and suppose $T : M \to M$ is a mapping for which T^N is a contraction mapping for some positive integer $N \geq 1$. Then T has a unique fixed point.*

It is not clear in general whether T has a fixed point whenever T^N has a fixed point. Note that fixed points of T^N are also known as periodic points of T.

1.3 Caristi–Ekeland Extension

This is one of the most interesting extensions of Banach Contraction Principle. In order to understand this extension, let us first go over Caristi's new proof of the Banach Contraction Principle. Indeed Caristi [27, 28] proved that if $T : M \to M$ is a contraction with Lipschitz constant $k < 1$, then we have

$$d(x, T(x)) \leq \varphi(x) - \varphi(T(x)), \quad \text{for all } x \in M,$$

where $\varphi : M \to \mathbb{R}^+ = [0, +\infty)$ is defined by

$$\varphi(x) = \frac{1}{1-k} d(T(x), x).$$

As a generalization, Caristi [28] and Ekeland [35] considered mappings $T : M \to M$ which satisfy the following property

$$d(x, T(x)) \leq \varphi(x) - \varphi(T(x)), \quad \text{for all } x \in M,$$

where $\varphi : M \to \mathbb{R}^+ = [0, +\infty)$. Both Caristi and Ekeland investigated this new class of mappings to find out when a fixed point exists. Recall that the function φ is said to be *lower semi-continuous* (l.s.c.), if for any sequence $\{x_n\} \subset M$, if $\lim_{n\to\infty} x_n = x$ and $\lim_{n\to\infty} \varphi(x_n) = r$, then $\varphi(x) \leq r$.

Theorem 1.3 (Ekeland Variational Principle, 1974). *Let (M, d) be a complete metric space and $\varphi : M \to \mathbb{R}^+$ l.s.c. Define:*

$$x \preceq y \Leftrightarrow d(x, y) \leq \varphi(x) - \varphi(y), \quad \textit{for all } x, y \in M.$$

Then, $(M, \preceq)$ has a maximal element.

Theorem 1.4 (Caristi Fixed Point Theorem, 1975). *Let (M, d) be a complete metric space and $\varphi : M \to \mathbb{R}^+$ l.s.c. Suppose $T : M \to M$ satisfies:*

$$d(x, T(x)) \leq \varphi(x) - \varphi(T(x)), \quad \textit{for all } x \in M.$$

Then, T has a fixed point.

Though both theorems have a different setting, in fact they are both equivalent [68]. The proof of Caristi–Ekeland's theorems is based on discrete technique: Zorn's lemma and Axiom of Choice applied to Bronsted partial order. There are some trials of finding a pure metric proof of Caristi's fixed point theorem (without success so far).

1.4 The Converse Problem

In other words, what kind of set M will have the conclusion of Banach Contraction Principle? The most elegant result in this direction is due to Bessaga [13].

Theorem 1.5. *Suppose that M is an arbitrary nonempty set and suppose that $T : M \to M$ has the property that T and each of its iterates T^n has a unique fixed point. Then, for each $\lambda \in (0,1)$, there exists a metric d_λ on M such that (M, d_λ) is complete and for which*

$$d_\lambda\left(T(x), T(y)\right) \leq \lambda\ d_\lambda(x,y), \quad \textit{for each } x,y \in M.$$

Example 1.1 (Ultrametric Spaces [70, 71]). Though the definition of ultrametric spaces is too strong, it is a very natural concept used in logic programming, for example [72]. Recall that a metric space (M,d) is *ultrametric* if and only if for each $x,y,z \in M$, we have

$$d(x,y) \leq \max\{d(x,z), d(y,z)\}.$$

It is immediate from the above definition that if $d(x,y) \neq d(y,z)$, then we have $d(x,z) = \max\{d(x,y), d(y,z)\}$, that is, each three points are vertices of an isosceles triangle. This leads to:

Lemma 1.1. *If $B(a,r_1)$ and $B(b,r_2)$ are two closed balls in an ultrametric space, with $r_1 \leq r_2$, then either $B(a,r_1) \cap B(b,r_2) = \emptyset$ or $B(a,r_1) \subset B(b,r_2)$.*

An ultrametric space M is said to be spherically complete if every descending sequence of closed balls in M has nonempty intersection. Thus, a spherically complete ultrametric space is complete. Some nice and immediate consequences of isosceles property are:

(a) If $\mathscr{F}$ is a family of closed balls in a spherically complete ultrametric space such that each two members of $\mathscr{F}$ intersect, then the family $\mathscr{F}$ has a nonempty intersection, that is, $\cap\mathscr{F} \neq \emptyset$.
(b) If $B(a,r_1)$ and $B(b,r_2)$ are two closed balls in an ultrametric space such that $B(a,r_1) \subset B(b,r_2)$, and if $b \notin B(a,r_1)$, then $d(b,a) = d(b,z)$ for any $z \in B(a,r_1)$.

Priess-Crampe [70] proved an analogue to Banach's Contraction Principle in ultrametric spaces.

Theorem 1.6. *An ultrametric space M is spherically complete if and only if every strictly contractive mapping $T : M \to M$ has a (unique) fixed point.*

Recall that $T : M \to M$ is strictly contractive if

$$d\left(T(x), T(y)\right) < d\left(x,y\right), \quad \text{for any } x,y \in M.$$

1.5 Some Applications

Many examples are known to use Banach's Contraction Principle. Here we will discuss two of them.

1.5.1 ODE and Integral Equations

Consider the integral equation

$$f(x) = \lambda \int_a^x K(x,t)f(t)dt + \phi(x),$$

for a fixed real number λ, where $K(x,t)$ is continuous on $[a,b] \times [a,b]$. Consider the metric space $\mathscr{C}[a,b]$ of continuous real-valued functions defined on $[a,b]$. Consider the map $T : \mathscr{C}[a,b] \to \mathscr{C}[a,b]$ defined by

$$(T(f))(x) = \lambda \int_a^x K(x,t)f(t)dt + \phi(x).$$

For $f_1, f_2 \in \mathscr{C}[a,b]$, we have

$$d\left(T^n(f_1), T^n(f_2)\right) \leq |\lambda|^n M^n \frac{(b-a)^n}{n!} d(f_1, f_2),$$

where

$$d(f_1, f_2) = \max\{|f_1(x) - f_2(x)|;\ x \in [a,b]\},$$

and

$$M = \max\left\{|K(x,t)|;\ (x,t) \in [a,b] \times [a,b]\right\}.$$

Clearly, there exists $n \geq 1$ such that T^n is a contraction, which implies that the above equation has a unique solution $f(x)$. In general, the map T may not be a contraction on $[a,b]$. Bielecki [14] discovered another way to remedy this "problem." Indeed, for $\lambda > 0$, set

$$\|f\|_\lambda = \max_{a \leq x \leq b} \left\{e^{-\lambda\,(x-a)} |f(x)|\right\},$$

it is now possible to prove that for any $f_1, f_2 \in \mathscr{C}[a,b]$, we have

$$d_\lambda\left(T(f_1), T(f_2)\right) = \|T(f_1) - T(f_2)\|_\lambda \leq \frac{M}{\lambda}\, \|f_1 - f_2\|_\lambda = \frac{M}{\lambda}\, d_\lambda(f_1, f_2),$$

where $M = \max\limits_{a \leq x,y \leq b} |K(x,y)|$ is as before. It is then clear that for λ sufficiently large T is a contraction for the new distance d_λ.

1.5.2 Cantor and Fractal Sets

Let (M,d) be a complete metric space, let $\mathscr{M}$ denote the family of all nonempty bounded closed subsets of M, and let $\mathscr{C}$ denote the subfamily of $\mathscr{M}$ consisting of all compact sets. For $A \in \mathscr{M}$ and $\varepsilon > 0$, define the ε-neighborhood of A to be the set

$$N_\varepsilon(A) = \{x \in M : dist(x,A) < \varepsilon\},$$

where $dist(x,A) = \inf_{y \in A} d(x,y)$. Now for $A, B \in \mathscr{M}$, set

$$H(A,B) = \inf\{\varepsilon > 0 : A \subseteq N_\varepsilon(B) \text{ and } B \subseteq N_\varepsilon(A)\}.$$

Then, $(\mathscr{M},H)$ is a metric space, and H is called the *Hausdorff metric* on $\mathscr{M}$. Notice that if (M,d) is complete, then $(\mathscr{M},H)$ is also complete. Let $T_i : M \to M$, $i = 1,\ldots,n$, be a family of contractions. Define the map $T^* : \mathscr{C} \to \mathscr{C}$ by

$$T^*(X) = \bigcup_{i=1}^{n} T_i(X).$$

Then, T^* is a contraction and its Lipschitz constant is smaller than the maximum of all Lipschitz constants of the mappings T_i, $i = 1,\ldots,n$. Then, Banach Contraction Principle implies the existence of a unique nonempty compact subset X of M such that

$$X = \bigcup_{i=1}^{n} T_i(X).$$

As an application of this, consider the real interval $[0,1]$ and the two contractions

$$T_1(x) = \frac{1}{3}x \quad \text{and} \quad T_2(x) = \frac{1}{3}x + \frac{2}{3}.$$

Then, the compact X which satisfies $X = T_1(X) \cup T_2(X)$ is the well-known Cantor set. Moreover this fixed point set is the limit of the iterate of $[0,1]$.

1.6 Historical Note

We all have learned that the origins of the metric contraction principles and, ergo, metric fixed point theory itself rest in the method of successive approximations for proving existence and uniqueness of solutions of differential equations. This method is associated with the names of such celebrated nineteenth century mathematicians as Cauchy, Liouville, Lipschitz, Peano, and specially Picard. In fact the iterative process used in the proof of the Contraction theorem bears the name of Picard iterates. It is quite interesting to know that in 1429 Al-Kashani [12] already published a book entitled: "The Calculator's Key," where he used Picard iterates. In fact, Al-Kashani set the stage to the so-called numerical techniques some 600 years ago. He was keen to develop ideas with practical matters, like the approximate values of $\sin(1^\circ)$ which enabled muslim scientists to come up with a very good approximations to the circumference of the earth.

1.7 Metric Fixed Point Theory in Banach Spaces

The formal definition of Banach spaces is due to Banach himself. But examples like the finite dimensional vector space $\mathbb{R}^n$ were studied prior to Banach's formal definition. In 1912, Brouwer [21] proved the following:

Theorem 1.7 (Brouwer Fixed Point Theorem). *Let B be a closed ball in $\mathbb{R}^n$. Then, any continuous mapping $T : B \to B$ has at least one fixed point.*

This theorem has a long history. The ideas used in its proof were known to Poincare as early as 1886. In 1909, Brouwer proved the theorem when $n = 3$. And in 1910 Hadamard gave the first proof for arbitrary n, and Brouwer gave another proof in 1912. All of these are older results than the Banach Contraction Principle. Though in nature the two theorems are different, they bare some similarities. A combination of the two led to the so-called metric fixed point theorem in Banach spaces. Indeed, in Brouwer's theorem the convexity, compactness, and the continuity of T are crucial, while the Lipshitz behavior of the contraction and completeness are crucial in Banach's fixed point theorem. In infinite normed linear vector spaces, we lose the compactness of the bounded closed convex sets (like closed balls). So if we assume completeness we get Banach spaces. On these, we have another natural topology (other than the norm topology), that is the weak-topology. So a weakly compact convex set need not be compact for the norm. The best example here is the Hilbert space. In it any bounded closed convex set is weakly-compact. The Lipschitz condition considered is when the Lipschitz constant is equal to 1. Such mappings are called *nonexpansive*. In other words, if (M,d) is a metric space, then $T : M \to M$ is *nonexpansive* if

$$d\,(T(x), T(y)) \leq d(x,y), \quad \text{for any } x, y \in M.$$

The metric fixed point problem in Banach spaces becomes:

> *Let X be a Banach space, and C a nonempty bounded closed convex subset of X. When does any nonexpansive mapping $T : C \to C$ have a fixed point?*

Other interesting problems closely related to this one are:

- The structure of the fixed points sets [23]
- The approximation of fixed points.

Recognition of fixed point theory for nonexpansive mappings as a noteworthy avenue of research almost surely dates from the 1965 publication of

1. *Browder–Gohde Theorem [22, 45].* If K is a bounded closed convex subset of a uniformly convex Banach space X and if $T : K \to K$ is nonexpansive, then T has a fixed point. Moreover the fixed point set of T is a closed convex subset of K.
2. *Kirk Theorem [58].* Let K be a weakly-compact convex subset of a Banach space X. Assume that K has the normal structure property, then any nonexpansive mapping $T : K \to K$ has a fixed point.

1.7.1 Classical Existence Results

While studying a paper by Brodskii and Milman [20], Kirk [58] was able to discover the above stated result. So the normal structure property is an old concept not directly related to nonexpansive mappings. Hilbert space and uniformly convex Banach spaces have the normal structure property. In fact, the proofs given independently by Browder [22] and Gohde [45] of their results do not use this property.

Let K be a bounded closed convex subset of a Banach space $(X, \|.\|)$ and $T : K \to K$ be a nonexpansive mapping. Let $\varepsilon \in (0,1)$ and $x_0 \in K$. Define T_ε by

$$T_\varepsilon(x) = \varepsilon x_0 + (1-\varepsilon)T(x), \quad \text{for any } x \in K.$$

It is easy to check that $T_\varepsilon(K) \subset K$ (since K is convex) and it is a contraction. The Banach Contraction Principle implies the existence of a unique point $x_\varepsilon \in K$ such that

$$x_\varepsilon = \varepsilon x_0 + (1-\varepsilon)T(x_\varepsilon).$$

which implies

$$\|x_\varepsilon - T(x_\varepsilon)\| = \varepsilon\ \|x_0 - T(x_\varepsilon)\| \leq \varepsilon\ \mathrm{diam}(K) = \varepsilon\ \sup\{\|x-y\|; x, y \in K\}.$$

Therefore, we have

$$\inf_{x\in K} \|x - T(x)\| = 0.$$

This property is known as the *approximate fixed point property*. So any nonexpansive mapping defined on a bounded closed convex subset of Banach space has this property. By taking a sequence $\{\varepsilon_n\}$ which goes to 0, we generate a sequence of points $\{x_n\}$ from K such that

$$\lim_{n\to\infty} \|x_n - T(x_n)\| = 0.$$

Such sequence is called an *approximate fixed point sequence* (a.f.p.s.).

In order to grasp the ideas behind the proofs of the existence of fixed point for nonexpansive mappings in the Hilbert and uniformly convex spaces, we need the concept of *asymptotic center* of a sequence discovered by Edelstein [34]. This concept is very useful whenever one deals with sequential approximations in Banach spaces.

Let $\{x_n\}$ be a bounded sequence in a Banach space X, and let C be a closed convex subset of X. Consider the functional $f : C \to [0,\infty)$ defined by

$$f(x) = \limsup_{n\to\infty} \|x_n - x\|.$$

Usually, we use the notation $f(x) = r(x,\{x_n\})$. The infimum of $f(x)$ over C is called the *asymptotic radius* of $\{x_n\}$ and denoted by $r(C,\{x_n\})$, that is,

$$r(C,\{x_n\}) = \inf_{x\in C} r(x,\{x_n\}) = \inf_{x\in C} \left(\limsup_{n\to\infty} \|x_n - x\| \right).$$

The set

$$A(C,\{x_n\}) = \{x \in C;\ r(x,\{x_n\}) = r(C,\{x_n\})\}$$

is the set of all asymptotic centers of $\{x_n\}$. If $\{x_n\}$ converges to $x \in C$, then $A(C,\{x_n\}) = \{x\}$. In general the set $A(C,\{x_n\})$ is not reduced to one point.

Theorem 1.8. *Every bounded sequence in a uniformly convex Banach space X has a unique asymptotic center with respect to any closed convex subset of X.*

In the Hilbert case, we get more:

Theorem 1.9. *In a Hilbert space H, the weak limit of a weakly convergent sequence coincides with its asymptotic center with respect to H.*

Note that in any Hilbert space H, the asymptotic center of a bounded sequence $\{x_n\}$ with respect to H belongs to the closed convex hull of $\{x_n\}$. But this does not necessarily hold in (even uniformly convex) Banach spaces.

Let C be a bounded closed convex subset of a uniformly convex Banach space X and $T : C \to C$ be nonexpansive.

(a) Let $\{x_n\}$ be an a.f.p.s. of T in C. Let $z \in C$ be its asymptotic center with respect to C, then it is quite easy to check that $T(z)$ is also an asymptotic center of $\{x_n\}$ with respect to C. By the uniqueness of the asymptotic center, we get $T(z) = z$.
(b) Let $x \in X$ and consider the orbit of x under T, that is, $\{T^n(x)\}$. Let $z \in C$ be its asymptotic center with respect to C, then again it is quite easy to check that $T(z)$ is also an asymptotic center of $\{T^n(x)\}$ with respect to C. By the uniqueness of the asymptotic center, we get $T(z) = z$.

In 1965, Browder [22] in fact discovered something truly amazing. Let C be a bounded closed convex subset of a uniformly convex Banach space X and $T : C \to C$ be nonexpansive. If $\{x_n\} \subset C$ converges weakly to x and $\{x_n - T(x_n)\}$ converges strongly (with respect to the norm) to 0, then we have

$$x - T(x) = 0.$$

This is known as the *Demi-closedness principle*.

1.7.2 The Normal Structure Property

The reason behind separating it from the other results is the important role it played during the first 20 years (since 1965) of the theory. In order to appreciate this property, let us give more information on nonexpansive mappings in Banach spaces. Indeed, let C be a weakly compact convex subset of a Banach space X. Let $T : C \to C$ be nonexpansive. The first pioneers of the metric fixed point theory were mostly concerned about the existence of fixed points. So assume that T fails to have a fixed point in C. Since C is weakly compact, there exists (by Zorn) a minimal nonempty closed convex subset K of C invariant under T, that is, $T(K) \subset K$. The central research of the metric fixed point theory in Banach spaces always centered around the discovery of new properties of such minimal convex sets. The first property is due to Kirk.

Theorem 1.10. *Under the above assumptions and notations, we have*

(a) $\overline{conv}(T(K)) = K$.
(b) $\sup_{x \in K} \|z - x\| = diam(K) > 0$, *for any* $z \in K$.

The normal structure property forbids the conclusion 2 to hold (see [11]).

Definition 1.2. A closed convex subset C of a Banach space X is said to have the *normal structure property* if any bounded convex subset K of C which contains more than one point contains a *nondiametral point*, that is, there exists a point $x_0 \in K$ such that

$$\sup_{x \in K} \|x_0 - x\| < \text{diam}(K).$$

We will also say that X has the normal structure property if any bounded closed convex subset has the normal structure property.

Throughout this section, we will use the following notations:

- $\text{diam}(C) = \sup_{x,y \in C} \|x - y\|$
- $r_z(C) = \sup_{x \in C} \|z - x\|$

- $R(C) = \inf_{x \in C} r_x(C)$
- $\mathscr{C}(C) = \{z \in C;\ r_z(C) = r(C)\}$

The number $R(C)$ and the set $\mathscr{C}(C)$ are called the *Chebyshev radius* and *Chebyshev center* of C, respectively.

Now we are ready to state Kirk's theorem [58].

Theorem 1.11 (Kirk Fixed Point Theorem). *Let X be a Banach space and suppose that C is a nonempty weakly compact convex subset of X which has the normal structure property. Then, any nonexpansive mapping $T : C \to C$ has a fixed point.*

For the following ten years most of the results were solely concentrated on the study of the normal structure property. For example, most of the classical known Banach spaces have normal structure property. So most of the fixed point existence theorems were in fact theorems about normal structure property. It got so bad that some thought that any reflexive (or superreflexive) Banach spaces have the normal structure property. It was not before 1972 that James [50] renormed the Hilbert space l_2 to get rid of this property. For any $\beta > 0$, define the new norm $\|.\|_\beta$ on l_2 by

$$\|x\|_\beta = \max\{\|x\|_2, \beta\ \|x\|_\infty\}.$$

Set $X_\beta = (l_2, \|.\|_\beta)$. James showed that the superreflexive Banach space $X_{\sqrt{2}}$ fails to have the normal structure property. This answered the question asked by the people working on the fixed point property in Banach spaces. Right after James' result was known, the natural question of whether $X_{\sqrt{2}}$ has the fixed point property arose. It was Karlovitz [50] who settled this question to the affirmative. We will discuss more Karlovitz ideas later on. Hence, we see that normal structure property (which is a geometric property) implies the fixed point property but they are not equivalent. Then people started to dissociate themselves from the normal structure property. In fact, Baillon and Schoneberg [7] studied the space $X_{\sqrt{2}}$ and proved that this space has a geometric property, known as *asymptotic normal structure property* which implies the fixed point property. In fact, they proved that X_β has this geometric property for $\beta < 2$. Then it was asked what happened to X_β for $\beta \geq 2$. No definitive answer was given to a point that some though again that may be these spaces may lead an example of a superreflexive space which fails the fixed point property. No luck again. This time it was Lin [64] who settled this question by proving that X_β has the fixed point property for any $\beta > 0$. After this, and other results some went too far to claim that no geometric property is equivalent to the fixed point property. This question is still open.

1.7.3 More on Normal Structure Property

Since Kirk publication of his paper, people intensified their investigation of this property. For example it was proved that uniformly convex Banach spaces have this property. In particular, let X be a Banach space. Set

$$\delta_X(\varepsilon) = \inf\left\{1 - \left\|\frac{x+y}{2}\right\| : \|x\| \leq 1, \|y\| \leq 1, \|x-y\| \geq \varepsilon,\ x,y \in X\right\}$$

defined for $\varepsilon \in [0,2]$. The characteristic (or coefficient) of convexity of the Banach space X is the number

$$\varepsilon_0(X) = \sup\{\varepsilon \geq 0 : \delta(\varepsilon) = 0\}.$$

X is uniformly convex if and only if $\varepsilon_0(X) = 0$, that is, $\delta(\varepsilon) > 0$, for any $\varepsilon \in [0,2]$. And X is said to be *uniformly non-square* if and only if $\varepsilon_0(X) < 2$. James [48] studied these spaces extensively (see also [38]). Assume that $\varepsilon_0(X) < 1$, and let $\varepsilon \in (\varepsilon_0(X), 1)$. Then, for any nonempty bounded closed convex subset C of X not reduced to one point, there exists a point $x \in C$ such that

$$r(x,C) \leq (1 - \delta(\varepsilon))\ \mathrm{diam}(C).$$

So, we have

$$N(X) = \sup\left\{\frac{r(C)}{\mathrm{diam}(C)}\right\} \leq (1 - \delta(\varepsilon_0(X))),$$

where the supremum is taken over all nonempty bounded closed convex subset of X with more than one point. The number $N(X)$, introduced by Bynum [26], is known as the *coefficient of uniform normal structure property*. So we have the normal structure property and even more: uniform normal structure property, that is, $N(X) < 1$. So uniformly convex Banach spaces have a stronger geometric property. In fact, a weaker version was discovered to still imply the normal structure property. It is known as *uniform convexity in every direction* (U.C.E.D.) [30]. Indeed, let $z \in X$ be a unit vector ($\|z\| = 1$). The modulus of uniform convexity $\delta_X(\varepsilon, z)$ of X in the direction z is defined by

$$\inf\left\{1 - \left\|\frac{x+y}{2}\right\| : \|x\| \leq 1, \|y\| \leq 1,\ x - y = \alpha z\,, \|x-y\| \geq \varepsilon\right\}.$$

It is not hard to show that if X is U.C.E.D., then for any nonempty bounded closed convex subset K of X, the Chebishev center $\mathscr{C}(K)$ of K has at most one point. So clearly if X is U.C.E.D., then it has the normal structure property. Spaces which are U.C.E.D. or equivalently renormed to be U.C.E.D. were extensively studied by Zizler [82].

There is a sequential characterization of the normal structure property discovered by Brodskii and Milman in their original work [20], which played a key role in studying this property.

Definition 1.3. A bounded sequence $\{x_n\}$ in a Banach space is said to be a *diametral sequence* if

$$\lim_{n\to\infty} dist(x_{n+1}, \overline{conv}\{x_1,x_2,\cdots,x_n\}) = \text{diam}\{x_1,x_2,\cdots\}.$$

We have

Theorem 1.12. *A bounded convex subset K of a Banach space has normal structure if and only if it does not contain a diametral sequence.*

1.7.4 Normal Structure and Smoothness

As with convexity, it is possible to scale Banach spaces with respect to their smoothness. Let X be a Banach space. Define the modulus of smoothness ρ_X of X by

$$\rho_X(\eta) = \sup\left\{\left\|\frac{x+\eta y}{2}\right\| + \left\|\frac{x-\eta y}{2}\right\| - 1\,;\|x\|\le 1,\ \|y\|\le 1\right\},$$

for any $\eta > 0$. X is said to be *uniformly smooth* (U.S.) if

$$\lim_{\eta\to 0}\frac{\rho_X(\eta)}{\eta} = 0.$$

Theorem 1.13. *Let X be a Banach space and X^* its dual. We have*

(a) $\rho_{X^*}(\eta) = \sup\left\{\frac{\eta\varepsilon}{2} - \delta_X(\varepsilon); 0\le\varepsilon\le 2\right\}$*, for any* $\eta > 0$*;*
(b) $\lim_{\eta\to 0}\frac{\rho_{X^*}(\eta)}{\eta} = \frac{\varepsilon_0(X)}{2}$*;*
(c) X is uniformly convex if and only if X^* is uniformly smooth.

Note that one may reverse the roles of X and X^* to obtain for example

$$\rho'_X(0) = \lim_{\eta\to 0}\frac{\rho_X(\eta)}{\eta} = \frac{\varepsilon_0(X^*)}{2}.$$

Using James non-squareness result, we see that X is *superreflexive* if and only if $\rho'_X(0) < \frac{1}{2}$. In [5] Baillon proved that any uniformly smooth space X has the fixed point property. In fact, he precisely proved this conclusion provided $\rho'_X(0) < \frac{1}{2}$.

In [79] Turett proved that the condition $\rho'_X(0) < \frac{1}{2}$ implies the normal structure property. Unfortunately Turett's proof is an exact reproduction of Baillon's original proof. Later on this conclusion was strengthened by Prus [73] and Khamsi [52] to show that the condition $\rho'_X(0) < \frac{1}{2}$ implies the super-normal structure property (and hence the uniform normal structure property) for both X and its dual X^*. The proof is based on nonstandard techniques.

1.7.5 Karlovitz–Goebel Lemma

As mentioned before, if K is a minimal convex associated with a nonexpansive mapping, then little is known about its properties except the two properties discovered by Kirk. Independently Karlovitz [50] and Goebel [40] proved the following:

Theorem 1.14. *Let K be a subset of a Banach space X which is minimal with respect to being nonempty, weakly compact, convex, and T-invariant for some nonexpansive mapping T, and suppose $\{x_n\} \subseteq K$ is an a.f.p.s., that is, $\lim_{n\to\infty} \|x_n - T(x_n)\| = 0$. Then for each $x \in K$, we have*

$$\lim_{n\to\infty} \|x - x_n\| = diam(K).$$

Recall again that it was this property that allowed Karlovitz to prove that the space $X_{\sqrt{2}}$ has the fixed point property. In this theorem a.f.p.s. assumption is crucial. It is this property that pushed Maurey [66] to use ultrapowers approach to study the fixed point property in Banach spaces.

From 1965 to 1980 most of the classical Banach spaces were investigated and proved to have the fixed point property. Missing to the call was c_0 and (general) lattice Banach spaces. During that period of time, it was strongly believed that any weakly compact convex subset of any Banach space has the fixed point property. It was Alspach [2] who disproved this claim by constructing an isometry on a weakly compact convex subset of $L^1[0,1]$ without a fixed point. This example set the stage to the second revolution in the theory (after the first one in 1965). Indeed, almost right after Alspach's example was made public, Maurey [66] published his famous results (using nonstandard techniques).

The Open Problem. *It is still unknown whether reflexive (or superreflexive) Banach spaces have the fixed point property.*

For more on reflexive and superreflexive Banach spaces the reader may consult [33, 49].

1.8 Nonstandard Techniques

In order to appreciate Maurey's ideas, we need to define the concept of ultrapower of a Banach space [1, 9, 46, 74]. Throughout this section $(X, \|.\|)$ denotes a Banach space and $\mathcal{U}$ a nontrivial ultrafilter over the positive integers.

Consider the vector space

$$\ell_\infty(X) = \left\{(x_n) \subset X : \sup_{1\leq n<\infty} \|x_n\| < \infty\right\}.$$

It is known (and easily checked) that $\ell_\infty(X)$ is a Banach space with the norm defined by

$$\|(x_n)\|_\infty = \sup_{1\le n<\infty} \|x_n\| \ , \quad \text{for } (x_n) \in \ell_\infty(X).$$

Set

$$\mathcal{N} = \left\{ (x_n) \in \ell_\infty(X);\ \lim_{\mathcal{U}} \|x_n\| = 0 \right\}.$$

The Banach space *ultrapower* $\tilde{X}$ of X (relative to the ultrafilter $\mathcal{U}$ is the quotient space $\ell_\infty(X)/\mathcal{N}$. Thus, the elements of $\tilde{X}$ are equivalence classes $[(x_n)]$ of bounded sequences $(x_n) \subset X$, where one agrees that two such sequences (x_n) and (y_n) are equivalent if and only if

$$\lim_{\mathcal{U}} \|x_n - y_n\| = 0.$$

The norm $\|\cdot\|_{\mathcal{U}}$ in $\tilde{X}$ is the usual quotient norm. Thus, for $\tilde{x} = [(x_n)] \in \tilde{X}$,

$$\|\tilde{x}\|_{\mathcal{U}} = \inf\{\|(x_n + y_n)\|_\infty : (y_n) \in \mathcal{N}\}.$$

We remark that there is another approach which leads to the same thing. Notice that since $\{\|x_n\|\}$ is bounded it lies in a compact subset of $\mathbb{R}$. Therefore, $\lim_{\mathcal{U}} \|x_n\|$ always exists. In fact one can prove that for $\tilde{x} = [(x_n)] \in \tilde{X}$,

$$\|\tilde{x}\|_{\mathcal{U}} = \lim_{\mathcal{U}} \|x_n\| .$$

For each $x \in X$ let (x_n) denote the sequence for which $x_n = x$, and let $\dot{x} = [(x_n)] \in \tilde{X}$. Then the subspace $\dot{X} = \{\dot{x} : x \in X\}$ is isometric to X via the mapping $x \to \dot{x}$ since

$$\|\dot{x}\|_{\mathcal{U}} = \lim_{\mathcal{U}} \|x_n\| = \lim_{\mathcal{U}} \|x\| = \|x\| .$$

Definition 1.4. We will say that the Banach space X has the property super-$\mathcal{P}$ if and only if its ultrapower $\tilde{X}$ has the property $\mathcal{P}$.

A precise definition using finite representability may be used but here we tried a simple approach. For example, we can talk of: superreflexive, super normal structure property, super fixed point property, etc.

1.8.1 Extending Mappings to Ultrapowers

Let $K \subseteq X$ be bounded closed and convex, and suppose $T : K \to K$ is nonexpansive. Letting

$$\tilde{K} = \left\{ [(x_n)] \in \tilde{X};\ x_n \in K \text{ for each } n \right\},$$

there is a canonical way to extend T to a mapping $\tilde{T} : \tilde{K} \to \tilde{K}$ by setting for $\tilde{x} = [(x_n)] \in \tilde{K}$,

$$\tilde{T}(\tilde{x}) = [(T(x_n))].$$

It is immediate that $\tilde{K}$ is a bounded closed and convex subset of $\tilde{X}$. It is quite straightforward to prove that if T is nonexpansive, then $\tilde{T}$ is nonexpansive as well. Moreover, since K is bounded and T is nonexpansive there always exist a.f.p. s. $(x_n) \subset K$, that is, $\lim\limits_{n\to\infty} \|x_n - T(x_n)\| = 0$. Set $\tilde{x} = [(x_n)]$. Then, $\tilde{T}(\tilde{x}) = \tilde{x}$, that is, $Fix(\tilde{T}) \neq \emptyset$. Not only does the mapping $\tilde{T}$ always have fixed points, but in fact we can prove that $Fix(\tilde{T})$ is metrically convex (Maurey [66]). When dealing with minimal invariant convex sets, more can be said. Indeed, let X be a Banach space and C be a nonempty, convex, weakly compact subset of X. Assume that there exists a nonexpansive mapping $T : C \to C$ whose fixed point set $Fix(T)$ is empty. Since C is weakly compact, there exists K a closed convex subset of C, which is T-invariant and minimal. Before, we have seen some of the properties of this minimal set. Define $\tilde{K}$ as usual to be $\tilde{K} = \big\{[(x_n)] \in \tilde{X} : x_n \in K \text{ for each } n\big\}$. Maurey proved some more properties of $\tilde{K}$:

(a) $\mathrm{diam}(K) = \mathrm{diam}(\tilde{K}) = \mathrm{diam}(Fix(\tilde{T}))$;
(b) For any $x \in K$, $dist(\dot{x}, Fix(\tilde{T})) = \mathrm{diam}(K)$;
(c) For any $\tilde{x} \in Fix(\tilde{T})$, $dist(\tilde{x}, \dot{K}) = \mathrm{diam}(K)$.

Using these properties Maurey proved the following:

Theorem 1.15 (Maurey Fixed Point Theorem). *Any reflexive subspace of $L^1[0,1]$) has the fixed point property as well as the Hardy space H^1. Moreover, if K is a weakly compact, convex nonempty subset of a superreflexive space X, then any isometric map $T : K \to K$ has a fixed point.*

Recall that up to 1980, it was still unknown whether the classical space c_0 has the fixed point property. Maurey, through his new ideas, was able to prove it by using the basic lattice properties of c_0 [36]. In fact, Borwein and Sims [18] did rewrite these ideas in general lattice Banach spaces to obtain similar conclusions.

Refining some of the properties satisfied by the minimal invariant sets (in the ultrapower language), Lin [64] proved the following result:

Theorem 1.16. *Let $\{\tilde{w}_n\}$ be an approximate fixed point sequence (a.f.p.s.) for $\tilde{T}$ in $\tilde{K}$, that is, $\lim\limits_{n\to\infty} \|\tilde{T}(\tilde{w}_n) - \tilde{w}_n\|_{\mathscr{U}} \to 0$. Then for any $x \in K$, we have*

$$\lim_{n\to\infty} \|\tilde{w}_n - \dot{x}\|_{\mathscr{U}} = diam(K).$$

This enabled him to prove one of the most elegant applications of Maurey's ideas:

Theorem 1.17 (Lin's Fixed Point Theorem [64]). *Let X be a Banach space with an unconditional Schauder basis. Assume that its constant of unconditionality λ satisfies*

$$\lambda < \frac{\sqrt{33}-3}{2} \approx 1.37$$

Then, X has the fixed point property.

Note that the canonical basis of the spaces X_β, that is, l_2 renormed with James' norm $\|.\|_\beta$, are 1-unconditional, then X_β has the fixed point property for any $\beta > 0$. This conclusion was unknown till Lin published the above theorem in 1985.

1.9 More on Metric Fixed Point Theory in Metric Spaces

1.9.1 Menger Convexity in Metric Spaces

The Banach Contraction Principle theorem is a metric result and does not depend on any linear structure. But Kirk's fixed point theorem is strongly connected to the linear convexity structure of linear spaces. As early as 1965, many have tried to weaken this tiding. Takahashi [77] was may be the first one to give a metric analogue to Kirk's theorem. His approach was based on defining a convexity in metric spaces extremely similar to the linear convexity also known as Menger convexity [15–17, 67].

Definition 1.5. Let (M,d) be a metric space with $I = [0,1]$. A mapping $W : M \times M \times I \to M$ is said to be a *convex structure on M* if for each $(x,y,\lambda) \in M \times M \times I$ and $z \in M$, we have

$$d(z, W(x,y,\lambda)) \leq \lambda\ d(z,x) + (1-\lambda)d(z,y).$$

Throughout let us write $W(x,y,\lambda) = \lambda x \oplus (1-\lambda)y$ whenever the choice of the convexity mapping W is irrelevant. Moreover, if we have

$$d\left(\frac{1}{2}p \oplus \frac{1}{2}x, \frac{1}{2}p \oplus \frac{1}{2}y\right) \leq \frac{1}{2}d(x,y), \quad \text{for all } p,x,y \in M,$$

then M is said to be a *hyperbolic metric space* (see [43]).

Obviously, normed linear spaces are hyperbolic spaces. As nonlinear examples, one can consider the Hadamard manifolds [25], the Hilbert open unit ball equipped with the hyperbolic metric [43], and the CAT(0) spaces [58–60, 63] (see Example 1.3).

Using the convexity structure W, one will easily define a convex subset of M and prove similar properties of convex sets in the linear case. It is not hard to check that balls in hyperbolic metric spaces are convex sets.

Example 1.2 ([43]). Let B be the open unit ball of the infinite Hilbert space H. On B, we consider the Poincaré hyperbolic metric ρ:

$$\rho(x,y) = \inf_{\gamma} \int_0^1 \alpha(\gamma(t), \gamma'(t))dt,$$

where the infimum is taken over all piece-wise differentiable curves such that $\gamma(0) = x$ and $\gamma(1) = y$, and where

$$\alpha(x,v) = \sup_{f \in \mathscr{F}} \|Df(x)(v)\|,$$

with $\mathscr{F} = \{f : B \to B; \text{holomorphic}\}$. (B,ρ) is a complete metric space (unbounded). Using the Mobius transformations, one can prove that

$$\rho(x,y) = \text{Argth}\Big(1 - \sigma(x,y)\Big)^{1/2} \quad \text{with } \sigma(x,y) = \frac{(1-|x|^2)(1-|y|^2)}{|1- <x,y>|^2}.$$

Using some complicated computations, one may prove that for any x,y in B, and $\lambda \in [0,1]$, there exists a unique $z \in B$ such that

$$\rho(z,w) \leq \lambda \rho(x,w) + (1-\lambda)\rho(y,w), \quad \text{for any } w \in B.$$

In other words, we have $z = W(x,y,\lambda) = \lambda x \oplus (1-\lambda)y$. In fact, the metric space (B,ρ) enjoys some geometric properties similar to uniformly convex Banach spaces. Moreover, we have

$$\begin{cases} \rho(a,x) \leq r \\ \rho(a,y) \leq r \\ \rho(x,y) \geq r\varepsilon, \end{cases} \Longrightarrow \rho\left(a, \frac{1}{2}x \oplus \frac{1}{2}y\right) \leq r(1 - \delta(r,\varepsilon)),$$

for any a,x,y in B and any positive r and ε, where

$$\delta(r,\varepsilon) = 1 - \frac{1}{r}\text{Argth}\left(\frac{\sinh(r(1+\varepsilon/2))\sinh(r(1-\varepsilon/2))}{\cosh(r)}\right)^{1/2}.$$

It is easy to check that for $r > 0$ and $\varepsilon > 0$, we have $\delta(r,\varepsilon) > 0$. This is the analogue of the uniform convexity in the nonlinear case. It is not hard to check that convex subsets enjoy the normal structure property and that the Chebishev center is reduced to one point. Note that holomorphic mappings are nonexpansive mappings for the Poincare distance ρ.

Example 1.3 ([19]). Let (X,d) be a metric space. A *geodesic* from x to y in X is a mapping c from a closed interval $[0,l] \subset \mathbb{R}$ to X such that $c(0)=x$, $c(l)=y$, and $d(c(t),c(t'))=|t-t'|$ for all $t,t' \in [0,l]$. In particular, c is an isometry and $d(x,y)=l$. The image α of c is called a geodesic (or metric) *segment* joining x and y. The space (X,d) is said to be a *geodesic space* if every two points of X are joined by a geodesic and X is said to be *uniquely geodesic* if there is exactly one geodesic joining x and y for each $x,y \in X$, which we will be denoted by $[x,y]$, and called the segment joining x to y.

A *geodesic triangle* $\Delta(x_1,x_2,x_3)$ in a geodesic metric space (X,d) consists of three points x_1,x_2,x_3 in X (the *vertices* of Δ) and a geodesic segment between each pair of vertices (the *edges* of Δ). A *comparison triangle* for geodesic triangle $\Delta(x_1,x_2,x_3)$ in (X,d) is a triangle $\overline{\Delta}(x_1,x_2,x_3) := \Delta(\bar{x}_1,\bar{x}_2,\bar{x}_3)$ in $\mathbb{R}^2$ such that $d_{\mathbb{R}^2}(\bar{x}_i,\bar{x}_j)=d(x_i,x_j)$ for $i,j \in \{1,2,3\}$. Such a triangle always exists (see [19]).

A geodesic metric space is said to be a *CAT* (0) space if all geodesic triangles of appropriate size satisfy the following *CAT* (0) comparison axiom:

Let Δ be a geodesic triangle in X and let $\overline{\Delta} \subset \mathbb{R}^2$ be a comparison triangle for Δ. Then Δ is said to satisfy the *CAT* (0) *inequality* if for all $x,y \in \Delta$ and all comparison points $\bar{x},\bar{y} \in \overline{\Delta}$,

$$d(x,y) \le d(\bar{x},\bar{y}).$$

Complete *CAT* (0) spaces are often called *Hadamard spaces (see [58])*. If x,y_1,y_2 are points of a *CAT* (0) space and y_0 is the midpoint of the segment $[y_1,y_2]$, that is, $y_0 = \frac{y_1 \oplus y_2}{2}$, then the *CAT* (0) inequality implies:

$$d^2\left(x, \frac{y_1 \oplus y_2}{2}\right) \le \frac{1}{2}d^2(x,y_1) + \frac{1}{2}d^2(x,y_2) - \frac{1}{4}d^2(y_1,y_2).$$

This inequality is the (CN) inequality of Bruhat and Tits [24].

The previous two examples suggest a thorough discussion of uniform convexity in metric spaces.

1.9.2 Uniformly Convex Metric Spaces

Let (M,d) be a metric space. Suppose that there exists a family $\mathscr{F}$ of metric segments such that any two points x,y in M are endpoints of a unique metric segment $[x,y] \in \mathscr{F}$ ($[x,y]$ is an isometric image of the real line interval $[0,d(x,y)]$). We will say that a subset C of a hyperbolic metric space M is convex if $[x,y] \subset C$ whenever x,y are in C.

Let τ be another topology on M that is, weaker than the metric topology. We will assume that τ is lower semi-continuous, that is,

$$d(x,y) \leq \liminf_{n\to\infty} d(x_n, y_n),$$

for every $\{x_n\}$ and $\{y_n\}$ in M which τ-convergent to x and y, respectively.

Definition 1.6. Let (M,d) be a hyperbolic metric space. We say that M is *uniformly convex* (in short, UC) if for any $a \in M$, for every $r > 0$, and for each $\varepsilon > 0$

$$\delta(r,\varepsilon) = \inf\left\{1 - \frac{1}{r}\, d\left(\frac{1}{2}x \oplus \frac{1}{2}y, a\right); d(x,a) \leq r, d(y,a) \leq r, d(x,y) \geq r\varepsilon\right\} > 0.$$

The definition of uniform convexity finds its origin in Banach spaces [29]. To the best of our knowledge, the first attempt to generalize this concept to metric spaces was done in [44]. The reader may also consult [43].

We have the following properties [56]:

(a) Let us observe that $\delta(r,0) = 0$, and $\delta(r,\varepsilon)$ is an increasing function of ε for every fixed r.
(b) For $r_1 \leq r_2$, there holds

$$1 - \frac{r_2}{r_1}\left(1 - \delta\left(r_2, \varepsilon\, \frac{r_1}{r_2}\right)\right) \leq \delta(r_1,\varepsilon).$$

(c) If (M,d) is uniformly convex, then (M,d) is strictly convex, that is, whenever

$$d\left(\frac{1}{2}x \oplus \frac{1}{2}y, a\right) = d(x,a) = d(y,a),$$

for any $x,y,a \in M$, then we must have $x = y$.

The next technical results will help establish a property in uniformly convex metric spaces equivalent to reflexivity in Banach spaces.

Lemma 1.2 ([56]). *Assume that (M,d) is uniformly convex. Let $\{C_n\} \subset M$ be a sequence of nonempty, nonincreasing, convex, bounded and closed sets. Let $x \in M$ be such that*

$$0 < d = \lim_{n\to\infty} d(x, C_n) < \infty.$$

Let $x_n \in C_n$ be such that $d(x,x_n) \to d$. Then, $\{x_n\}$ is a Cauchy sequence.

Recall that a hyperbolic metric space (M,d) is said to have the property (R) [53] if any nonincreasing sequence of nonempty, convex, bounded and closed sets has a nonempty intersection. Our next result deals with the existence and the uniqueness of the best approximants of convex, closed and bounded sets in a uniformly convex metric space. This result is of interest by itself as uniform convexity implies the property (R), which reduces to reflexivity in the linear case.

Theorem 1.18. *Assume that (M,d) is complete and uniformly convex. Let $C \subset M$ be nonempty, convex and closed. Let $x \in M$ be such that $d(x,C) < \infty$. Then there exists a unique best approximant of x in C, that is, there exists a unique $x_0 \in C$ such that*

$$d(x,x_0) = d(x,C).$$

The following result gives the analogue result to the well-known theorem that states any uniformly convex Banach space is reflexive. For a reference the reader may read Theorem 2.1 in [43].

Theorem 1.19. *If (M,d) is complete and uniformly convex, then (M,d) has the property (R).*

Remark 1.1. Note that any hyperbolic metric space M which satisfies the property (R) is complete. Indeed, let $\{x_n\}$ be a Cauchy sequence in M. Denote

$$\varepsilon_n = \sup\{d(x_m,x_s);\ m,s \geq n\},\ n = 1,\cdots.$$

Our assumption implies that $\lim_{n\to\infty} \varepsilon_n = 0$. In hyperbolic metric spaces, closed balls are convex. Therefore, the property (R) implies that $\bigcap_{n\geq 1} B(x_n,\varepsilon_n) \neq \emptyset$. It is easy to check that this intersection is reduced to one point which is the limit of $\{x_n\}$.

The following technical lemma is useful to establish a metric version of the main results in [80] proved in the setting of Banach spaces.

Lemma 1.3 ([56]). *Let (M,d) be a uniformly convex metric space. Assume that there exists $R \in [0,+\infty)$ such that*

$$\limsup_{n\to\infty} d(x_n,a) \leq R,\ \limsup_{n\to\infty} d(y_n,a) \leq R,\ \textit{and}\ \lim_{n\to\infty} d\left(a,\frac{1}{2}x_n \oplus \frac{1}{2}y_n\right) = R.$$

Then, $\lim_{n\to\infty} d(x_n,y_n) = 0$.

A metric version of the parallelogram identity goes as follows (see [10,43,80]).

Theorem 1.20 (Parallelogram Inequality [56]). *Let (M,d) be uniformly convex. Fix $a \in M$. For each $0 < r$ and for each $\varepsilon > 0$ denote*

$$\Psi(r,\varepsilon) = \inf\left\{\frac{1}{2}d^2(a,x) + \frac{1}{2}d^2(a,y) - d^2\left(a,\frac{1}{2}x \oplus \frac{1}{2}y\right)\right\},$$

where the infimum is taken over all $x,y \in M$ such that $d(a,x) \leq r$, $d(a,y) \leq r$, and $d(x,y) \geq r\varepsilon$. Then, $\Psi(r,\varepsilon) > 0$ for any $0 < r$ and for each $\varepsilon > 0$. Moreover, for a fixed $r > 0$, we have

(a) $\Psi(r,0) = 0$;
(b) $\Psi(r,\varepsilon)$ *is a nondecreasing function of* ε;
(c) if $\lim_{n\to\infty} \Psi(r,t_n) = 0$, *then* $\lim_{n\to\infty} t_n = 0$.

The concept of p-uniform convexity was used extensively by Xu [80] (see also [9] p. 310); its nonlinear version for $p = 2$ is given below [56].

Definition 1.7. We will say that (M,d) is *2-uniformly convex* if

$$c_M = \inf\left\{\frac{\Psi(r,\varepsilon)}{r^2\varepsilon^2}; r > 0\ ,\ \varepsilon > 0\right\} > 0.$$

Note that (M,d) is 2-uniformly convex if and only if

$$\inf\left\{\frac{\delta(r,\varepsilon)}{\varepsilon^2}; r > 0\ ,\ \varepsilon > 0\right\} > 0.$$

Example 1.4. Let (M,d) be $CAT(0)$ geodesic metric space. As for the Hilbert space, the (CN) inequality satisfied by M implies that M is uniformly convex with

$$\delta(r,\varepsilon) = 1 - \sqrt{1 - \frac{\varepsilon^2}{4}}.$$

The (CN) inequality also implies that

$$\Psi(r,\varepsilon) = \frac{r^2\varepsilon^2}{4}.$$

This clearly implies that any CAT(0) space is 2-uniformly convex with $c_M = \frac{1}{4}$ [56].

1.9.3 Fixed Point Property in Uniformly Convex Metric Spaces

As an application of the previous technical results, we discuss the existence of fixed points of uniformly Lipschitzian mappings [41].

Definition 1.8. A mapping $T : C \to C$ (a subset of M) is said to be *uniformly Lipschitzian* if there exists a nonnegative number k such that

$$d(T^n(x), T^n(y)) \leq k\, d(x,y), \quad \text{for all } x,y \in C, \text{ and } n \geq 1.$$

The smallest such constant k will be denoted by $\lambda(T)$.

It is well known that if a mapping is uniformly Lipschitzian, then one may find an equivalent distance for which the mapping is nonexpansive (see [43, pp. 34–38]). Indeed, let $T : C \to C$ be uniformly Lipschitzian. Denote

$$\rho(x,y) = \sup\{d(T^n x, T^n y) : n = 0,1,2,\ldots\}, \quad \text{for all } x,y \in C.$$

Then, ρ is an equivalent distance to the distance d and relative to which T is nonexpansive. In this context, it is natural to ask the question: if a set C has the fixed point property (fpp) for nonexpansive mappings with respect to the metric d, then does C also have (fpp) for mappings which are nonexpansive relative to an equivalent metric? This is known as the stability of (fpp). The first result in this direction is due to Goebel and Kirk [41]. Motivated by such questions, many investigated the fixed point property of uniformly Lipschitzian mappings in uniformly convex Banach spaces and hyperbolic metric spaces.

Recall that the normal structure coefficient $N(M)$ of the hyperbolic metric space M is defined (see [26]):

$$N(M) = \inf\left\{\frac{\operatorname{diam}(C)}{R(C)};\ C \text{ bounded convex subset of } M \text{ with } \operatorname{diam}(C) > 0\right\},$$

where $\operatorname{diam}(C) = \sup\{d(x,y);\ x,y \in C\}$ is the diameter of C, and $R(C) = \inf\{\sup_{y\in C} d(x,y);\ x \in C\}$ is the *Chebyshev radius* of C.

For further development, we will need the following technical lemma [56]:

Lemma 1.4. *Let (M,d) be hyperbolic metric space and let C be a nonempty, closed and convex subset of M. Assume that M is 2-uniformly convex. Let $\{x_n\}$ be a bounded sequence in C. Then, there exists a unique point $z \in C$ such that*

$$\limsup_{n\to\infty} d^2(x_n,z) + 2c_M\, d^2(z,x) \le \limsup_{n\to\infty} d^2(x_n,x), \quad \textit{for any } x \in C.$$

The next result is the nonlinear version of Theorem 3 of [80].

Theorem 1.21 ([56]). *Let (M,d) be hyperbolic metric space which is 2-uniformly convex. Let C be a nonempty, closed, convex and bounded subset of M. Let $T : C \to C$ be uniformly Lipschitzian with*

$$\lambda(T) < \left(\frac{1+\sqrt{1+8c_M N(M)^2}}{2}\right)^{1/2}.$$

Then, T has a fixed point in C.

1.10 The Convexity Structures

In 1977, Penot [69] was successful in giving an abstract formulation of Kirk's theorem via the concept of *Convexity Structures*.

Definition 1.9. Let M be an abstract set. A family Σ of subsets of M is called a *convexity structure* if

(a) the empty set $\emptyset \in \Sigma$;
(b) $M \in \Sigma$;
(c) Σ is closed under arbitrary intersections.

The convex subsets of M are the elements of Σ. If M is a metric space, we will always assume that closed balls are convex. The smallest convexity structure which contains the closed balls is $\mathscr{A}(M)$ the family of admissible subsets of M. Recall that A is an admissible subset of M if it is an intersection of closed balls. Since Kirk's theorem involves some kind of compactness and the normal structure property, it was of no surprise that the generalized attempts did define these two concepts in metric spaces. Takahashi [77] in his attempt considered compact metric spaces, which was very restrictive. Penot [69], on the other hand, defined compactness for convexity structures which leads to weak compactness in the linear case. Indeed, a convexity structure Σ is said to be compact if and only if every family of subsets of Σ which has the finite intersection property has a nonempty intersection, that is, if $(A_i)_{i\in I}$, with $A_i \in \Sigma$, then

$$\bigcap_{i\in I} A_i \neq \emptyset$$

provided $\bigcap_{i\in I_f} A_i \neq \emptyset$ for any finite subset I_f of I. Since the normal structure property is a metric notion, then it was not that difficult to extend it to convexity structures. Indeed, the convexity structure Σ is said to be normal if and only if for any nonempty and bounded $A \in \Sigma$ not reduced to one point, there exists $a \in A$ such that

$$\sup\{d(a,x);\ x \in A\} < \sup\{d(x,y);\ x,y \in A\} = diam(A).$$

Penot's formulation becomes

Theorem 1.22. *Let (M,d) be a nonempty bounded metric space which possesses a convexity structure which is compact and normal. Then, every nonexpansive mapping $T : M \to M$ has a fixed point.*

In the general Kirk's theorem, the Banach space is supposed to have normal structure which means that the family of all convex sets is normal. But this family is a large one and contains the admissible sets. For example, the Banach space l^∞ is a wonderful example which illustrates the power behind Penot's formulation. Indeed, l^∞ fails to have the normal structure property but $\mathscr{A}(l^\infty)$ is compact and normal which implies the following theorem discovered separately by Sine [75] and Soardi [76] in 1979.

Theorem 1.23. *Let A be a nonempty admissible subset of l^∞. Then, every nonexpansive mapping $T : A \to A$ has a fixed point.*

Remark 1.2. In the original proof of Kirk's theorem, the weak compactness is used to prove the existence of a minimal invariant set via Zorn's lemma. Gillespie and Williams [39] showed that a constructive proof may be found which uses only countable compactness. In other words, the convexity structure Σ is assumed to satisfy a countable intersection property, that is, for any $(A_n)_{1 \leq n}$, with $A_n \in \Sigma$, then

$$\bigcap_{n=1}^{\infty} A_n \neq \emptyset,$$

provided $\bigcap_{n=1}^{m} A_n \neq \emptyset$ for any $m \geq 1$. This weakening is very important since in many practical cases, we do not have a compactness generated by a topology but a compactness defined sequentially. The latest usually generates some kind of countable compactness. Note also that if the convexity structure Σ is uniformly normal, then it is countably compact [4, 53]. This is an amazing metric translation of a well-known similar result in Banach spaces due to Maluta [65]. It is natural to ask whether a convexity structure which is countably compact is basically compact. The answer is yes if we are dealing with uniform normal structure [61]. In fact, if $\mathscr{A}(M)$ is countably compact and normal, then $\mathscr{A}(M)$ is compact.

The special role played by l^∞ above is not unique. In fact it is a special case of a more general theory which we will introduce in the next section.

1.10.1 Hyperconvex Metric Spaces

The notion of hyperconvexity is due to Aronszajn and Panitchpakdi [3] (see also [37]) who discovered it when investigating an extension of Hahn–Banach theorem to metric spaces. The corresponding linear theory is well developed and associated with the names of Gleason, Goodner, Kelley, and Nachbin (see [51, 62]). The nonlinear theory is still developing. The recent interest into these spaces goes back to the results of Sine and Soardi [75, 76] who proved independently that the fixed point property for nonexpansive mappings holds in bounded hyperconvex spaces.

Recall the Hahn–Banach theorem:

Theorem 1.24 (Hahn–Banach Theorem). *Let X be a real vector space, Y be a linear subspace of X, and ρ a seminorm on X. Let f be a linear functional defined on Y such that $f(y) \leq \rho(y)$, for all $y \in Y$. Then there exists a linear functional g defined on X, which is an extension of f (that is, $g(y) = f(y)$, for all $y \in Y$), which satisfies $g(x) \leq \rho(x)$, for all $x \in X$.*

The proof is based on the following well-known fundamental property of the real line $\mathbb{R}$:

> *"If $\{I_\alpha\}_{\alpha \in \Gamma}$ is a collection of intervals such that $I_\alpha \cap I_\beta \neq \emptyset$, for any $\alpha, \beta \in \Gamma$, then we have $\bigcap_{\alpha \in \Gamma} I_\alpha \neq \emptyset$."*

It is this property that is at the heart of the new concept discovered by Aronszajn and Panitchpakdi [3]. Note that an interval may also be seen on the real line as a closed ball. So the above intersection property may also be seen as a ball intersection property. Recall that M is metrically convex (in the sense of Menger) if

$$B(x,\alpha)\cap B(y,\beta)\neq\emptyset \text{ if and only if } d(x,y)\leq\alpha+\beta,$$

for any points $x,y\in M$ and positive numbers α and β.

Therefore, the Hahn–Banach extension theorem is closely related to an intersection property of the closed balls combined with some kind of metric convexity. Hyperconvexity captures these ideas.

Definition 1.10. The metric space M is said to be *hyperconvex* if for any collection of points $\{x_\alpha\}_{\alpha\in\Gamma}$ in M and positive numbers $\{r_\alpha\}_{\alpha\in\Gamma}$ such that $d(x_\alpha,x_\beta)\leq r_\alpha+r_\beta$ for any α and β in Γ, we must have

$$\bigcap_{\alpha\in\Gamma} B(x_\alpha,r_\alpha)\neq\emptyset\,.$$

Clearly the real line $\mathbb{R}$ is hyperconvex. In fact, we can easily prove that the infinite dimensional Banach space l_∞ is also hyperconvex, while the Hilbert space l_2 is not hyperconvex. In the next result, we see how one can generate hyperconvex metric spaces from known ones.

Theorem 1.25. *Let $(M_\alpha,d_\alpha)_{\alpha\in\Gamma}$ be a collection of hyperconvex metric spaces. Consider the product space $\mathscr{M}=\prod\limits_{\alpha\in\Gamma} M_\alpha$. Fix $a=(a_\alpha)\in\mathscr{M}$ and consider the subset M of $\mathscr{M}$ defined by*

$$M=\left\{(x_\alpha)\in\mathscr{M};\ \sup_{\alpha\in\Gamma} d_\alpha(x_\alpha,a_\alpha)<\infty\right\}.$$

Then, (M,d_∞) is a hyperconvex metric space where d_∞ is defined by

$$d_\infty\Big((x_\alpha),(y_\alpha)\Big)=\sup_{\alpha\in\Gamma} d_\alpha(x_\alpha,y_\alpha),\quad \textit{for any } (x_\alpha),(y_\alpha)\in M.$$

As we have seen earlier, it is clear that hyperconvex metric spaces are complete. We have the following:

Theorem 1.26. *Suppose A is a bounded subset of a hyperconvex metric space M. Set $cov(A)=\bigcap\{B : B \text{ is a ball and } B\supseteq A\}$. Then:*

(a) $cov(A)=\bigcap\{B(x,r_x(A)) : x\in M\}$.
(b) $r_x(cov(A))=r_x(A)$, for any $x\in M$.
(c) $R(cov(A))=R(A)=\dfrac{1}{2}diam(A)$.
(d) $diam(cov(A))=diam(A)$.

Let A be an admissible subset of M, i.e. $A \in \mathscr{A}(M)$. We have

$$C(A) = \bigcap_{a \in A} B\Big(a, R(A)\Big) \bigcap A \in \mathscr{A}(M).$$

Moreover, $\mathrm{diam}(C(A)) \leq \mathrm{diam}(A)/2$. So we have $A = C(A)$ if and only if $A \in \mathscr{A}(M)$ and $\mathrm{diam}(A) = 0$, that is, A is reduced to one point.

Next we discuss another fundamental property satisfied by hyperconvex metric spaces discovered by Aronszajn and Panitchpakdi. First recall that a metric space M is said to be *injective* if it has the following extension property: *whenever Y is a subspace of X and $f : Y \to M$ is nonexpansive, then f has a nonexpansive extension $\tilde{f} : X \to M$. This fact has several nice consequences.*

Theorem 1.27. *Let H be a metric space. The following statements are equivalent:*

(a) H is hyperconvex;
(b) H is injective.

This fundamental result may be stated in terms of retractions as follows:

Theorem 1.28. *Let H be a metric space. The following statements are equivalent:*

(a) H is hyperconvex;
(b) for every metric space M which contains H metrically, there exists a nonexpansive retraction $R : M \to H$;
(c) for any point ω not in H, there exists a nonexpansive retraction $R : H \cup \{\omega\} \to H$.

Remark 1.3. Note that statement (b) is also known as an absolute retract property. This is why hyperconvex metric spaces are also called absolute nonexpansive retract (or in short ANR).

Using the statement (c), Khamsi [54] introduced a new concept called 1-*local retract*.

Definition 1.11. Let M be a metric space. A subset $N \subset M$ is called a *1-local retract of M* if for any point $x \in M \setminus N$, there exists a nonexpansive retraction $R : M \cup \{x\} \to N$.

If we take in this definition the set M to be any metric space which contains N metrically, the 1-local retract property becomes *absolute 1-local retract* property. Note that absolute 1-local retracts are absolute nonexpansive retract, i.e. hyperconvex. Also it is easy to check that a nonexpansive retract of a hyperconvex metric space is also hyperconvex.

The most beautiful result proved in hyperconvex metric spaces is may be the one discovered by Baillon [6].

Theorem 1.29. *Let M be a bounded metric space. Let $(H_\beta)_{\beta\in\Gamma}$ be a decreasing family of nonempty hyperconvex subsets of M, where Γ is totally ordered. Then, $\bigcap_{\beta\in\Gamma} H_\beta$ is not empty and is hyperconvex.*

The proof of this theorem is highly technical. It is still open whether a simpler proof exists. One of the implications of Baillon's Theorem is the existence of hyperconvex closures. Indeed, let M be a metric space and consider the family $\mathscr{H}(M) = \{H; H \text{ is hyperconvex and } M \subset H\}$. In view of what we said previously, the family $\mathscr{H}(M)$ is not empty. Using Baillon's result, any descending chain of elements of $\mathscr{H}(M)$ has a nonempty intersection. Therefore one may use Zorn's lemma which will insure us of the existence of minimal elements. These minimal hyperconvex sets are called *hyperconvex hulls*. Isbell [47] was among the first to investigate the properties of the hyperconvex hulls. In fact he was the first one to give a concrete construction of a hyperconvex hull.

It is clear that hyperconvex hulls are not unique. But they do enjoy some kind of uniqueness. Indeed, we have:

Proposition 1.1. *Let M be a metric space. Assume that H_1 and H_2 are two hyperconvex hulls of M. Then, H_1 and H_2 are isometric.*

Let us finish this section by discussion the fixed point property in hyperconvex metric spaces. Indeed both Sine and Soardi showed that nonexpansive mappings defined on a bounded hyperconvex metric space have fixed points. Their results were stated in different context but the underlying spaces are simply hyperconvex spaces.

Theorem 1.30. *Let H be a bounded hyperconvex metric space. Any nonexpansive map $T : H \to H$ has a fixed point. Moreover, the fixed point set of T, $Fix(T)$, is hyperconvex.*

Using Baillon's theorem, we get the following:

Theorem 1.31. *Let H be a bounded hyperconvex metric space. Any commuting family of nonexpansive maps $\{T_i\}_{i\in I}$, with $T_i : H \to H$, has a common fixed point. Moreover, the common fixed point set $\bigcap_{i\in I} Fix(T_i)$ is hyperconvex.*

Remark 1.4. Baillon asked whether boundedness may be relaxed. He precisely asked whether the conclusion holds if the nonexpansive map has a bounded orbit. In the classical Kirk's fixed point theorem, having a bounded orbit implies the existence of a fixed point. Prus [57] answered this question in the negative. Indeed, consider the hyperconvex Banach space $H = l_\infty$ and the map $T : H \to H$ defined by

$$T\Big((x_n)\Big) = (1 + n\lim_{\mathscr{U}} x_n, x_1, x_2, \ldots),$$

where $\mathscr{U}$ is a nontrivial ultrafilter on the set of positive integers. We may also take a Banach limit instead of a limit over an ultrafilter. The map T is an isometry and has no fixed point. On the other hand, we have

$$T^n(0) = (1,1,\ldots,1,0,0,\ldots),$$

where the first block of length n has all its entries equal to 1 and 0 after that. So T has bounded orbits.

References

1. Aksoy, A.G., Khamsi, M.A.: Nonstandard Methods in Fixed Point Theory. Springer, New York, Berlin (1990)
2. Alspach, D.E.: A fixed point free nonexpansive map. Proc. Am. Math. Soc. **82**, 423–424 (1981)
3. Aronszajn N., Panitchpakdi, P.: Extensions of uniformly continuous transformations and hyperconvex metric spaces. Pac. J. Math. **6**, 405–439 (1956)
4. Bae, J.S., Park, M.S.: Fixed point theorems in metric Spaces with uniform normal structure. J. Kor. Math. Soc. **30**, 51–62 (1993)
5. Baillon, J.B.: Quelques aspects de la théorie des points fixes dans les espaces de Banach I. *Séminaire d' Analyse Fonctionnelle de l'École Polytechnique VII* (1978–1979). École Polytechnique, Palaiseau (1979) (in French)
6. Baillon, J.B.: Nonexpansive mapping and hyperconvex spaces. *Fixed Point Theory and its Applications.* In: Brown, R.F. (ed.) Contemporary Math., vol. 72, pp. 11–19. American Mathematical Society, Providence (1988)
7. Baillon, J.B., Schoneberg, R.: Asymptotic normal structure and fixed points of nonexpansive mappings. Proc. Am. Math. Soc. **81**, 257–264 (1981)
8. Banach, S.: Sur les opérations dans les ensembles abstraits et leurs applications. Fund. Math. **3**, 133–181 (1922)
9. Beauzamy, B.: Introduction to Banach Spaces and Their Geometry. North-Holland, Amsterdam (1985)
10. Beg, I.: Inequalities in metric spaces with applications. Top. Meth. Nonlinear Anal. **17**, 183–190 (2001)
11. Belluce, L.P., Kirk, W.A., Steiner, E.F.: Normal structure in Banach spaces. Pac. J. Math. **26**, 433–440 (1968)
12. Berggren, J.L.: Episodes in the Mathematics of Medieval Islam. Springer, New York (1986)
13. Bessaga, C.: On the converse of the Banach fixed-point principle. Colloq. Math. **7**, 41–43 (1959)
14. Bielecki, A.: Une remarque sur application de la méthode de Banach-Cacciopoli-Tikhonov dans la théorie de equation $s = f(x,y,z,p,q)$. Bull. Acad. Polon. Sci. Ser. Sci. Math. Astronom. Phys. **4**, 265–268 (1956)
15. Blumenthal, L.M.: Distance Geometries: A Study of the Development of Abstract Metrics. University of Missouri Studies, vol. 13, 1938
16. Blumenthal, L.M.: Theory and Applications of Distance Geometry. The Clarendon Press, Oxford (1953)
17. Blumenthal, L.M., Menger, K.: Studies in Geometry. W. H. Freeman, San Francisco (1970)
18. Borwein, J.M., Sims, B.: Non-expansive mappings on Banach lattices and related topics. Houston J. Math. **10**, 339–356 (1984)
19. Bridson M., Haefliger, A.: Metric Spaces of Non-positive Curvature. Springer, Berlin, Heidelberg, New York (1999)
20. Brodskii, M.S., Milman, D.P.: On the center of a convex set. Dokl. Akad. Nauk SSSR **59**, 837–840 (1948) (Russian)

21. Brouwer, L.E.J.: Über Abbildungen von Mannigfaltigkeiten. Math. Ann. **71**, 97–115 (1912)
22. Browder, F.E.: Nonexpansive nonlinear operators in a Banach space. Proc. Natl. Acad. Sci. USA **54**, 1041–1044 (1965)
23. Bruck, R.E.: Properties of fixed-point sets of nonexpansive mappings in Banach spaces. Trans. Am. Math. Soc. **179**, 251–262 (1973)
24. Bruhat, F., Tits, J.: Groupes réductifs sur un corps local. I. Données radicielles valuées. Inst. Hautes Études Sci. Publ. Math. **41**, 5–251 (1972)
25. Busemann, H.: Spaces with non-positive curvature. Acta. Math. **80**, 259–310 (1948)
26. Bynum, W.L.: Normal structure coefficients for nnormal structure for Banach spaces. Pac. J. Math. **86**, 427–436 (1980)
27. Caristi, J.: The Fixed Point Theory for Mappings Satisfying Inwardness Conditions. Ph.D. Dissertation, University of Iowa (May 1975)
28. Caristi, J.: Fixed point theorems for mappings satisfying inwardness conditions. Trans. Am. Math. Soc. **215**, 241–251 (1976)
29. Clarkson, J.A.: Uniformly convex spaces. Trans. Am. Math. Soc. **40**, 396–414 (1936)
30. Day, M.M., James, R.C., Swaminathan, S.: Normed linear spaces that are uniformly convex in every direction. Canad. J. Math. **23**, 1051–1059 (1971)
31. Diestel, J.: Geometry of Banach spaces: Selected Topics. Springer Lecture Notes in Math, vol. 485, Springer, Berlin, New York (1975)
32. Dugundji, J., Granas, A.: Fixed Point Theory. Polska Akademia Nauk, Instytut Matematyczny, PWN-Polish Scientific Publ., Warszawa (1982)
33. van Dulst, D.: Reflexive and Super-reflexive Banach Spaces. Mathematical Centre Tracts, vol. 102. Mathematisch Centrum, Amsterdam (1978)
34. Edelstein, M.: The construction of an asymptotic center with a fixed-point property. Bull. Am. Math. Soc. **78**, 206–208 (1972)
35. Ekeland, I.: On the variational principle. J. Math. Anal. Appl. **47**, 324–353 (1974)
36. Elton, J., Lin, P.K., Odell, E., Szarek, S.: Remarks on the fixed point problem for nonexpansive maps. In: Sine, R. (ed.) Fixed Points and Nonexpansive Maps, Contemporary Math., vol. 18, pp. 87–120. American Mathematical Society, Princeton (1983)
37. Espínola, R., Khamsi, M.A.: Introduction to hyperconvex spaces. In: Kirk, W.A., Sims, B. (eds.) Handbook of Metric Fixed Point Theory, pp. 391–435. Kluwer Academic, Dordrecht (2001)
38. Garcia-Falset, J.: The fixed point property in Banach spaces whose characteristic of uniform convexity is less than 2. J. Aust. Math. Soc. **54**, 169–173 (1993)
39. Gillespie, A., Williams, B.: Fixed point theorem for nonexpansive mappings on Banach spaces. Appl. Anal. **9**, 121–124 (1979)
40. Goebel, K.: On the structure of minimal invariant sets for nonexpansive mappings. Ann. Univ. Mariae Curie-Skłodowski **29**, 73–77 (1975)
41. Goebel, K., Kirk, W.A.: A fixed point theorem for transformations whose iterates have uniform Lipschitz constant. Stud. Math. **47**, 135–140 (1973)
42. Goebel, K., Kirk, W.A.: Topics in Metric Fixed Point Theory. Cambridge University Press, Cambridge (1990)
43. Goebel, K., Reich, S.: Uniform Convexity, Hyperbolic Geometry, and Nonexpansive Mappings. Series of Monographs and Textbooks in Pure and Applied Mathematics, vol. 83. Marcel Dekker, New York (1984)
44. Goebel, K., Sekowski, T., Stachura, A.: Uniform convexity of the hyperbolic metric and fixed points of holomorphic mappings in the Hilbert ball. Nonlinear Anal. **4**, 1011–1021 (1980)
45. Göhde, D.: Zum Prinzip der kontraktiven Abbildung. Math. Nachr. **30**, 251–258 (1965)
46. Heinrich, S.: Ultraproducts in Banach space theory. J. Reine Angew. Math. **313**, 72–104 (1980)
47. Isbell, J.R.: Six theorems about injective metric spaces. Comment. Math. Helv. **39**, 439–447 (1964)
48. James, R.C.: Uniformly non-square Banach spaces. Ann. Math. **80**, 542–550 (1964)
49. James, R.C.: Super-reflexive Banach spaces. Canad. J. Math. **24**, 896–904 (1972)

50. Karlovitz, L.: Existence of fixed points for nonexpansive mappings in spaces without normal structure. Pac. J. Math. **66**, 153–156 (1976)
51. Kelley, J.L.: Banach spaces with the extension property. Trans. Am. Math. Soc. **72**, 323–326 (1952)
52. Khamsi, M.A.: Uniform smoothness implies super-normal structure property. Nonlinear Anal. **19**, 1063–1069 (1992)
53. Khamsi, M.A.: On metric spaces with uniform normal structure. Proc. Am. Math. Soc. **106**, 723–726 (1989)
54. Khamsi, M.A.: One-local retract and common fixed point for commuting mappings in metric spaces. Nonlinear Anal. **27**, 1307–1313 (1996)
55. Khamsi, M.A., Kirk, W.A.: An Introduction to Metric Spaces and Fixed Point Theory. Wiley, New York (2001)
56. Khamsi, M.A., Khan, A.R.: Inequalities in metric spaces with applications. Nonlinear Anal. **74**, 4036–4045 (2011)
57. Khamsi, M.A., Lin, M., Sine, R.: On the fixed points of commuting nonexpansive maps in hyperconvex spaces. J. Math. Anal. Appl. **168**, 372–380 (1992)
58. Kirk, W.A.: A fixed point theorem for mappings which do not increase distances. Am. Math. Mon. **72**, 1004–1006 (1965)
59. Kirk, W.A.: Fixed point theory for nonexpansive mappings, I and II. Lecture Notes in Mathematics, vol. 886, pp. 485–505. Springer, Berlin (1981)
60. Kirk, W.A.: A fixed point theorem in CAT(0) spaces and R-trees. Fixed Point Theory Appl. **2004**, 309–316 (2004)
61. Kulesza, J., Lim, T.C.: On weak compactness and countable weak compactness in fixed point theory. Proc. Am. Math. Soc. **124**, 3345–3349 (1996)
62. Lacey, H.E.: The Isometric Theory of Classical Banach Spaces. Springer, Berlin, Heidelberg, New York (1974)
63. Leustean, L.: A quadratic rate of asymptotic regularity for CAT(0)-spaces. J. Math. Ana. Appl. **325**, 386–399 (2007)
64. Lin, P.K.: Unconditional bases and fixed points of nonexpansive mappings. Houston J. Math. **116**, 69–76 (1985)
65. Maluta, E.: Uniformly normal structure and related coefficients. Pac. J. Math. **111**, 357–369 (1984)
66. Maurey, B.: Points fixes des contractions sur un convexe fermé de L_1. Seminaire d'Analyse Fonctionelle, vol. 80–81. Ecole Polytechnique, Palaiseau (1981)
67. Menger, K.: Untersuchungen über allgemeine Metrik. Math. Ann. **100**, 75–163 (1928)
68. Oettli, W., Théra, M.: Equivalents of Ekeland's principle. Bull. Aust. Math. Soc. **48**, 385–392 (1993)
69. Penot, J.P.: Fixed point theorem without convexity. Bull. Soc. Math. France Mem. **60**, 129–152 (1979)
70. Prieß-Crampe, S.: Der Banachsche Fixpunkstaz für ultrametrische Raüme. Results Math. **18**, 178–186 (1990)
71. Prieß-Crampe, S.: Some results of functional analysis for ultrametric spaces and valued vector spaces. Geometriae Dedicata **58**, 79–90 (1995)
72. Prieß-Crampe, S., Ribenboimb, P.: Ultrametric spaces and logic programming. J. Logic Program **42**, 59–70 (2000)
73. Prus, S.: Nearly uniformly smooth Banach spaces. Boll. Un. Mat. Ital. **7**, 507–521 (1989)
74. Sims, B.: Ultra-techniques in Banach Space Theory. Queen's Papers in Pure and Applied Mathematics, vol. 160. Queen's University, Kingston, Ontario (1982)
75. Sine, R.: On nonlinear contractions in sup-norm spaces. Nonlinear Anal. **3**, 885–890 (1979)
76. Soardi, P.: Existence of fixed points of nonexpansive mappings in certain Banach lattices. Proc. Am. Math. Soc. **73**, 25–29 (1979)
77. Takahashi, W.: A convexity in metric space and nonexpansive mappings. I. Kodai Math. Sem. Rep. **22**, 142–149 (1970)

78. Tarski, A.: A lattice theoretical fixpoint theorem and its applications. Pac. J. Math. **5**, 285–309 (1955)
79. Turett, B.: A dual view of a theorem of Baillon. In: Singh, S.P., Burry, J.H. (eds.) Nonlinear Analysis and Applications. Lecture Notes in Pure and Applied Mathematics, vol. 80, pp. 279–286. Marcel Dekker, New York (1982)
80. Xu, H.K.: Inequalities in Banach spaces with applications. Nonlinear Anal. **16**, 1127–1138 (1991)
81. Zeidler, E.: Nonlinear Functional Analysis and its Applications I: Fixed-Point Theorems. Springer, New York, Berlin, Heidelberg, Tokyo (1986)
82. Zizler, V.: On some rotundity and smoothness properties of Banach spaces. Dissertationes Math. (Rozprawy Mat.) **87**, 415–440 (1971)

Chapter 2
Banach Contraction Principle and Its Generalizations

Abdul Latif

2.1 Introduction

In 1922, the Polish mathematician Stefan Banach established a remarkable fixed point theorem known as the "Banach Contraction Principle" (BCP) which is one of the most important results of analysis and considered as the main source of metric fixed point theory. It is the most widely applied fixed point result in many branches of mathematics because it requires the structure of complete metric space with contractive condition on the map which is easy to test in this setting. The BCP has been generalized in many different directions. In fact, there is vast amount of literature dealing with extensions/generalizations of this remarkable theorem. In this chapter, it is impossible to cover all the known extensions / generalizations of the BCP. However, an attempt is made to present some extensions of the BCP in which the conclusion is obtained under mild modified conditions and which play important role in the development of metric fixed point theory.

2.2 Contractions: Definition and Examples

Throughout this paper, we denote by $\mathbb{N}$ the set of all positive integers and by $\mathbb{R}$ the set of all real numbers.

Definition 2.1. Let (X,d) be a metric space and let $f : X \to X$ be a mapping.

(a) A point $x \in X$ is called a *fixed point* of f if $x = f(x)$.
(b) f is called *contraction* if there exists a fixed constant $h < 1$ such that

A. Latif (✉)
Department of Mathematics, King Abdulaziz University, Jeddah, 21589, Saudi Arabia
e-mail: alatif@kau.edu.sa

S. Almezel et al. (eds.), *Topics in Fixed Point Theory*, DOI 10.1007/978-3-319-01586-6_2,

$$d(f(x), f(y)) \leq hd(x,y), \quad \text{for all } x, y \in X. \tag{2.1}$$

A contraction mapping is also known as Banach contraction. If we replace the inequality (2.1) with strict inequality and $h = 1$, then f is called *contractive* (or *strict contractive*). If (2.1) holds for $h = 1$, then f is called *nonexpansive*; and if (2.1) holds for fixed $h < \infty$, then f is called *Lipschitz continuous* . Clearly, for the mapping f, the following obvious implications hold:

contraction $\Rightarrow$ contractive $\Rightarrow$ nonexpansive $\Rightarrow$ Lipschitz continuous

Example 2.1. (a) Consider the usual metric space $(\mathbb{R}, d)$. Define

$$f(x) = \frac{x}{a} + b, \quad \text{for all } x \in \mathbb{R}.$$

Then, f is contraction on $\mathbb{R}$ if $a > 1$ and the solution of the equation $x - f(x) = 0$ is $x = \frac{ab}{a-1}$.

(b) Consider the Euclidean metric space $(\mathbb{R}^2, d)$. Define

$$f(x,y) = \left(\frac{x}{a} + b, \frac{y}{c} + b\right), \quad \text{for all } (x,y) \in \mathbb{R}^2.$$

Then, f is contraction on $\mathbb{R}^2$ if $a, c > 1$. Now, solving the equation $f(x,y) = (x,y)$ for a fixed point, we get $x = \frac{ab}{a-1}$ and $y = \frac{cb}{c-1}$.

Using induction, one can easily get the following concerning iterates of a contraction mapping.

If f is a contraction mapping on a metric space (X, d) with contraction constant h, then f^n (where the superscript represents the nth iterate of f) is also a contraction on X with constant h^n.

Consider then the situation in which $f : X \to X$ is not necessarily a contraction mapping, but f^n is a contraction for some n.

Example 2.2. (a) Let $f : [0,2] \to [0,2]$ be defined by

$$f(x) = \begin{cases} 0, & x \in [0,1], \\ 1, & x \in (1,2]. \end{cases}$$

Then, $f^2(x) = 0$ for all $x \in [0,2]$, and so, f^2 is a contraction on $[0,2]$. Note that f is not continuous and thus not a contraction map.

(b) Define $f(x) = \cos x$ on $\mathbb{R}$. Then, f is not a contraction on $\mathbb{R}$. Indeed, suppose there exists $h \in (0,1)$ such that

$$\left|\frac{\cos x - \cos y}{x - y}\right| \leq h, \quad \text{for all } x \neq y.$$

Letting $y \to x$, we get $|\sin x| \leq h$ for all x, which is false.

Note that the iterated function $f^2(x) = \cos(\cos x)$ satisfies

$$\left|\frac{d}{dx}(\cos(\cos x))\right| = |\sin(\cos x)\sin(x)| < \sin 1 < 1,$$

and thus, by the mean-value theorem, f^2 is a contraction on $\mathbb{R}$.

(c) Define $f(x) = \exp(-x)$ on $\mathbb{R}$. Then, f is not a contraction on $\mathbb{R}$. But, $f^2(x) = \exp(-\exp(-x))$ is a contraction on $\mathbb{R}$. Indeed,

$$\left|\frac{d}{dx}(\exp(-\exp(-x)))\right| = |\exp(-x-\exp(-x))| \leq e^{-1} < 1,$$

and thus, the conclusion follows by the mean-value theorem.

For non-contractions, there are examples where f has a unique fixed point but an iterate of f does not.

Example 2.3. Define

$$f(x) = 1 - x, \quad \text{for all } x \in \mathbb{R}.$$

Then, f is not a contraction, has a unique fixed point, but note that

$$f^2(x) = x, \quad \text{for all } x \in \mathbb{R}$$

is rich with fixed points.

2.3 The Banach Contraction Principle with Some Applications

In this section, we will discuss the most basic fixed point theorem in analysis, known as the Banach Contraction Principle (BCP). It is due to Banach [11] and appeared in his Ph.D. thesis (1920, published in 1922). The BCP was first stated and proved by Banach for the contraction maps in the setting of complete normed linear spaces. At about the same time the concept of an abstract metric space was introduced by Hausdorff, which then provided the general framework for the principle for contraction mappings in a complete metric space. The BCP can be applied to mappings which are differentiable, or more generally, Lipschitz continuous. A number of articles with applications on the topic can also be found in [1, 7, 8, 29, 34, 46, 103].

Theorem 2.1 (Banach Contraction Principle). *Let (X,d) be a complete metric space, then each contraction map $f : X \to X$ has a unique fixed point.*

Proof. Let h be a contraction constant of the mapping f. We will explicitly construct a sequence converging to the fixed point. Let x_0 be an arbitrary but fixed element in X. Define a sequence of iterates $\{x_n\}$ in X by

$$x_n = f(x_{n-1}) \quad (= f^n(x_0)), \quad \text{for all } n \geq 1. \tag{2.2}$$

Since f is a contraction, we have

$$d(x_n, x_{n+1}) = d(f(x_{n-1}), f(x_n)) \leq hd(x_{n-1}, x_n), \quad \text{for any } n \geq 1.$$

Thus, we obtain

$$d(x_n, x_{n+1}) \leq h^n d(x_0, x_1), \quad \text{for all } n \geq 1.$$

Hence, for any $m > n$, we have

$$d(x_n, x_m) \leq \left(h^n + h^{n+1} + \cdots + h^{m-1}\right) d(x_0, x_1) \leq \frac{h^n}{1-h} d(x_0, x_1).$$

We deduce that $\{x_n\}$ is Cauchy sequence in a complete space X. Let $x_n \to p \in X$. Now using the continuity of the map f, we get

$$p = \lim_{n\to\infty} x_n = \lim_{n\to\infty} f(x_{n-1}) = f(p).$$

Finally, to show f has at most one fixed point in X, let p and q be fixed points of f. Then,

$$d(p,q) = d(f(p), f(q)) \leq hd(p,q).$$

Since $h < 1$, we must have $p = q$. □

The proof of the BCP yields the following useful information about the rate of convergence towards the fixed point.

Corollary 2.1. *Let f be a contraction mapping on a complete metric space (X,d) with contraction constant h and fixed point p. For any $x_0 \in X$, with f-iterates $\{x_n\}$, we have the estimates*

$$d(x_n, p) \leq \frac{h^n}{1-h} d(x_0, f(x_0)), \tag{2.3}$$

$$d(x_n, p) \leq hd(x_n, p), \tag{2.4}$$

$$d(x_n, p) \leq \frac{h}{1-h} d(x_{n-1}, x_n). \tag{2.5}$$

In fact, the three inequalities in Corollary 2.1 serve different purposes. The inequality (2.3) tells us, in terms of the distance between x_0 and $f(x_0) = x_1$, how many times we need to iterate f starting from x_0 to be certain that we are within a specified distance from the fixed point. This is an upper bound on how long we need to compute. It is called an a priori estimate. Inequality (2.4) shows that once we find a term by iteration within some desired distance of the fixed point, all further iterates will be within that distance. However, (2.4) is not so useful as an error estimate since both sides of (2.4) involve the unknown fixed point. The inequality (2.5) tells us, after each computation, how much closer we are to the fixed point in terms of the previous two iterations. This kind of estimate, called an a posteriori estimate, is very important because if two successive iterations are nearly equal, (2.5) guarantees that we are very close to the fixed point.

Corollary 2.2. *Let $f : X \to X$ be a contraction mapping on a complete metric space and $M \subseteq X$ be a closed subset such that $f(M) \subseteq M$. Then, the unique fixed point of f is in M.*

It may be the case that $f : X \to X$ is not a contraction on the whole space X, but rather a contraction on some neighborhood of a given point. In this case we have the following result:

Theorem 2.2. *Let (X,d) be a complete metric space and let $B_r(y) = \{x \in X : d(x,y) < r\}$, where $y \in X$ and $r > 0$. Let $f : B_r(y) \to X$ be a contraction map with contraction constant $h < 1$. Further, assume that*

$$d(y, f(y)) < r(1-h).$$

Then, f has a unique fixed point in $B_r(y)$.

Proof. Note that the uniform continuity of f allows us to extend f to a mapping defined on $\overline{B_r(y)}$ which is a contraction map having the same Lipschitz constant as the original map. We show that $f(\overline{B_r(y)}) \subseteq B_r(y)$. Let $x \in \overline{B_r(y)}$, then

$$d(y, f(x)) \leq d(y, f(y)) + d(f(y), f(x)) < r(1-h) + hr = r,$$

and hence, $f : \overline{B_r(y)} \to B_r(y)$. Since $\overline{B_r(y)}$ is a complete metric space, using Theorem 2.1, f has a unique fixed point $p \in \overline{B_r(y)}$. Thus, $p \in B_r(y)$ because $p = f(p) \in B_r(y)$. □

Remark 2.1. If f is a contraction map on a complete metric space (X,d) with contraction constant h, then f^n is also a contraction on X with constant h^n and the unique fixed point of f is also the unique fixed point of any f^n.

We observed in Example 2.2 that in some situations a function is not a contraction but its iterate is a contraction map. To get the conclusion of the BCP for the original function, the following early example of an extension of BCP is due Caccioppoli [18].

Theorem 2.3. *Let (X,d) be a complete metric space and let $f : X \to X$ be a mapping such that for each $n \geq 1$, there exists a constant c_n such that*

$$d(f^n(x), f^n(y)) \leq c_n d(x,y), \quad \textit{for all } x,y \in X,$$

where $\sum_{n=1}^{\infty} c_n < \infty$. *Then, f has a unique fixed point.*

While, Bryant [17] extended BCP as follows.

Theorem 2.4. *Let (X,d) be a complete metric space and let $f : X \to X$ be a mapping such that for some positive integer n, f^n is contraction on X. Then, f has a unique fixed point.*

Proof. By BCP, f^n has a unique fixed point, say $x \in X$ with $f^n(x) = x$. Since

$$f^{n+1}(x) = f(f^n(x)) = f(x),$$

it follows that $f(x)$ is a fixed point of f^n, and thus, by the uniqueness of x, we have $f(x) = x$, that is, f has a fixed point. Since, the fixed point of f is necessarily a fixed point of f^n, so is unique. □

Remark 2.2. (a) Theorem 2.4 has importance in the scene that the mapping f is not even assumed to be continuous while the same result was appeared in the literature with continuity assumption on the mapping f, see, for example, [15, 30].
(b) In some applications, it is the case that the mapping f is a Lipschitz which is not necessarily a contraction, whereas some power of f is a contraction mapping.

Example 2.4. Consider the metric space $X = C[a,b]$, of continuous real-valued functions defined on the compact interval $[a,b]$. This is a Banach space with respect to the sup norm. Define $f : X \to X$ by

$$f(u)(t) = \int_a^t u(s)ds.$$

Then,

$$|f(u) - f(v)| \leq (b-a)\,|u-v|.$$

Now, we compute

$$f^2(u)(t) = \int_a^t (t-s)u(s)ds,$$

and inductively,

$$f^n(u)(t) = \frac{1}{(n-1)!}\int_a^t (t-s)^{n-1}u(s)ds.$$

Thus, we get

$$|f^n(u) - f^n(v)| \leq \frac{(b-a)^n}{n!}\,|u-v|.$$

Hence, f^n is a contraction map for all values of n for which $\frac{(b-a)^n}{n!} < 1$. It follows that f has the unique fixed point $u = 0$.

Remark 2.3. It was important in the proof of BCP that the contraction constant h be strictly less than 1. That gave us control over the rate of convergence of $f^n(x_0)$ to the fixed point since $h^n \to 0$ as $n \to \infty$. If we consider f is contractive mapping instead of a contraction, then we lose that control and indeed a fixed point need not exist.

Example 2.5. Let I be a closed interval in $\mathbb{R}$ and $f : I \to I$ be differentiable with $|f'(t)| < 1$ for all t. Then, the mean-value theorem implies $|f(x) - f(y)| < |x - y|$ for all $x \neq y$ in I. The following three functions satisfy this condition, where $I = [1, \infty)$ in the first case and $I = \mathbb{R}$ in the second and third cases:

$$f(x) = x + \frac{1}{x}; \quad f(x) = \sqrt{x^2 + 1}; \quad f(x) = \log(1 + exp(x)).$$

In each case, $f(x) > x$, so none of these functions has a fixed point in I.

Despite such examples, there is a fixed point theorem of Edelstein [30] when $d(f(x), f(y)) < d(x, y)$ for all $x \neq y$ provided the space is compact, which is not the case in the previous example.

Similarly, a nonexpansive mapping on a complete metric space need not has a fixed point. For instance, consider the translation operator by a nonzero vector in a Banach space, which is clearly a nonexpansive fixed point free mapping. On the other hand, a fixed point of a nonexpansive map need not be unique. For instance, consider an identity operator, which is obviously nonexpansive and rich with fixed points. Thus the fixed point theory of nonexpansive mappings is fundamentally different from that of the contraction mapping, and thus we shall not discuss in this chapter.

Remark 2.4. It is worth to mention that on the one hand, the BCP is very forceful and simple, and it became a classical tool in nonlinear analysis. But, on the other hand, Connell [26] gave an example of a metric space X such that X is not complete and every contraction on X has a fixed point. Thus, Theorem 2.1 cannot characterize the metric completeness of X which means the notion of contractions is too strong from this point of view.

A mapping f on a metric space (X, d) is called *Kannan mapping* if there exists $h \in [0, \frac{1}{2})$ such that

$$d(f(x), f(y)) \leq h \, \{d(x, Tx) + d(y, Ty)\}, \quad \text{for all } x, y \in X.$$

Kannan [44] proved that if X is complete, then every Kannan mapping has a fixed point. We note that Kannan fixed point theorem is not an extension of the BCP. In our opinion, Kannan theorem is also very important because Subrahmanyam [85] proved that Kannan theorem characterizes the metric completeness. That is, a metric space X is complete if and only if every Kannan mapping on X has a fixed point.

2.4 Some Other Extensions of BCP for Single-Valued Mappings

There have been numerous extensions of a milder form of BCP. In this section we present some of these.

The first generalization in this direction which received a significant importance is the following result of Rakotch [76].

Theorem 2.5. *Let (X,d) be a complete metric space, and suppose that $f : X \to X$ satisfies*

$$d(f(x), f(y)) \leq \eta(d(x,y))d(x,y), \quad \textit{for all } x,y \in X,$$

where η is a decreasing function on $\mathbb{R}^+$ to $[0,1)$. Then, f has a unique fixed point.

A variant of Rokotch's theorem has been given by Geraghty [33], in which the function η satisfies the simple condition that $\eta(t_n) \to 1 \Rightarrow t_n \to 0$.

In [14], Boyd and Wong obtained a more general result as follows.

Theorem 2.6 (Boyd–Wong Theorem). *Let (X,d) be a complete metric space, and suppose that $f : X \to X$ satisfies*

$$d(f(x), f(y)) \leq \psi(d(x,y)), \quad \textit{for all } x,y \in X,$$

where $\psi : \mathbb{R} \to [0,\infty)$ is upper semicontinuous from the right (that is, for any sequence $t_n \downarrow t \geq 0 \Rightarrow \limsup_{n\to\infty} \psi(t_n) \leq \psi(t)$) and satisfies $0 \leq \psi(t) < t$ for $t > 0$. Then, f has a unique fixed point.

Proof. Let x_0 be an arbitrary but fixed element in X. Define a sequence of iterates $\{x_n\}$ in X by

$$x_n = f(x_{n-1}) \ \ (= f^n(x_0)), \quad \text{for all } n \geq 1.$$

Set $a_n = d(x_{n-1}, x_n)$. Note that the sequence $\{a_n\}$ is monotone decreasing and bounded below, thus it is convergent and we let $\lim_{n\to\infty} a_n = a$. If $a > 0$, we obtain a contradiction. Indeed,

$$a_{n+1} \leq \psi(a_n),$$

and thus, $a \leq \psi(a)$, which contradicts to the property of ψ, and therefore, $\{a_n\}$ converges to 0. Now, we show that $\{x_n\}$ is a Cauchy sequence. Suppose that this is not so. Then, there exists $\varepsilon > 0$ such that for any $k \in \mathbb{N}$, there is $m_k > n_k \geq k$ such that we have the relation

$$d_k = d(x_{m_k}, x_{n_k}) \geq \varepsilon.$$

Of course we can assume that $d(x_m, x_{m-1}) \leq \varepsilon$ and thus we have

$$d_k < d(x_{m_k}, x_{m_k-1}) + d(x_{m_k-1}, x_{n_k}) \leq a_{m_k} + \varepsilon \leq a_k + \varepsilon,$$

which implies that $d_k \to 0$. On the other hand

$$\begin{aligned} d_k &\leq d(x_{m_k}, x_{m_k+1}) + d(x_{m_k+1}, x_{n_k+1}) + d(x_{n_k+1}, x_{n_k}) \\ &\leq 2a_k + \psi(d_k). \end{aligned}$$

It follows that $\varepsilon \leq \psi(\varepsilon)$, a contradiction. Hence $\{x_n\}$ is a Cauchy sequence in X. Since X is complete, $x_n \to z \in X$. Thus by the continuity of f, we get $f(z) = z$. Uniqueness of z follows from the contractive condition. □

Remark 2.5. In fact, two key steps involved in proving the existence of fixed point in each of the above results, showing that for given $x_0 \in X$, the iterative sequence $x_n = f^n(x_0) = f(x_{n-1})$ is a Cauchy sequence in the underlying spaces and then the continuity of the mapping guarantees the required fixed point.

Remark 2.6. In [14], Boyd and Wong observed that the upper semicontinuity condition of ψ can be dropped if the space X is metrically convex. While in [62], Matkowski further extended this result by assuming that ψ is continuous at 0 and that there exists a sequence $t_n \downarrow 0$ for which $\psi(t_n) < t_n$.

The following variant result is due to Matkowski [63], where the continuity condition on ψ is replaced with another suitable assumption.

Theorem 2.7. *Let (X,d) be a complete metric space, and suppose that $f : X \to X$ satisfies*

$$d(f(x), f(y)) \leq \psi(d(x,y)), \quad \textit{for all } x, y \in X,$$

where $\psi : (0,\infty) \to (0,\infty)$ is monotone nondecreasing and satisfies $\lim_{n\to\infty} \psi^n(t) = 0$ for all $t > 0$. Then, f has a unique fixed point.

Proof. Let x_0 be an arbitrary but fixed element in X. Define a sequence of iterates $\{x_n\}$ in X by

$$x_n = f(x_{n-1}) \ \ (= f^n(x_0)), \quad \text{for all } n \geq 1.$$

Set $a_n = d(x_{n+1}, x_n)$. Note that

$$0 \leq \limsup_{n\to\infty} a_n \leq \limsup_{n\to\infty} \psi^n(d(x_1, x_0)) = 0.$$

Thus, $\lim_{n\to\infty} a_n = 0$. Now, we show that $\{x_n\}$ is a Cauchy sequence. Also note that for any $\varepsilon > 0$, $\psi(\varepsilon) < \varepsilon$. And since $\lim_{n\to\infty} a_n = 0$, so for $\varepsilon > 0$, we can choose n such that $a_n \leq \varepsilon - \psi(\varepsilon)$. Now, define

$$M = \{x \in X : d(x, x_n) \leq \varepsilon\}.$$

Then for any $y \in M$, we have

$$\begin{aligned} d(f(y), x_n) &\leq d(f(y), f(x_n)) + d(f(x_n), x_n) \\ &\leq \psi(d(y, x_n)) + d(x_{n+1}, x_n) \\ &\leq \psi(\varepsilon) + \varepsilon - \psi(\varepsilon) = \varepsilon. \end{aligned}$$

Thus $f(y) \in M$, that is; $f(M) \subseteq M$. It follows that $d(x_m, x_n) \leq \varepsilon \ \forall m \geq n$. We obtain the conclusion by following the rest of the proof as in Theorem 2.6 □

Using somewhat different approach Meir and Keeler [64] extended Theorem 2.6 as follows.

Theorem 2.8. *Let (X, d) be a complete metric space, and suppose that $f : X \to X$ satisfies the condition: for each $\varepsilon > 0$, there exists $\delta > 0$ such that for all $x, y \in X$,*

$$\varepsilon \leq d(x, y) \leq \varepsilon + \delta \Rightarrow d(f(x), f(y)) \leq \varepsilon. \tag{2.6}$$

Then, f has a unique fixed point.

Clearly, the condition (2.6) implies that the mapping f is contractive. Thus, f is continuous and has a unique fixed point if it exists. Further, the condition (2.6) implies $d(x_n, x_{n+1})$ decreasing to zero. Finally, it is easy to show that $\{x_n\}$ is a Cauchy sequence in a complete metric space X by using the contrary technique.

In [21], Ćirić has obtained the following generalization of the BCP.

Theorem 2.9. *Let (X, d) be a complete metric space, and let $f : X \to X$ be a quasi-contraction, that is, for a fixed constant $h < 1$*

$$d(f(x), f(y)) \leq h\max\{d(x, y), d(x, fx), d(y, fy), d(x, fy), d(y, fx)\}, \textit{for all } x, y \in X. \tag{2.7}$$

Then, f has a unique fixed point.

It has been observed in [79] that Theorem 2.9 is also true if we replace (2.7) with the following equivalent contractive condition.

$$d(f(x), f(y)) \leq h \max\left\{d(x, y), d(x, f(x)), d(y, f(y)), \frac{1}{2}[d(x, f(y)) + d(y, f(x))]\right\}. \tag{2.8}$$

Several such type of contractive conditions have been studied by Rhoades [79], Jachymaski [39], and Ćirić [22].

Asymptotic fixed point theory involves assumptions about the iterates of the mapping in question. In fact the concept of asymptotic contractions is suggested in Theorem 2.3 which is the earliest extension of Banach contraction principle, (also see Theorem 2.4). For further historical comments, see, for example, [51].

Let Ψ denote the class of all mappings $\psi : [0,\infty) \to [0,\infty)$ such that ψ is continuous, and $\psi(t) < t$ for all $t > 0$.

Note that if (X,d) is any complete metric space and $f : X \to X$ is any mapping satisfying

$$d(f(x), f(y)) \leq \psi(d(x,y)), \quad \text{for all } x, y \in X,$$

for any $\psi \in \Psi$, then by Theorem 2.6, f has a unique fixed point.

Using the Cantor's intersection theorem, Kirk [49] obtained the following asymptotic version of Theorem 2.6.

Theorem 2.10. *Let (X,d) be a complete metric space, and suppose that $f : X \to X$ satisfies*

$$d(f^n(x), f^n(y)) \leq \psi_n(d(x,y)), \quad \textit{for all } x, y \in X,$$

where $\psi_n : [0,\infty) \to [0,\infty)$ are continuous and $\psi_n \to \psi \in \Psi$ uniformly. Further, assume that some orbit of f is bounded. Then, f has a unique fixed point.

See also [9, 41, 52] which are dealing with asymptotic version of Theorem 2.6. In [91] Suzuki obtained a result for asymptotic contractions of final type and claims his result is the final generalization in some sense.

In [4], Alber et al. suggested a generalization of BCP in the setting of Hilbert spaces and subsequently Rhoades [80] extended and improved their result to metric spaces.

Theorem 2.11. *Let (X,d) be a complete metric space, and suppose that $f : X \to X$ satisfies the following inequality*

$$d(f(x), f(y)) \leq d(x,y) - \psi(d(x,y)), \quad \textit{for all } x, y \in X, \tag{2.9}$$

where $\psi : [0,\infty) \to [0,\infty)$ is a continuous and nondecreasing function such that $\psi(t) = 0$ if and only if $t = 0$. Then, f has unique fixed point.

Note that if one takes $\psi(t) = (1-h)t$ where $0 < h < 1$, then the inequality (2.9) reduces to the inequality (2.1).

Remark 2.7. In the literature, a map $f : X \to X$ with inequality (2.9) is known as weakly contractive map. The function ψ involved in Theorem 2.11 known as alternating distance (also called control function), which was initially used in metric fixed point theory by Khan et al. [47]. This function and its generalizations have been used in fixed point problems in metric and probabilistic metric spaces, see, for example, [65, 67, 81, 82].

In [28], the following generalization has been appeared.

Theorem 2.12. *Let* (X,d) *be a complete metric space, and suppose that* $f : X \to X$ *satisfies the following inequality*

$$\phi(d(f(x), f(y))) \leq \phi(d(x,y)) - \psi(d(x,y)), \quad \textit{for all } x,y \in X,$$

where both the functions $\phi, \psi : [0,\infty) \to [0,\infty)$ *are continuous and nondecreasing such that* $\psi(t) = 0 = \phi(t)$ *if and only if* $t = 0$. *Then,* f *has unique fixed point.*

Recently, more general results in this direction have been appeared. For example, one of the results in [20] as follows.

Theorem 2.13. *Let* (X,d) *be a complete metric space, and suppose that* $f : X \to X$ *satisfies the following inequality*

$$\phi\left(d(f(x), f(y))\right) \leq \phi(m(x,y)) - \psi\left(\max\{d(x,y), d(y, f(y))\}\right),$$

where

$$m(x,y) = \max\left\{d(x,y), d(x, f(x)), d(y, f(y)), \frac{1}{2}\left[d(x, f(y)) + d(y, f(x))\right]\right\},$$

for all $x,y \in X$, *and* $\phi, \psi : [0,\infty) \to [0,\infty)$ *are functions such that* ϕ *is alternating distance and* ψ *is continuous with* $\psi(t) = 0$ *if and only if* $t = 0$. *Then,* f *has unique fixed point.*

A direct consequence is the following result.

Corollary 2.3. *Let* (X,d) *be a complete metric space, and suppose that* $f : X \to X$ *satisfies the following inequality for all* $x,y \in X$,

$$\begin{aligned}
&\phi(d(f^n(x), f^n(y))) \\
&\leq \phi\left(\max\left\{d(x,y), d(x, f^n(x)), d(y, f^n(y)), \frac{1}{2}\left[d(x, f^n(y)) + d(y, f^n(x))\right]\right\}\right) \\
&\quad - \psi(\max\{d(x,y), d(y, f^n(y))\}),
\end{aligned}$$

where n *is a positive integer and* $\phi, \psi : [0,\infty) \to [0,\infty)$ *are functions such that* ϕ *is alternating distance and* ψ *is continuous with* $\psi(t) = 0$ *if and only if* $t = 0$. *Then,* f *has unique fixed point.*

Example 2.6. Let $X = \{0,1,2,3,\ldots\}$. Let $d : X \times X \to \mathbb{R}$ be given as

$$d(x,y) = \begin{cases} x+y, & \text{if } x \neq y, \\ 0, & \text{if } x = y. \end{cases}$$

Then, (X,d) is a complete metric space. Define $\phi, \psi : [0,\infty) \to [0,\infty)$ by $\phi(t) = t^2$ for all t, and

$$\psi(t) = \begin{cases} \frac{t^2}{2}, & \text{if } t \leq 1, \\ \frac{1}{2}, & \text{if } t > 1. \end{cases}$$

Let $f : X \to X$ be defined by

$$f(x) = \begin{cases} x - 1, & \text{if } x \neq 0, \\ 0, & \text{if } x = 0. \end{cases}$$

Then, ϕ, ψ and f satisfy all the conditions of Theorem 2.13 and clearly 0 is the unique fixed of f.

Recently, Suzuki [88] established a new type of generalization of the BCP and characterizes the metric completeness of the underlying space.

Theorem 2.14. *Let (X,d) be a complete metric space, and suppose that $f : X \to X$. Define a nonincreasing function $\psi : [0,1) \to (1/2, 1]$ by*

$$\psi(h) = \begin{cases} 1, & \text{if } 0 \leq h \leq \frac{1}{2}(\sqrt{5}-1), \\ \frac{1-h}{h^2}, & \text{if } \frac{1}{2}(\sqrt{5}-1) < h < \frac{1}{\sqrt{2}}, \\ \frac{1}{1+h}, & \text{if } \frac{1}{\sqrt{2}} \leq h < 1. \end{cases}$$

Assume that there exists $h \in [0,1)$ such that

$$\psi(h) d(x, f(x)) \leq d(x,y) \Rightarrow d(f(x), f(y)) \leq h d(x,y), \quad \text{for all } x, y \in X.$$

Then, f has a unique fixed point.

Proof. Since $\psi(h) \leq 1$, we get $\psi(h) d(x, f(x)) \leq d(x, f(x))$ for every $x \in X$. Note that

$$d(f(x), f^2(x)) \leq h d(x, f(x)) \ \forall x \in X.$$

Let x_0 be an arbitrary but fixed element in X. Define a sequence of iterates $\{x_n\}$ in X by

$$x_n = f(x_{n-1}) \ (= f^n(x_0)), \quad \text{for all } n \geq 1.$$

Then, we have $d(x_n, x_{n+1}) \leq h^n d(x_0, f(x_0))$, and thus $\sum_{n=1}^{\infty} d(x_n, x_{n+1}) < \infty$. By the usual arguments, we get $\{x_n\}$ is a Cauchy sequence. Since X is complete, $\{x_n\}$ converges to some point $z \in X$. Then, for any $x \in X \backslash \{z\}$, there exists $n_0 \in \mathbb{N}$ such that $d(x_n, z) \leq \frac{1}{3} d(x,z)$ for all $n \in \mathbb{N}$ with $n \geq n_0$. Now, we have

$$\begin{aligned}\psi(h)d(x_n, f(x_n)) &\le d(x_n, f(x_n) = d(x_n, x_{n+1}) \\ &\le d(x_n, z) + d(x_{n+1}, z) \\ &\le (2/3)d(x,z) = d(x,z) - d(x,z)/3 \\ &\le d(x,z) - d(x_n, z) \le d(x_n, x).\end{aligned}$$

Thus, by hypothesis

$$d(x_{n+1}, f(x)) \le hd(x_n, x), \quad n \ge n_0,$$

and hence, we get

$$d(f(x), z) \le hd(x, z), \quad \text{for all } x \in X \backslash \{z\}.$$

Now, for using the contrary arguments, we assume that $f^k(z) \neq z$ for all $k \in \mathbb{N}$. Note that

$$d(f^{k+1}(z), z) \le h^k d(f(z), z), \quad k \in \mathbb{N}.$$

In this situation, each of the following three cases of ψ yields contradiction.

- $0 \le h \le \frac{1}{2}(\sqrt{5} - 1)$,
- $\frac{1}{2}(\sqrt{5} - 1) < h < \frac{1}{\sqrt{2}}$,
- $\frac{1}{\sqrt{2}} \le h < 1$

(see for detail [88]). Therefore, in all the cases, there exists some $k \in \mathbb{N}$ such that $f^k(z) = z$. Since $\{f^n(z)\}$ is a Cauchy sequence, we get $f(z) = z$. The uniqueness of the fixed point follows from the fact that

$$d(f(x), z) \le hd(x, z), \quad \text{for all } x \in X \backslash \{z\}.$$

Indeed, if $f(w) = w \neq z$, then $d(f(w), z) \le hd(w, z)$ implies $d(w, z) < d(w, z)$, which is not possible. □

Obviously, the class of contraction mappings given in Theorem 2.14 contains the class of usual contractions. However, it has been observed in [88] that Suzuki's contractions and Kannan's contractions are independent but both types of contractions characterize the metric completeness of the underlying spaces.

Remark 2.8. Recently, a number of results appeared on the existence of a unique fixed point for a single-valued mapping f of a metric space (X, d) endowed with a partial order $\preceq$. Indeed, almost all such results deal with monotone mapping satisfying Banach contractive condition (2.1) with some restriction and for fixed $x_0 \in X$, $x_0 \preceq f(x_0)$ (or $f(x_0) \preceq x_0$). The first result in this direction was proved by Ran and Reurings [77], which is an analogue of the BCP in partially ordered sets. They also presented several applications to linear and nonlinear matrix equations. Subsequently, many fixed point results with interesting applications have appeared in this direction, see, for example, [35–38, 69–71, 74, 97] and others.

2.5 Caristi's Fixed Point Theorem

In 1976, Caristi [19] proved a wonderful fixed point theorem on complete metric spaces, which is related to the BCP and is equivalent to Ekland variational principle [31]. The Caristi's fixed point theorem has found many applications in nonlinear analysis, see [10, 16, 40, 45, 52, 93] for detail discussion.

Definition 2.2. (a) A real-valued function φ defined on X is said to be *lower semicontinuous* at x if for any sequence $\{x_n\} \subset X$, we have

$$x_n \to x \in X \quad \Rightarrow \quad \varphi(x) \leq \liminf_{n\to\infty} \varphi(x_n).$$

(b) A single-valued self-mapping f on a metric space (X,d) is said to be *Caristi mapping* if there exists a lower semicontinuous function $\varphi : X \to \mathbb{R}^+$ such that

$$d(x, f(x)) \leq \varphi(x) - \varphi(f(x)), \quad \text{for all } x \in X. \tag{2.10}$$

Example 2.7. Each Banach contraction mapping f on a metric space (X,d) is a Caristi mapping with a function

$$\varphi(x) = \frac{1}{1-h} d(x, f(x)), \quad \text{for all } x \in X,$$

where h is a contraction constant.

Clearly, φ is a continuous real valued function on X and

$$\varphi(fx) \leq \frac{h}{1-h} d(x, fx) = h\varphi(x).$$

Note that for all $x \in X$,

$$d(x, fx) = (1-h)\varphi(x) = \varphi(x) - h\varphi(x) \leq \varphi(x) - \varphi(fx),$$

that is, f is a Caristi mapping.

Remark 2.9. In fact, the class of single-valued Caristi mappings is very large, including at least usual contractions, Ćirić contractive mappings and in particular Kannan mapping.

Theorem 2.15 (Caristi's Fixed Point Theorem). *Let (X,d) be a complete metric space. Then, each Caristi map $f : X \to X$ has a fixed point.*

Remark 2.10. The original proof of this results involved transfinite induction arguments. But, after the appearance of this remarkable theorem of Caristi, numerous papers were published on various proofs of this result. For example, see Wong [96], Penot [75], Siegel [83], and others. An elegant and direct proof of the Caristi's fixed point theorem is given in Deimling [29]. An elementary and straightforward approach is due to Brezis-Browder [16].

Remark 2.11. The key relation between Caristi's fixed point theorem and BCP was noted in Example 2.7. From the Caristi's fixed point theorem one cannot expect the all conclusions of the BCP. In the Caristi's fixed point theorem, the fixed point need not be unique and the sequence $\{f^n(x_0)\}$ need not even converge to a fixed point of f. Secondly, the map f satisfying (2.1) is continuous while the map f satisfying (2.10) is not necessarily continuous.

Remark 2.12. It is well known that the fixed point property for contraction mappings does not characterize metric completeness, see, for example, Suzuki and Takahashi [87]. However, some characterizations of metric completeness have been discussed by several authors. For example, Kirk [52] and Weston [95] proved that a metric space is complete if and only if it has the fixed point property for Caristi mappings. Moreover, Shiojiet et al. [84] proved that a metric space is complete if and only if it has the fixed point property for Kannan mappings. Thus, Kannan mappings and Caristi maps characterize metric completeness, while contraction mappings do not.

Regarding the problem of characterizations of metric completeness by means of contraction mappings, Suzuki and Takahashi [87] and independently Anisiu and Anisiu [5] proved that a convex subset Y of a normed space is complete if and only if every contraction $f : Y \to Y$ has a fixed point in Y. The most elegant result in this direction is due to Bessage [12] which states that if any mapping f on an arbitrary set X and each of its iterates f^n has a unique fixed point, then for each $h \in (0,1)$ there exists a metric d_h on X for which X is complete and f is a contraction mapping with contraction constant h. See [86] for more on Ekeland's variational principle and the equivalence between the Caristi's fixed point result and the completeness of metric spaces. Also, see [2,3,6–8,61,73].

2.6 Some Extensions of BCP Under Generalized Distances

In recent years, distances in metric have been introduced which generalize metrics and which have applications to obtaining the solutions of several new important problems in nonlinear analysis. The pioneering effort in this direction is papers of Kada et al. [43], Suzuki and Takahashi [87], Suzuki [89, 90], Lin and Du [59,60], and Ume [99] in metric spaces. In these papers, among other things, various distances are introduced, and relations between these distances with applications are established.

In [43], Kada et al. introduced a notion of w-distance on a metric space and using this notion, they improved the Caristi's fixed point theorem, Ekland variational principle, and Takahashi minimization theorem. Using the notion of w-distance, Suzuki and Takahashi [87] have introduced notions of single-valued and multivalued weakly contractive (in short, w-contractive) mappings and proved fixed point results for such mappings. Consequently, they generalized the Banach Contraction principle and Nadler's fixed point result [68].

Definition 2.3. A function $p : X \times X \to \mathbb{R}^+$ is called a *w-distance* on X if it satisfies the following for any $x,\ y,\ z \in X$:

(w_1) $p(x,z) \leq p(x,y) + p(y,z)$;
(w_2) a map $p(x,.) : X \to [0,\infty)$ is lower semicontinuous;
(w_3) for any $\varepsilon > 0$, there exists $\delta > 0$ such that $p(z,x) \leq \delta$ and $p(z,y) \leq \delta$ imply $d(x,y) \leq \varepsilon$.

Note that, in general for $x,y \in X$, $p(x,y) \neq p(y,x)$ and not either of the implications $p(x,y) = 0 \Leftrightarrow x = y$ necessarily hold.

Example 2.8. (a) The metric d is a w-distance on X.
(b) Let $(Y,\|.\|)$ be a normed space. Then, the functions $p_1, p_2 : Y \times Y \to \mathbb{R}^+$ defined by

$$p_1(x,y) = \|y\| \quad \text{and} \quad p_2(x,y) = \|x\| + \|y\|, \quad \text{for all } x,y \in Y,$$

are w-distances.

For other examples and related results, see [43]. Here we state two useful lemmas. For further details, see [43, 60, 98].

Lemma 2.1. *Let (X,d) be a metric space and let p be a w-distance on X. Let $\{x_n\}$ and $\{y_n\}$ be sequences in X and let $\{\alpha_n\}$ and $\{\beta_n\}$ be sequences in $[0,\infty)$ converging to 0. Then, the following statements hold for every $x,y,z \in X$:*

(a) If $p(x_n,y) \leq \alpha_n$ and $p(x_n,z) \leq \beta_n$ for any $n \in \mathbb{N}$, then $y = z$, in particular, if $p(x,y) = 0$ and $p(x,z) = 0$, then $y = z$;
(b) If $p(x_n,y_n) \leq \alpha_n$ and $p(x_n,z) \leq \beta_n$ for any $n \in \mathbb{N}$, then $\{y_n\}$ converges to z;
(c) If $p(x_n,x_m) \leq \alpha_n$ for any $n,m \in \mathbb{N}$ with $m > n$, then $\{x_n\}$ is a Cauchy sequence;
(d) If $p(y,x_n) \leq \alpha_n$ for any $n \in \mathbb{N}$, then $\{x_n\}$ is a Cauchy sequence.

Lemma 2.2. *Let (X,d) be a metric space and let p be a w-distance on X. Let K be a closed subset of X. Suppose that there exists $u \in X$ such that $p(u,u) = 0$. Then, $p(u,K) = 0$ if and only if $u \in K$, where $p(u,K) = \inf_{y \in K} p(u,y)$.*

Using the concept of w-distance, Kada et al. [43] improved Caristi's fixed point theorem as follows:

Theorem 2.16. *Let (X,d) be a complete metric space and p be a w-distance on X. Then, each Caristi mapping f on X with respect to p has a fixed point $x_0 \in X$ and $p(x_0,x_0) = 0$.*

While Suzuki and Takahashi [87] improved the BCP as follows.

Theorem 2.17. *Let (X,d) be a complete metric space and p be a w-distance on X. Then, each contraction mapping f on X with respect to p has a unique fixed point $x_0 \in X$ and $p(x_0,x_0) = 0$.*

In [87], they also obtained a result on characterization of metric completeness.

Theorem 2.18. *Let (X,d) be a metric space. Then, X is complete if and only if every contraction mapping f on X with respect to p has a fixed point in X.*

Among others results, Ume [100] improved Theorem 2.9 for w-distance.

Theorem 2.19. *Let (X,d) be a complete metric space and let p be a w-distance on X. Let $f : X \to X$ be a mapping such that for a fixed constant $h < 1$ and for all $x, y \in X$,*

$$p(f(x), f(y)) \leq h \max\{p(x,y), p(x,fx), p(y,fy), p(x,fy), p(y,fx)\},$$

and $\inf\{p(x,u) + p(x,fx) : x \in X\} > 0$ *for every $u \in X$ with $u \neq f(u)$. Then, f has a unique fixed point $x_0 \in X$ and $p(x_0,x_0) = 0$.*

A number of fixed point results w. r. t w-distance have been appeared in the literature.

Generalizing the concept of w-distance, Suzuki [89] introduced the following notion of τ-distance on metric spaces.

Definition 2.4. A function $p : X \times X \to \mathbb{R}^+$ is said to be a *τ-distance* on X if it satisfies the following conditions for any x, y, $z \in X$:

(τ_1) $p(x,z) \leq p(x,y) + p(y,z)$;
(τ_2) $\eta(x,0) = 0$ and $\eta(x,t) \geq t$ for all $x \in X$ and $t \geq 0$, and η is concave and continuous in its second variable;
(τ_3) $\lim_n x_n = x$ and $\limsup_n \{\eta(z_n, p(z_n,x_m)) : m \geq n\} = 0$ imply $p(u,x) \leq \liminf_n p(u,x_n)$ for all $u \in X$;
(τ_4) $\limsup_n \{p(x_n,y_m) : m \geq n\} = 0$ and $\lim_n \eta(x_n,t_n) = 0$ imply $\lim_n \eta(y_n,t_n) = 0$;
(τ_5) $\lim_n \eta(z_n, p(z_n,x_n)) = 0$ and $\lim_n \eta(z_n, p(z_n,y_n)) = 0$ imply $\lim_n d(x_n,y_n) = 0$.

It has been observed in [89] that (τ_2) can be replaced by

(τ_2)′ $\inf\{\eta(x,t) : t \geq 0\} = 0$ for all $x \in X$, and η is nondecreasing in its second variable.

In general, a τ-distance p does not necessarily satisfy $p(x,x) = 0$. The metric d is a τ-distance on X. Each w-distance on a metric space X is also a τ- distance on X. Other examples and properties of τ-distance are given in [89].

Using the concept of τ-distance, Susuki [89] improved the BCP and Caristi's fixed point theorem as under:

Theorem 2.20. *Let (X,d) be a complete metric space and p be a τ-distance on X. Then, each contraction mapping f on X with respect to p has a unique fixed point $x_0 \in X$ and $p(x_0,x_0) = 0$.*

Theorem 2.21. *Let (X,d) be a complete metric space and p be a τ-distance on X. Then, each Caristi's mapping f on X with respect to p has a fixed point $x_0 \in X$ and $p(x_0,x_0) = 0$.*

Using this result, Suzuki [93] obtained generalized Caristi's fixed point theorem as follows.

Theorem 2.22. *Let (X,d) be a complete metric space, p be a τ-distance on X and let $g : X \to (0,\infty)$ be a function such that for some $r > 0$*

$$\sup\{g(x) : x \in X, \psi(x) \leq \inf_{z \in X} \psi(z) + r\} < \infty,$$

where $\psi : X \to (0,\infty)$ is a lower semicontinuous function. Let $f : X \to X$ be a map such that for each $x \in X$,

$$p(x, f(x)) \leq g(x)(\psi(x) - \psi(f(x))).$$

Then, there exists $x_o \in X$ such that $f(x_o) = x_o$ and $p(x_o, x_o) = 0$.

See also [90] for further results in this direction. Recently, Ume [99] introduced a new concept of a distance called u-distance, which generalizes w-distance, TataruŠs distance [94], and τ-distance. Some interesting fixed point results including BCP with respect to u-distance appeared in [13,99]. In the literature, some other distances have introduced, and among others results the BCP has been also studied with respect to these generalized distances see, for example, [3, 59, 72, 101].

2.7 Multivalued Versions of BCP

Investigations on the existence of fixed points for multivalued contraction mappings in the setting of metric spaces were initiated by Nadler in 1979. Using the concept of Hausdorff metric, he established multivalued version of the Banach contraction principle.

Let (X,d) be a metric space. We denote by 2^X the collection of all nonempty subsets of X, $Cl(X)$ the collection of all nonempty closed subsets of X, $CB(X)$ the collection of all nonempty closed bounded subsets of X, and H the *Hausdorff metric* on $CB(X)$, that is,

$$H(A,B) = \max\{\sup_{x \in A} d(x,B), \sup_{y \in B} d(y,A)\}, \quad \text{for all } A, B \in CB(X),$$

where $d(x,A) = \inf_{y \in A} d(x,y)$. In the metric space $(CB(X), H)$, $\lim_{n\to\infty} A_n = A$ means that $\lim_{n\to\infty} H(A_n, A) = 0$. Let $A_1, A_2 \in CB(X)$. Then, for each $x \in A_1$ and $\varepsilon > 0$, there is $y \in A_2$ such that

$$d(x,y) \leq H(A_1, A_2) + \varepsilon.$$

Example 2.9. (a) Let $X = \mathbb{R}, A = [0,1], B = [2,4]$. Then,

$$\sup_{a\in A} d(a,B) = 2, \quad \sup_{b\in B} d(b,A) = 3, \quad \text{and} \quad H(A,B) = 3.$$

(b) Let $A = B_r(a)$, $B = B_s(b)$, $a,b \in (X,d)$, $0 < r \le s$. Then, $H(A,B) = d(a,b) + s - r$.

Definition 2.5. Let (X,d) be a metric space and let $T : X \to 2^X$.

(a) An element $x \in X$ is called a *fixed point* of a multivalued mapping T if $x \in T(x)$. We denote $Fix(T) = \{x \in X : x \in T(x)\}$.
(b) A sequence $\{x_n\}$ in X is said to be an *iterative sequence* of T at $x_0 \in X$ if $x_n \in T(x_{n-1})$ for all $n \in \mathbb{N}$.
(c) T is said to be a *contraction* [68] if for a fixed constant $h < 1$ and for each $x, y \in X$,

$$H(T(x), T(y)) \le h\, d(x,y). \tag{2.11}$$

Such a mapping T is also known as Nadler contraction.

Using the concept of Hausdorff metric, Nadler [68] proved the following theorem on the existence of fixed points for multivalued mappings, known as Nadler contraction principle (NCP).

Theorem 2.23 (Nadler' Fixed Point Theorem). *Let (X,d) be a complete metric space. Then, each contraction mapping $T : X \to CB(X)$ has a fixed point.*

Proof. Let $x_0 \in X$ be an arbitrary fixed and let $x_1 \in T(x_0)$. Then, there exists $x_2 \in T(x_1)$ such that

$$d(x_1, x_2) \le H(T(x_0), T(x_1)) + h,$$

where $h < 1$ is a contraction constant. Continuing this iterative process, in general, there exists $x_{n+1} \in T(x_n)$ for each $n \in \mathbb{N}$ such that

$$d(x_n, x_{n+1}) \le H(T(x_{n-1}), T(x_n)) + h^n \le \cdots \le h^n d(x_0, x_1) + n\, h^n.$$

Thus, we have

$$\sum_{n=0}^{\infty} d(x_n, x_{n+1})) \le d(x_0, x_1) \left(\sum_{n=0}^{\infty} h^n \right) + \sum_{n=0}^{\infty} n\, h^n < \infty.$$

Hence, $\{x_n\}$ is a Cauchy sequence, and thus, there exists some $x \in X$ such that $\lim_{n\to\infty} x_n = x$. Now the continuity of T implies $\lim_{n\to\infty} H(T(x_n), T(x)) = 0$. Since $x_n \in T(x_{n-1})$, we have $\lim_{n\to\infty} d(x_n, T(x)) = 0$, which implies $x \in T(x)$ because $T(x)$ is closed. □

Remark 2.13. (a) We observed in the above proof that using the property of the Hausdorff metric, we get an iterative sequence which is a Cauchy and converges to the fixed point of T.

(b) In contrast to its single-valued counterpart, fixed point in Theorem 2.23 need not be unique. Indeed, if X is bounded, then the map $T(x) = X$, for all $x \in X$ satisfies the conditions of Theorem 2.23.

Without using an iterative methods, a beautiful proof of Theorem 2.23 is given in [38] by Jachymski. The proof depends on the Axiom of choice and the Caristi's fixed point theorem.

Theorem 2.24. *Caristi's fixed point result (Theorem 2.15) yields NCP (Theorem 2.23).*

Proof. Let $T : X \to CB(X)$ be a Nadler contraction mapping with contraction constant h. Choose real α such that $h < \alpha < 1$. Let $x \in X$, then

$$\{y \in T(x) : \alpha d(x,y) \leq d(x, T(x))\} \neq \emptyset.$$

By the axiom of choice, there is a map $f : X \to X$ such that $f(x) \in T(x)$ and $\alpha d(x, f(x)) \leq d(x, T(x))$. Thus, we have

$$d(f(x), T(f(x))) \leq H(T(x), T(f(X))) \leq h(d(x, f(x))).$$

Note that

$$\begin{aligned} d(x, f(x)) &= \frac{1}{\alpha - h}(\alpha d(x, f(x)) - hd(x, f(x))) \\ &\leq \frac{1}{\alpha - h}(d(x, T(x)) - d(f(x), T(f(x)))). \end{aligned}$$

Set $\varphi(x) = \frac{1}{\alpha - h} d(x, T(x))$. Then, we have

$$d(x, f(x)) \leq \varphi(x) - \varphi(f(x)).$$

Also, note that for each $x, y \in X$, we get

$$|\varphi(x) - \varphi(y)| \leq \frac{1}{\alpha - h}\{d(x,y) + H(T(x), T(y))\} \leq \frac{h+1}{\alpha - h} d(x,y),$$

and thus, φ is continuous. Hence, by Theorem 2.15, there exists a fixed point in X. □

At the same time another proof of Theorem 2.23 has appeared in [42] without using an iterative technique.

In [66], Mizoguchi and Takahashi generalized Nadler's fixed point theorem as follows (which is also a partial answer to the problem proposed by Reich [78]).

Theorem 2.25. *Let (X,d) be a complete metric space and let $T: X \to CB(X)$ be a mapping such that for each $x, y \in X$,*

$$H(T(x), T(y)) \leq k(d(x,y))d(x,y),$$

where k is a function from $[0,\infty)$ to $[0,1)$ satisfying $\limsup_{r \to t^+} k(r) < 1$, *for every* $t \geq 0$. *Then, T has a fixed point.*

Remark 2.14. In the original statement, the domain of the function k is $(0,\infty)$. However both are equivalent because $d(x,y) = 0 \Rightarrow H(T(x),T(y)) = 0$. Also, note that the stronger condition assumed on k implies that $k(t) < h$ for some $0 < h < 1$. Thus with this condition, one may get that the map T is a contraction over a region for which $d(x,y)$ is sufficiently small.

Remark 2.15. In fact, the original proof of Theorem 2.25 is not simple. Alternative proofs appeared in [27, 81, 102]. The simplest alternative proof given in [92].

Proof. Define a function $\beta : [0,\infty) \to [0,1)$ by $\beta(t) = (k(t)+1)/2$. Then, we have

$$\limsup_{r \to t+0} \beta(r) < 1, \quad \text{for all } t \geq 0,$$

and for all $x, y \in X$ and $u \in T(x)$, there exists an element $v \in T(x)$ such that

$$d(u,v) \leq \beta(d(x,y))\, d(x,y).$$

Thus, we can define a sequence $\{x_n\}$in X such that for all integer $n \geq 1$, $x_{n+1} \in T(x_n)$ and

$$d(x_{n+1}, x_{n+2}) \leq \beta(d(x_n, x_{n+1}))\, d(x_n, x_{n+1}).$$

For convenience, we put $a_n = d(x_n, x_{n+1})$. Hence, the sequence of nonnegative real numbers $\{a_n\}$ is non-increasing and thus converges to some nonnegative real number α. Note that there exist some $b \in [0,1)$ and $\varepsilon > 0$ such that $\beta(r) \leq b$ for all $r \in [\alpha, \alpha+\varepsilon]$. Now we can choose some integer $m \geq 1$ such that $m \leq a_n \leq \alpha + \varepsilon$ with $n \geq m$. Note that

$$a_{n+1} \leq \beta(a_n)\, a_n \leq b\, a_n,$$

and thus, we have

$$\sum_{n=1}^{\infty} a_n < \infty.$$

Hence, $\{x_n\}$ is a Cauchy sequence in the complete space X. Let $\{x_n\}$ converges to some $z \in X$. Note that

$$\begin{aligned} d(z,T(z)) &= \lim_{n\to\infty} d(x_{n+1},T(z)) \\ &\le \lim_{n\to\infty} H(T(x_n),T(z)) \\ &\le \lim_{n\to\infty} \beta(d(x_n,z))d(x_n,z) \\ &\le \lim_{n\to\infty} d(x_n,z) = 0. \end{aligned}$$

Since $T(z)$ is closed, we get $z \in T(z)$. □

Recently, Kikkawa and Suzuki [48] generalized Nadler's fixed point theorem and Theorem 2.14.

Theorem 2.26. *Define a strictly increasing function η from $[0,1)$ onto $(\frac{1}{2},1]$ by*

$$\eta(h) = \frac{1}{1+h}.$$

Let (X,d) be a complete metric space and let $T : X \to CB(X)$. Assume that there exists fixed $h \in [0,1)$ such that

$$\eta(h)d(x,T(x)) \le d(x,y) \Rightarrow H(T(x),T(y)) \le hd(x,y), \quad \text{for all } x,y \in X.$$

Then, T has a fixed point.

Proof. Take a real number k with $0 < h < k < 1$. Let $x_0 \in X$ be an arbitrary fixed and let $x_1 \in T(x_0)$. Then we have

$$\eta(h)d(x_0,T(x_0)) \le \eta(h)d(x_0,x_1) \le d(x_0,x_1).$$

From the hypothesis, it follows that

$$d(x_1,T(x_1)) \le H(T(x_0),T(x_1)) \le hd(x_0,x_1).$$

So, there exits $x_2 \in T(x_1)$ such that $d(x_1,x_2) \le kd(x_0,x_1)$. Continuing this process, we get a sequence $\{x_n\}$ in X such that $x_n \in T(x_{n-1})$ and $d(x_{n-1},x_n) \le kd(x_{n-2},x_{n-1})$. Thus, we have

$$\sum_{n=1}^{\infty} d(x_{n-1},x_n) \le \sum_{n=1}^{\infty} k^{n-1}d(x_0,x_n) < \infty,$$

and hence $\{x_n\}$ is a Cauchy sequence. Since X is complete, $\{x_n\}$ converges to some point $z \in X$. We next show that

$$d(z,T(x)) \le hd(z,x), \quad \text{for all } x \in X \setminus \{z\}.$$

Since X is complete, $\{x_n\}$ converges to some point $z \in X$. Then, for any $x \in X \backslash z$, there exists $n_0 \in \mathbb{N}$ such that $d(x_n, z) \leq \frac{1}{3} d(x, z)$ for all $n \in \mathbb{N}$ with $n \geq n_0$. Now we have

$$\begin{aligned}
\eta(h) d(x_n, T(x_n)) &\leq d(x_n, T(x_n)) \leq d(x_n, x_{n+1}) \\
&\leq d(x_n, z) + d(x_{n+1}, z) \\
&\leq (2/3) d(x, z) = d(x, z) - d(x, z)/3 \\
&\leq d(x, z) - d(x_n, z) \leq d(x_n, x).
\end{aligned}$$

Hence, $H(T(x_n, T(x)) \leq hd(x_n, x)$, it follows that $d(T(x), x_{n+1}) \leq hd(x, x_n)$ for all $n \in \mathbb{N}$ with $n \geq n_0$. Thus, we get

$$d(T(x), z) \leq hd(x, z), \quad \text{for all } x \in X \backslash \{z\}.$$

We next prove that $H(T(x), T(z)) \leq hd(x, z)$ for all $x \in X$. If $d(x, z) = 0$, then we are done. So we assume that $d(x, z) > 0$. Then for every $n \in \mathbb{N}$, there exists $y_n \in T(x)$ such that

$$d(z, y_n) \leq d(z, T(x)) + \frac{1}{n} d(x, z).$$

For $n \in \mathbb{N}$, We have

$$\begin{aligned}
d(x, Tx) &\leq d(x, y_n) \leq d(x, z) + d(z, y_n) \\
&\leq d(x, z) + d(z, T(x)) + \frac{1}{n} d(x, z) \\
&\leq d(x, z) + hd(x, z) + \frac{1}{n} d(x, z) \\
&= \left(1 + h + \frac{1}{n}\right) d(x, z),
\end{aligned}$$

and hence, $\frac{1}{1+h} d(x, T(x)) \leq d(x, z)$. From the hypothesis, we have $H(T(x), T(z)) \leq hd(x, z)$. Finally, note that

$$d(z, T(z)) = \lim_{n \to \infty} d(x_{n+1}, T(z)) \leq \lim_{n \to \infty} H(T(x_n), T(z)) \leq \lim_{n \to \infty} hd(x_n, z) = 0,$$

and since $T(z)$ is closed, we get $z \in T(z)$. □

Many modifications and generalizations of Nadler's Theorem have been developed in successive years. In most cases, the nature of the Hausdorff metric is not used and the existence part of results can be proved without using the concept of a Hausdorff metric. For instance, recently, Feng and Liu [32] obtained some interesting fixed point results for multivalued mappings and extended Nadler's result as follows.

Theorem 2.27. *Let (X,d) be a complete metric space and let $T : X \to Cl(X)$ be a mapping such that for any fixed constants $h, b \in (0,1)$, $h < b$, and for each $x \in X$, there is $y \in T(x)$ satisfying the following conditions:*

$$b\, d(x,y) \leq d(x,T(x)),$$

and

$$d(y,T(y)) \leq h\, d(x,y).$$

Then, $Fix(T) \neq \emptyset$ provided the real-valued function g on X, $g(x) = d(x,T(x))$ is lower semicontinuous.

Proof. Let $x_0 \in X$ be an arbitrary fixed. Then by hypothesis, there exists $x_1 \in T(x_0)$ such that

$$bd(x_o,x_1) \leq d(x_0,T(x_0)) \quad \textit{and} \quad d(x_1,T(x_1)) \leq hd(x_o,x_1).$$

Similarly, there is $x_2 \in T(x_1)$ satisfying

$$b\, d(x_1,x_2) \leq d(x_1,T(x_1)) \quad \textit{and} \quad d(x_2,T(x_2)) \leq h\, d(x_1,x_2).$$

Continuing this process, we get a sequence $\{x_n\}$ in X satisfying $x_{n+1} \in T(x_n)$,

$$bd(x_n,x_{n+1}) \leq d(x_n,T(x_n)) \quad \textit{and} \quad d(x_{n+1},T(x_{n+1})) \leq hd(x_n,x_{n+1}),$$

for all $n = 0,1,2,\ldots$. Thus, from the last two inequalities, we have for all $n = 0,1,2,\ldots$,

$$d(x_{n+1},x_{n+2}) \leq \frac{h}{b} d(x_n,x_{n+1}), \tag{2.12}$$

and

$$d(x_{n+1},T(x_{n+1})) \leq \frac{h}{b} d(x_n,T(x_n)). \tag{2.13}$$

Note that

$$d(x_{n+1},T(x_{n+1})) \leq d(x_n,T(x_n)).$$

Thus, $\{d(x_n,T(x_n))\}$ is a decreasing sequence. Further, for each $n \in \{0,1,2,\ldots\}$ we have

$$d(x_n,x_{n+1}) \leq \frac{h^n}{b^n} d(x_o,x_1),$$

and

$$d(x_n, T(x_n)) \le \frac{h^n}{b^n} d(x_o, T(x_o)).$$

Since $h < b$, so we get that $d(x_n, T(x_n)) \to 0$ as $n \to \infty$. Set $a = \frac{h}{b}$. Then, for $m, n \in \mathbb{N}$ with $m > n$, we have

$$\begin{aligned} d(x_m, x_n) &\le d(x_m, x_{m-1}) + d(x_{m-1}, x_{m-2}) + \cdots + d(x_{n+1}, x_n) \\ &\le a^{m-1} d(x_o, x_1) + a^{m-2} d(x_o, x_1) + \cdots + a^n d(x_o, x_1) \\ &\le \frac{a^n}{1-a} d(x_o, x_1), \end{aligned}$$

Due to $h < b$, we have $a^n \to 0$ as $n \to \infty$, and hence $\{x_n\}$ is a Cauchy sequence. Since X is complete, $\{x_n\}$ converges to some point $x \in X$. We assert that x is a fixed point of T. Note that the sequence of nonnegative terms $\{f(x_n)\} = \{d(x_n, T(x_n))\}$ is decreasing to 0, and since f is lower semicontinous, we have

$$0 \le f(x) \le \liminf_{n \to \infty} f(x_n) = 0,$$

which implies $f(x) = 0$. Hence, the closeness of $T(x)$ implies $x \in T(x)$. □

Remark 2.16. Theorem 2.27 generalizes Nadler's fixed point result. Indeed, if T satisfies the condition of Nadler's result, then the lower semicontinuity of function $f(x) = d(x, T(x))$ follows from the contraction condition. Further, since $T(x)$ is closed and bounded, so there exists $y \in T(x)$ such that

$$bd(x, y) \le d(x, T(x)), \quad \text{for } b \in (0, 1).$$

Also,

$$d(y, T(y)) \le H(T(x), T(y)) \le hd(x, y), \quad \text{for } h \in (0, 1).$$

Thus, the existence of fixed point of T follows from the Theorem 2.27.

Klim and Wardowski [53] generalized Theorem 2.23 and Theorem 2.27. While in [23,24], Ćirić generalized all the above-mentioned fixed point results of this section.

On the other hand, mutivalued versions of the BCP with respect to generalized distances have appeared. Using w-distance, Suzuki and Takahashi [87] obtained multivalued version of the BCP which is an improved version of the Nadler's fixed point theorem (Theorem 2.23).

Theorem 2.28. *Let (X, d) be a complete metric space and let p be a w-distance on X. Let $T : X \to Cl(X)$ be a mapping such that for a fixed constant $h \in [0, 1)$ and for any $x, y \in X$, $u \in T(x)$, there is $v \in T(y)$ satisfying*

$$p(u,v) \leq h\,p(x,y).$$

Then, there exists $x_0 \in X$ *such that* $x_0 \in T(x_0)$ *and* $p(x_0,x_0) = 0$.

Proof. Let $u_o \in X$ be fixed and let $u_1 \in T(u_o)$. Then, there exists $u_2 \in T(u_1)$ such that

$$p(u_1,u_2) \leq hp(u_o,u_1).$$

Continuing this process, we have a sequence $\{u_n\}$ in X such that $u_{n+1} \in T(u_n)$ and

$$p(u_n,u_{n+1}) \leq hp(u_{n-1},u_n), \quad \text{for every } n \in \mathbb{N}.$$

Thus, we have

$$p(u_n,u_{n+1}) \leq hp(u_{n-1},u_n) \leq h^2 p(u_{n-2},u_{n-1}) \leq \cdots \leq h^n p(u_o,u_1),$$

and hence, for any $n,m \in \mathbb{N}$ with $m > n$,

$$\begin{aligned} p(u_n,u_m) &\leq p(u_n,u_{n+1}) + p(u_{n+1},u_{n+2}) + \cdots + p(u_{m-1},u_m) \\ &\leq \{h^n + h^{n+1} + \cdots + h^{m-1}\} p(u_o,u_1) \\ &\leq \frac{h^n}{1-h} p(u_o,u_1) \to 0. \end{aligned}$$

By Lemma 2.1, $\{u_n\}$ is a Cauchy sequence . Thus, $\{u_n\}$ converges to some $v_o \in X$. Fix $n \in \mathbb{N}$. Since $\{u_m\}$ converges to v_o and $p(u_n,.)$ is lower semicontinuous, we have

$$p(u_n,v_o) \leq \liminf_{m\to\infty} p(u_n,u_m) \leq \frac{h^n}{1-h} p(u_o,u_1).$$

Since $u_n \in T(u_{n-1})$ and $v_o \in X$, by hypothesis there is $w_n \in T(v_o)$ such that

$$p(u_n,w_n) \leq hp(u_{n-1},v_o) \leq \frac{h^n}{1-h} p(u_o,u_1).$$

By Lemma 2.1, $\{w_n\}$ converges to v_o. Since $T(v_o)$ is closed, we have $v_o \in T(v_o)$. For such v_o, there exists $v_1 \in T(v_o)$ such that

$$p(v_o,v_1) \leq hp(v_o,v_o).$$

Thus, we also have a sequence $\{v_n\}$ in X such that $v_{n+1} \in T(v_n)$ and

$$p(v_o,v_{n+1}) \leq hp(v_o,v_n), \quad \text{for every } n \in \mathbb{N}.$$

Thus, we have

$$p(v_o,v_n) \le hp(v_o,v_{n-1}) \le \cdots \le h^n p(v_o,v_o) \to 0 \text{ as } n \to \infty.$$

By Lemma 2.1, $\{v_n\}$ is a Cauchy sequence. Then, $\{v_n\}$ converges to a point $x_o \in X$. Since $p(v_o,.)$ is lower semicontinuous, we have

$$p(v_o,x_o) \le \liminf_{n\to\infty} p(v_o,v_n) \le 0,$$

and hence, $p(v_o,x_o) = 0$. Thus, for any $n \in \mathbb{N}$, we have

$$\begin{aligned} p(u_n,x_o) &\le p(u_n,v_o) + p(v_o,x_o) \\ &\le \frac{h^n}{1-h} p(u_o,u_1) \to 0 \text{ as } n \to \infty. \end{aligned}$$

So, by Lemma 2.1, we obtain $v_o = x_o$, and hence, $p(v_o,v_o) = 0$. □

The following result is a generalization of the Banach contraction principle.

Corollary 2.4. *Let (X,d) be a complete metric space and p be a w-distance on X. Then, each contraction mapping f on X with respect to p has a unique fixed point $x_0 \in X$ and $p(x_0,x_0) = 0$.*

Proof. From Theorem 2.28, there exists $x_o \in X$ with $f(x_o) = x_o$ and $p(x_o,x_o) = 0$. If $y_o = f(y_o)$, then we have

$$p(x_o,y_o) = p(f(x_o), f(y_o)) \le hp(x_o,y_o).$$

Since $h \in (0,1)$, we have $p(x_o,y_o) = 0$. So, by $p(x_o,x_o) = 0$ and by Lemma 2.1, we have $x_o = y_o$. □

In [89], Suzuki established the following multivalued version of the BCP with respect to τ-distance, which is a generalization of Theorem 2.28.

Theorem 2.29. *Let (X,d) be a complete metric space and let p be a τ-distance on X. Let $T : X \to Cl(X)$ be a mapping such that for a fixed constant $h \in [0,1)$ and for any $x,y \in X$, $u \in T(x)$, there is $v \in T(y)$ satisfying*

$$p(u,v) \le h\,p(x,y).$$

Then, T has a fixed point.

Many other fixed point results with respect to τ-distance and other generalized distances have appeared in the literature, see, for example, [13, 25, 54–58, 89] and the references therein.

References

1. Agarwal, R.P.: Contraction and approximate contraction with an application to multi-point boundary value problems. J. Comput. Appl. Math. **9**, 315–325 (1983)
2. Agarwal, R.P., Khamsi, M.A.: Extension of CaristiŠs fixed point theorem to vector valued metric spaces. Nonlinear Anal. **74**, 141–145 (2011)
3. Al-Homidan, S., Ansari, Q.H., Yao, J.-C.: Some generalizations of Ekeland-type variational principle with applications to equilibrium problems and fixed point theory. Nonlinear Anal. **69**, 126–139 (2008)
4. Alber, Ya.I., Guerre-Delabriere, S.: Principles of weakly contractive maps in Hilbert spaces. In: Goldberg, I., Lyubich, Yu. (eds.) New Results in Operator Theory. Advances and Applications, vol. 98, pp. 7–22. Birkhäuser, Basel (1997)
5. Anisiu, M.C., Anisiu, V.: On the closedness of sets with the fixed point property for contractions. Rev. Anal. Numér. Théor. Approx. **26**, 13–17 (1997)
6. Ansari, Q.H.: Vectorial form of Ekeland-type variational principle with applications to vector equilibrium problems and fixed point theory. J. Math. Anal. Appl. **334**, 561–575 (2007)
7. Ansari, Q.H.: Metric Spaces: Including Fixed Point Theory and Set-valued Maps. Narosa Publishing House, New Delhi (2010)
8. Ansari, Q.H.: Topics in Nonlinear Analysis and Optimization. World Education, Delhi (2012)
9. Arandelović, I.D.: On a fixed point theorem of Kirk. J. Math. Anal. Appl. **301**, 384–385 (2005)
10. Bae, J.S., Cho, E.W., Yeom, S.H.: A generalization of the Caristi-Kirk fixed point theorem and its application to mapping theorems. J. Kor. Math. Soc. **31**, 29–48 (1994)
11. Banach, S.: Sur les opérations dans les ensembles abstraits et leur applications aux équations intégrales. Fund. Math. **3**, 133–181 (1922)
12. Bessage, C.: On the converse of the Banch fixed point principle. Colloq. Math. **7**, 41–43 (1959)
13. Dehaish, B.A.B. ,Latif, A.: Fixed point theorems for generalized contractive type multivalued maps. Fixed Point Theor. Appl. **2012**, Article No. 135 (2012)
14. Boyd, D.W., Wong, J.S.W.: Nonlinear contractions. Proc. Am. Math. Soc. **20**, 458–464 (1969)
15. Bonsall, F.F.: Lectures on Some Fixed Point Theorems of Functional Analysis. Tata Institute of Fundamental Research, Bombay (1962)
16. Breziz, H., Browder, F.E.: A general principle on order sets in nonlinear functional analysis. Adv. Math. **21**, 355–364 (1978)
17. Bryant, V.W.: A remark on a fixed point theorem for iterated mappings. Am. Math. Mon **75**, 399–400 (1968)
18. Caccioppoli, R.: Un teorema generale sullésistenza de elemente uniti in una transformazione funzionale. Rend. Acad. Naz. Linzei. **11**, 31–49 (1930)
19. Caristi, J.: Fixed point theorem for mapping satisfying inwardness conditions. Trans. Am. Math. Soc. **215**, 241–251 (1976)
20. Choudhurya, B.S., Konarb, P., Rhoades, B.E., Metiya, N.: Fixed point theorems for generalized weakly contractive mappings. Nonlinear Anal. **74**, 2116–2126 (2011)
21. Ćirić, L.B.: A generalization of Banach's contraction principle. Proc. Am. Math. Soc. **45**, 267–273 (1974)
22. Ćirić, L.B.: Fixed Point Theory: Contraction Mapping Principle. Faculty of Mechanical Engneering, University of Belgrade, Serbia (2003)
23. Ćirić, L.B.: Fixed point theorems for multi-valued contractions in metric spaces. J. Math. Anal. Appl. **348**, 499–507 (2008)
24. Ćirić, L.B.: Multivalued nonlinear contraction mappings. Nonlinear Anal. **71**, 2716–2723 (2009)
25. Cho, Y.J., Hirunworakit, S., Petrot, N.: Set-valued fixed point theorems for generalized contractive mappings without the Hausdorff metric. Appl. Math. Lett. **24**, 1959–1967 (2011)
26. Connell, E.H.: Properties of fixed point spaces. Proc. Am. Math. Soc. **10**, 974–979 (1959)

27. Daffer, P.Z., Kaneko, H.: Fixed points of generalized contractive multi-valued mappings. J. Math. Anal. Appl. **192**, 655–666 (1995)
28. Dutta, P.N., Choudhury, B.S.: A generalisation of contraction principle in metric spaces. Fixed Point Theor. Appl. **2008**, Article ID 406368 (2008)
29. Deimling, K.: Multivalued Differential Equations. Walter de Gruyter, Berlin, New york (1992)
30. Edelstein, M.: On fixed and periodic points under contractive mappings. J. Lond. Math. Soc. **37**, 74–79 (1962)
31. Ekeland, I.: Nonconvex minimization problems. Bull. Am. Math. Soc. **1**, 443–474 (1979)
32. Feng, Y., Liu, S.: Fixed point theorems for multivalued contractive mappings and multivalued Caristi type mappings. J. Math. Anal. Appl. **317**, 103–112 (2006)
33. Geraghty, M.: On contraction mappings. Proc. Am. Math. Soc. **40**, 604–608 (1973)
34. Goebel, K., Kirk, W.A.: Topics in Metric Fixed Point Theory. Cambridge University Press, Cambridge (1990)
35. Gordji, M.E., Ramezani, M.: A generalization of Mizoguchi and Takahashi's theorem for single-valued mappings in partially ordered metric spaces. Nonlinear Anal. **74**, 4544–4549 (2011)
36. Amini-Harandi, A., Emami, H.: A fixed point theorem for contraction type maps in partially ordered metric spaces and application to ordinary differential equations. Nonlinear Anal. **72**, 2238–2242 (2010)
37. Harjani, J., Sadarangani, K.: Fixed point theorems for weakly contractive mappings in partially ordered sets. Nonlinear Anal. **71**, 3403–3410 (2009)
38. Jachymski, J.R.: The contraction principle for mappings on a metric space with graph. Proc. Am. Math. Soc. **136**, 1359–1373 (2008)
39. Jachymski, J.R.: Equivalence of some contractivity properties over metrical structure. Proc. Am. Math. Soc. **125**, 2327–2335 (1997)
40. Jachymski, J.R.: Caristi's fixed point theorem and selection of set-valued contractions. J. Math. Anal. Appl. **227**, 55–67 (1998)
41. Jachymski, J.R., Jóźwik, I.: On Kirk's asymptotic contractions. J. Math. Anal. Appl. **300**, 147–159 (2004)
42. Jachymski, J.R.: Fixed point theorems in metric and uniform spaces via the Knaster-Tarski principle. Nonlinear Anal. **32**, 225–233 (1998)
43. Kada, O., Susuki, T., Takahashi, W.: Nonconvex minimization theorems and fixed point theorems in complete metric spaces. Math. Jpn. **44**, 381–391 (1996)
44. Kannan, R.: Some results on fixed points Ű II. Am. Math. Mon. **76**, 405–408 (1969)
45. Khamsi, M.A.: Remarks on Caristi's fixed point theorem. Nonlinear Anal. **71**, 227–231 (2009)
46. Khamsi, M.A., Kirk, W.A.: An Introduction to Metric Spaces and Fixed Point Theory. Wiley, New York (2001)
47. Khan, M.S., Sessa, S.: Fixed points theorems by altering distances between the points. Bull. Aust. Math. Soc. **30**, 1–9 (1984)
48. Kikkawaa, M., Suzuk, T.: Three fixed point theorems for generalized contractions with constants in complete metric spaces. Nonlinear Anal. **69**, 2942–2949 (2008)
49. Kirk, W.A.: Fixed points of asymptotic contractions. J. Math. Anal. Appl. **277**, 645–650 (2003)
50. Kirk, W.A.: Asymptotic pointwise contractions. Nonlinear Anal. **69**, 4706–4712 (2008)
51. Kirk, W.A.: Contraction mappings and extensions. In: Kirk, W.A., Sims, B. (eds.) A Handbook of Metric Fixed Point Theory, pp. 1–34. Kluwer Academic, Dordrecht (2001)
52. Kirk, W.A.: Caristi's fixed point theorem and metric convexity. Colloq. Math. **36**, 81–86 (1976)
53. Klim, D., Wardowski, D.: Fixed point theorems for set-valued contractions in complete metric spaces. J. Math. Anal. Appl. **334**, 132–139 (2007)
54. Latif, A.: Generalized Caristi's fixed points theorems. Fixed Point Theor. Appl. **2009** Article ID 170140, 7 (2009)
55. Latif, A., Abdou, A.A.N.: Some new weakly contractive type multimaps and fixed point results in metric spaces. Fixed Point Theor. Appl. **2009** Article ID 412898, 12 (2009)

56. Latif, A., Abdou, A.A.N.: Multivalued generalized nonlinear contractive maps and fixed points. Nonlinear Anal. **74**, 1436–1444 (2011)
57. Latif, A., Al-Mezel, S.A.: Fixed point results in quasimetric spaces. Fixed Point Theor. Appl. **2011** Article ID 178306, 8 (2011)
58. Lin, L.-J., Chuang, C.-S: Some new fixed point theorems of generalized nonlinear contractive multivalued maps in complete metric spaces. Comput. Math. Appl. **62**, 3555–3566 (2011)
59. Lin, L.-J., Du, W.-S.: Ekeland's variational principle, minimax theorems and existence of nonconvex equilibria in complete metric spaces. J. Math. Anal. Appl. **323**, 360–370 (2006)
60. Lin, L.-J., Du, W.-S.: Some equivalent formulations of the generalized Ekeland's variational principle and their applications. Nonlinear Anal. **67**, 187–199 (2007)
61. Lin, L.-J., Wang, S.-Y., Ansari, Q.H.: Critical point theorems and Ekeland type varitational principle with applications. Fixed Point Theor. Appl. **2011** Article ID 914624, (2011)
62. Matkowski, J.: Nonlinear contractions in metrically convex spaces. Publ. Math. Debrecen **45**, 103–114 (1994)
63. Matkowski, J.: Integrable solutions of functional equations. Diss. Math. **127**, (1975)
64. Meir, A., Keeler, E.: A theorem on contraction mappings. J. Math. Anal. Appl. **28**, 326–329 (1969)
65. Mihep, D.: Altering distances in probabilistic Menger spaces. Nonlinear Anal. **71**, 2734–2738 (2009)
66. Mizoguchi, N., Takahashi, W.: Fixed points theorems for multivalued mappings on complete metric spaces. J. Math. Anal. Appl. **141**, 177–188 (1989)
67. Naidu, S.V.R.: Some fixed point theorems in Metric spaces by altering distances. Czech. Math. J. **53**, 205–212 (2003)
68. Nadler, S.B.: Multivalued contraction mappings. Pac. J. Math. **30**, 475–488 (1969)
69. Nieto, J.J., Pouso, R.L., Rodrígues-López, R.: Fixed point theorems in ordered abstract sets. Proc. Am. Math. Soc. **135**, 2505–2517 (2007)
70. Nieto, J.J., Rodrígues-López, R.: Contractive mapping theorems in partially ordered sets and applications to ordinary differential equations. Order **22**, 223–239 (2005)
71. Nieto, J.J., Rodrígues-López, R.: Existence and uniqueness of fixed point in partially ordered sets and applications to ordinary differential equations. Acta Math. Sin. **23**, 2205–2212 (2007)
72. Park, S.: On generalizations of the Ekland-type variational principles. Nonlinear Anal. **39**, 881–889 (2000)
73. Park, S.: Characterizations of metric completeness. Colloq. Math. **49**, 21–26 (1984)
74. O'Regan, D., Petruél, A.: Fixed point theorems for generalized contractions in ordered metric spaces. J. Math. Anal. Appl. **341**, 1241–1252 (2008)
75. Penot, J.P.: A short constractive proof of Caristi's fixed point theorem. Publ. Math. Univ. Paris **10**, 1–3 (1976)
76. Rakotch, E.: A note on contractive mappings. Proc. Am. Math. Soc. **13**, 459–465 (1962)
77. Ran, A.C.M., Reurings, M.C.B.: A fixed point theorem in partially ordered sets and some applications to matrix equations. Proc. Am. Math. Soc. **132**, 1435–1443 (2004)
78. Reich, S.: Some problems and results in fixed point theory. Contemp. Math. **21**, 179–187 (1983)
79. Rhoades, B.E.: A comparison of various definations of contractive mappings. Proc. Am. Math. Soc. **226**, 257–290 (1977)
80. Rhoades, B.E.: Some theorems on weakly contractive maps. Nonlinear Anal. **47**, 2683–2693 (2001)
81. Sastry, K.P.R., Babu, G.V.R.: Convergence of Ishikawa iterates for a multi-valued mapping with a fixed point. Czech. Math. J. **55**, 817–826 (2005)
82. Sastry, K.P.R., Babu, G.V.R.: Some fixed point theorems by altering distances between the points. Indian J. Pure Appl. Math. **30**, 641–647 (1999)
83. Siegel, J.: A new proof of Caristi's fixed point theorem. Proc. Am. Math. Soc. **66**, 54–56 (1977)
84. Shioji, N., Suzuki, T., Takahashi, W.: Contractive mappings, Kannan mappings and metric completeness, Proc. Am. Math. Soc. **126**, 3117–3124 (1998)

85. Subrahmanyam, V.: Completeness and fixed-points. Monatsh. Math. **80**, 325–330 (1975)
86. Sullivan, F.: A characterization of complete metric spaces. Proc. Am. Math. Soc. **85**, 345–346 (1981)
87. Suzuki, T., Takahashi, W.: Fixed point Theorems and characterizations of metric completeness. Topol. Meth. Nonlinear Anal. **8**, 371–382 (1996)
88. Suzuki, T.: A generalized Banach contraction principle that characterizes metric completeness. Proc. Am. Math. Soc. **136**, 1861–1869 (2008)
89. Suzuki, T.: Generalized distance and existence theorems in complete metric spaces. J. Math. Anal. Appl. **253**, 440–458 (2001)
90. Suzuki, T.: Several fixed point theorems concerning τ-distance. Fixed Point Theor. Appl. **3**, 195–209 (2004)
91. Suzuki, T.: A definitive result on asymptotic contractions. J. Math. Anal. Appl. **335**, 707–715 (2007)
92. Suzuki, T.: Mizoguchi-Takahashi's fixed point theorem is a real generalization of Nadler's. J. Math. Anal. Appl. **340**, 752–755 (2008)
93. Suzuki, T.: Generalized Caristi's fixed point theorem by Bae and others. J. Math. Anal. Appl. **302**, 502–508 (2005)
94. Tataru, D.: Viscosity solutions of Hamilton-Jacobi equations with unbounded nonlinear terms. J. Math. Anal. Appl. **163**, 345–392 (1992)
95. Weston, J.D.: A characterization of metric completeness. Proc. Am. Math. Soc. **64**, 186–188 (1977)
96. Wong, C.S.: On a fixed point theorem of contractive type. Proc. Am. Math. Soc. **57**, 283–284 (1976)
97. Yan, F., Su, Y., Feng, Q.: A new contraction mapping principle in partially ordered metric spaces and applications to ordinary differential equations. Fixed Point Theor. Appl. **2012** Article No. 152 (2012)
98. Takahashi, W.: Nonlinear Functional Analysis: Fixed Point Theory and Its Applications. Yokohama Publishers, Yokohama (2000)
99. Ume, J.S.: Existence theorems for generalized distance on complete metric space. Fixed Point Theor. Appl. **2010** Article ID 397150 (2010)
100. Ume, J.S.: Fixed point theorems related to Ćirić's contraction principle, J. Math. Anal. Appl. **225**, 630–640 (1998)
101. Włodarczyk, K., Plebaniak, R.: Contractions of Banach, Tarafdar, Meir-Keeler, Ćirić-Jachymski-Matkowski and Suzuki types and fixed points in uniform spaces with generalized pseudodistances. J. Math. Anal. Appl. **404**, 338–350 (2013)
102. Xu, H.K.: Metric fixed point theory for multivalued mappings. Diss. Math. **389**, 39 (2000)
103. Zeidler, E.: Nonlinear Functional Analysis and Its Applications I: Fixed-Point Theorems. Springer, New York, Berlin, Heidelberg, Tokyo (1986)

Chapter 3
Ekeland's Variational Principle and Its Extensions with Applications

Qamrul Hasan Ansari

3.1 Introduction

In 1972, Ekeland [35] (see also, [36, 37]) established a theorem on the existence of an approximate minimizer of a bounded below and lower semicontinuous function. This theorem is known as Ekeland's variational principle (in short, EVP). It is one of the most applicable results from nonlinear analysis and used as a tool to study the problems from fixed point theory, optimization, optimal control theory, game theory, nonlinear equations, dynamical systems, etc; see, for example, [7–9, 19, 20, 34–38, 46, 55, 60, 67, 72, 85] and the references therein. Later, it was found that several well-known results, namely, Caristi–Kirk fixed point theorem [24, 25], Takahashi's minimization theorem [84], the Petal theorem [72], and the Daneš drop theorem [32] from nonlinear analysis are equivalent to the Ekeland's variational principle. Since the discovery of EVP, there have also appeared many extensions or equivalent formulations of EVP; see, for example, [1, 9–11, 16, 21, 31, 41, 42, 45, 48–54, 57, 59, 65, 67, 70–72, 76–82, 86–88, 90, 91] and the references therein. Sullivan [76] established that the conditions of EVP on a metric space (X,d) imply the completeness of the metric space (X,d). In 1982, McLinden [59] showed that how EVP, or more precisely the augmented form of it provided by Rockafellar [73], can be adapted to extremum problems of minimax type. For further detail on the equivalence of these results, we refer to [1, 31, 33, 45, 48, 60, 70, 80, 81, 83] and the references therein.

In 1993, Oettli and Théra [67] extended the EVP for bifunctions and studied the existence theory of equilibrium problems in the setting of nonconvex metric spaces. Such extended form of EVP is known as extended Ekeland's variational principle (in short, EEVP). It is also shown that the existence result for a solution of an equilibrium problem is equivalent to the Ekeland-type variational principle

Q.H. Ansari (✉)
Department of Mathematics, Aligarh Muslim University, Aligarh 202 002, India
e-mail: qhansari@gmail.com

S. Almezel et al. (eds.), *Topics in Fixed Point Theory*, DOI 10.1007/978-3-319-01586-6_3,

for bifunctions, Caristi–Kirk fixed point theorem for set-valued maps [25], and a maximal element theorem. It is further studied in [1–3, 5, 14, 18, 56, 58, 67, 71] and the references therein. Hamel [50] and Park [71] studied such kind of variational principle in a more general setting.

In this chapter, we present several forms of Ekeland's variational principle, its equivalence to Takahashi's minimization theorem, and some applications to fixed point theory and weak sharp minima. We discuss equilibrium problem which is a unified model of several problems, namely, minimization problem, saddle point problem, Nash equilibrium problem, fixed point problem, variational inequality problem, etc. We present equilibrium version of EVP and prove the equivalence to some other problems. We use such EVP to establish the existence of a solution of an equilibrium problem without convexity assumption on the underlying set.

3.2 Ekeland's Variational Principle in Complete Metric Spaces

Let X be a nonempty set. A function $f : X \to \mathbb{R} \cup \{+\infty\}$ is said to be *proper* if $f(x) \neq +\infty$ for all $x \in X$. The domain of f, denoted by $dom(f)$, is defined by

$$dom(f) = \{x \in X : f(x) < +\infty\}.$$

Definition 3.1. Let X be a metric space. A function $f : X \to \mathbb{R}$ is said to be *lower semicontinuous* at a point $x \in X$ if $f(x) \leq \liminf_{n\to\infty} f(x_n)$ whenever $x_n \to x$ as $n \to \infty$. f is said to be *lower semicontinuous on* X if it is lower semicontinuous at each point of X.

A function $f : X \to \mathbb{R}$ is said to be *upper semicontinuous at a point* $x \in X$ if $f(x) \geq \limsup_{n\to\infty} f(x_n)$ whenever $x_n \to x$ as $n \to \infty$. f is said to be *upper semicontinuous on* X if it is upper semicontinuous at each point of X.

Example 3.1. Let $f : \mathbb{R} \to \mathbb{R}$ be defined as

$$f(x) = \begin{cases} -1, & \text{if } x < 0, \\ 1, & \text{if } x \geq 0. \end{cases}$$

Then, f is upper semicontinuous at $x = 0$ but not lower semicontinuous at $x = 0$.

Example 3.2. Let $f : \mathbb{R} \to \mathbb{R}$ be defined as

$$f(x) = \begin{cases} -1, & \text{if } x \leq 0, \\ 1, & \text{if } x > 0. \end{cases}$$

Then, f is lower semicontinuous at $x = 0$ but not upper semicontinuous at $x = 0$.

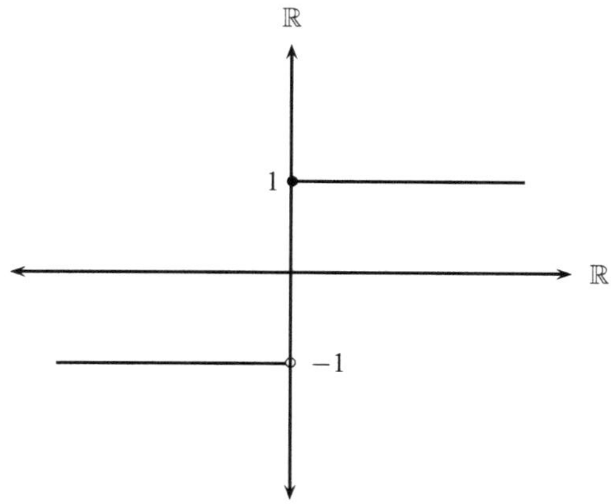

An upper semicontinuous function

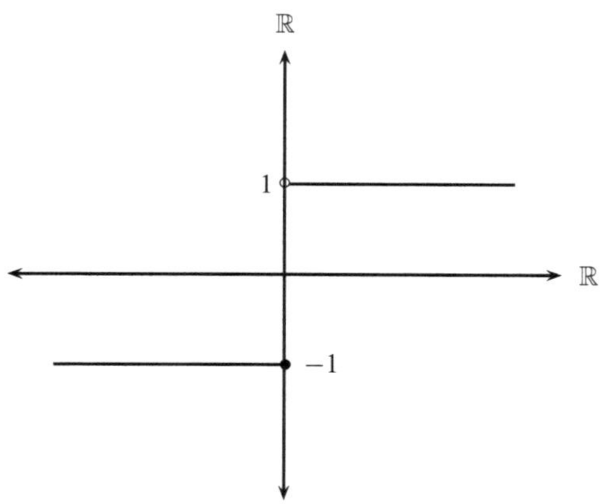

A lower semicontinuous function

Remark 3.1. A function $f : X \to \mathbb{R}$ is lower (respectively, upper) semicontinuous on X if and only if the set $\{x \in X : f(x) \leq \alpha\}$ (respectively, $\{x \in X : f(x) \geq \alpha\}$) is closed in X for all $\alpha \in \mathbb{R}$.

Let (X,d) be a metric space, K a nonempty subset of X, and $f : X \to \mathbb{R} \cup \{+\infty\}$ a proper functional. The well-known Weierstrass's theorem states that if K is compact and f is lower semicontinuous, then the following constrained minimization problem (in short, CMP)

$$\inf_{x \in K} f(x)$$

has a solution. Not only this, the solution set of CMP is compact.

Definition 3.2. For a given $\varepsilon > 0$, an element x_ε is said to be an *approximate ε-solution* of the following minimization problem (in short, MP)

$$\inf_{x \in X} f(x),$$

if

$$\inf_X f \leq f(x_\varepsilon) \leq \inf_X f + \varepsilon,$$

where $\inf_X f := \inf_{x \in X} f(x)$.

The following Ekeland's variational principle is the basic tool to establish the existence of solutions, or approximate ε-solutions, for minimization problems who fails to satisfy the compactness requirement of Weierstrass's theorem.

Theorem 3.1 (Strong Form of Ekeland's Variational Principle). *Let (X,d) be a complete metric space and $f : X \to \mathbb{R} \cup \{+\infty\}$ be a proper, bounded below and lower semicontinuous functional. Let $\varepsilon > 0$ and $\hat{x} \in X$ be given such that*

$$f(\hat{x}) \leq \inf_X f + \varepsilon.$$

Then, for a given $\lambda > 0$, there exists $\bar{x} \in X$ such that

(a) $f(\bar{x}) \leq f(\hat{x})$
(b) $d(\hat{x}, \bar{x}) \leq \lambda$
(c) $f(\bar{x}) < f(x) + \frac{\varepsilon}{\lambda} d(x, \bar{x})$, *for all* $x \in X \setminus \{\bar{x}\}$.

Proof. For the sake of convenience, we set $d_\lambda(u,v) = (1/\lambda) d(u,v)$. Then, d_λ is equivalent to d and (X, d_λ) is complete. Let us define a partial ordering $\preccurlyeq$ on X by

$$x \preccurlyeq y \quad \text{if and only if} \quad f(x) \leq f(y) - \varepsilon d_\lambda(x,y).$$

It is easy to see that this ordering is (i) reflexive, that is, for all $x \in X$, $x \preccurlyeq x$; (ii) antisymmetric, that is, for all $x, y \in X$, $x \preccurlyeq y$ and $y \preccurlyeq x$ imply $x = y$; (iii) transitive, that is, for all $x, y, z \in X$, $x \preccurlyeq y$ and $y \preccurlyeq z$ imply $x \preccurlyeq z$.

We define a sequence $\{S_n\}$ of subsets of X as follows: Start with $x_1 = \hat{x}$ ($\hat{x}$ is the same as given in the statement of the theorem) and define

$$S_1 = \{x \in X : x \preccurlyeq x_1\}; \quad x_2 \in S_1 \text{ such that } f(x_2) \leq \inf_{S_1} f + \frac{\varepsilon}{2},$$

$$S_2 = \{x \in X : x \preccurlyeq x_2\}; \quad x_3 \in S_2 \text{ such that } f(x_3) \leq \inf_{S_2} f + \frac{\varepsilon}{2^2},$$

and inductively

$$S_n = \{x \in X : x \preccurlyeq x_n\}; \quad x_{n+1} \in S_n \text{ such that } f(x_{n+1}) \le \inf_{S_n} f + \frac{\varepsilon}{2^n}.$$

Clearly, $S_1 \supset S_2 \supset S_3 \supset \ldots$. We claim that each S_n is closed.

Indeed, let $u_j \in S_n$ with $u_j \to u \in X$. Then $u_j \preccurlyeq x_n$ and so $f(u_j) \le f(x_n) - \varepsilon d_\lambda(u_j, x_n)$. By taking limit and using the lower semicontinuity of f and the continuity of d and so the continuity of d_λ, we conclude that $u \in S_n$.

Now, we prove that the diameter of these sets S_n, $\delta(S_n) \to 0$ as $n \to \infty$.

Indeed, take an arbitrary point $u \in S_n$, On the one hand, $u \preccurlyeq x_n$ implies that

$$f(u) \le f(x_n) - \varepsilon d_\lambda(u, x_n). \tag{3.1}$$

On the other hand, we observe that u also belongs to S_{n-1}. So it is one of the points which entered in the competition when we picked x_n. Therefore,

$$f(x_n) \le f(u) + \frac{\varepsilon}{2^{n-1}}. \tag{3.2}$$

From (3.1) and (3.2), we obtain

$$d_\lambda(u, x_n) \le 2^{-n+1}, \quad \text{for all } u \in S_n$$

which gives $\delta(S_n) \le 2^{-n}$, and hence, $\delta(S_n) \to 0$ as $n \to \infty$.

Since (X, d_λ) is complete and $\{S_n\}$ is a decreasing sequence of closed sets, by Cantor's Intersection Theorem 3.18, we infer that

$$\bigcap_{n=1}^{\infty} S_n = \{\bar{x}\}.$$

We still have to prove that this unique point $\bar{x}$ satisfies conditions (a)–(c).

Since $\bar{x} \in S_1$, we have

$$\bar{x} \preccurlyeq x_1 = \hat{x} \quad \text{if and only if} \quad f(\bar{x}) \le f(\hat{x}) - \varepsilon d_\lambda(\bar{x}, \hat{x})$$

and so, $f(\bar{x}) \le f(\hat{x})$. Hence (a) is proved.

Now let $x \ne \bar{x}$. We cannot have $x \preccurlyeq \bar{x}$, because otherwise x would belong to $\bigcap_{n=1}^{\infty} S_n$. So $x \not\preccurlyeq \bar{x}$, which means that

$$f(x) > f(\bar{x}) - \varepsilon d_\lambda(x, \bar{x}),$$

and hence, (c) is proved.

Finally, by writing

$$d_\lambda(\hat{x}, x_n) = d_\lambda(x_1, x_n) \leq \sum_{j=1}^{n-1} d_\lambda(x_j, x_{j+1}) \leq \sum_{j=1}^{n-1} 2^{-j},$$

and taking limit as $n \to \infty$, we obtain $d_\lambda(\hat{x}, \bar{x}) \leq 1$ and so $d(\hat{x}, \bar{x}) \leq \lambda$. This proves (b). □

Remark 3.2. Let $\mathbb{B}_r[x]$ be the closed sphere with center at x and radius $r > 0$. In this context, condition (b) of Theorem 3.1 can be written as $\bar{x} \in \mathbb{B}_\lambda[\hat{x}]$.

Remark 3.3. Strong form of Ekeland's variational principle says that for $\lambda, \varepsilon > 0$ and $\hat{x}$ an ε-approximate solution of an optimization problem, there exists a new point $\bar{x}$ that is not worse than $\hat{x}$ and belongs to a λ-neighborhood of $\hat{x}$, and especially, $\bar{x}$ satisfies (c). Relation (c) says, in fact, that $\bar{x}$ minimizes globally $f(\cdot) + \left(\frac{\varepsilon}{\lambda}\right) d(\cdot, \bar{x})$, which is a Lipschitz perturbation of f.

Aubin and Frankowska [9] established the following form of Ekeland's variational principle which is equivalent to Theorem 3.1.

Theorem 3.2 (Strong Form of Ekeland's Variational Principle). *Let (X,d) be a complete metric space and $f : X \to \mathbb{R} \cup \{+\infty\}$ be a proper, bounded below and lower semicontinuous functional. Let $\hat{x} \in dom(f)$ and $e > 0$ be fixed. Then, there exists $\bar{x} \in X$ such that*

(aa) $f(\bar{x}) - f(\hat{x}) + ed(\hat{x}, \bar{x}) \leq 0$
(bb) $f(\bar{x}) < f(x) + ed(x, \bar{x})$, *for all* $x \in X \setminus \{\bar{x}\}$.

Proof. For the sake of convenience, we set $d_e(u,v) = ed(u,v)$. Then, d_e is equivalent to d and (X, d_e) is complete. For all $x \in X$, define

$$S(x) = \{y \in X : f(y) - f(x) + d_e(x,y) \leq 0\}.$$

We assume that $f(x) \neq +\infty$ for all $x \in X$. For all $x \in X$, we have $x \in S(x)$, and therefore, $S(x)$ is nonempty. So, we can let $y \in S(x)$. Then,

$$f(y) - f(x) + d_e(x,y) \leq 0, \tag{3.3}$$

and also let $z \in S(y)$, then

$$f(z) - f(y) + d_e(y,z) \leq 0. \tag{3.4}$$

Adding (3.3) and (3.4), we obtain

$$f(z) - f(x) + d_e(x,z) \leq 0.$$

Therefore, $z \in S(x)$ which implies that $S(y) \subseteq S(x)$.

We claim that for every $x \in X$, $S(x)$ is closed. Indeed, let $\{x_n\}$ be a sequence in $S(x)$ such that $x_n \to x^* \in X$. Then,

$$f(x_n) - f(x) + d_e(x, x_n) \leq 0.$$

By using the lower semicontinuity of f and continuity of d and hence continuity of d_e, we have

$$f(x^*) - f(x) + d_e(x, x^*) \leq \lim_{n\to\infty} f(x_n) - f(x) + \lim_{n\to\infty} d_e(x, x_n) \leq 0.$$

Therefore, $x^* \in S(x)$ and so $S(x)$ is closed for every $x \in X$.

For all $x \in \operatorname{dom}(f)$, define $\mathscr{V}(x)$ by

$$\mathscr{V}(x) := \inf_{z \in S(x)} f(z).$$

For every $z \in S(x)$,

$$d_e(x,z) \leq f(x) - f(z) \leq f(x) - \mathscr{V}(x),$$

so that the diameter of $S(x)$, $\delta(S(x)) = \sup_{y,z \in S(x)} d_e(y,z)$, is not greater than $2\big(f(x) - \mathscr{V}(x)\big)$.

Define a sequence in the following manner: Fix $x_0 = \hat{x} \in X$; take $x_1 \in S(x_0)$ such that

$$f(x_1) \leq \mathscr{V}(x_0) + 2^{-1}.$$

Denote by x_2 any point in $S(x_1)$ such that

$$f(x_2) \leq \mathscr{V}(x_1) + 2^{-2}.$$

Continue in this manner, we obtain a sequence $\{x_n\}$ such that

$$x_{n+1} \in S(x_n) \quad \text{and} \quad f(x_{n+1}) \leq \mathscr{V}(x_n) + 2^{-(n+1)}.$$

Since $S(x_{n+1}) \subseteq S(x_n)$, we deduce that

$$\mathscr{V}(x_{n+1}) = \inf_{z \in S(x_{n+1})} f(z) \geq \inf_{z \in S(x_n)} f(z) = \mathscr{V}(x_n),$$

and thus,

$$\mathscr{V}(x_n) \leq \mathscr{V}(x_{n+1}).$$

On the other hand, inequality $\mathscr{V}(y) \leq f(y)$ implies that

$$\mathscr{V}(x_{n+1}) \leq f(x_{n+1}) \leq \mathscr{V}(x_n) + 2^{-(n+1)} \leq \mathscr{V}(x_{n+1}) + 2^{-(n+1)},$$

and therefore,

$$0 \leq 2\left(f(x_{n+1}) - \mathscr{V}(x_{n+1})\right) \leq 2^{-n}.$$

It follows that

$$\delta\left(S(x_n)\right) \leq 2^{-n} \to 0 \quad \text{as } n \to \infty.$$

By Cantor's Intersection Theorem 3.18, there exists exactly one point $\bar{x} \in X$ such that

$$\bigcap_{n=0}^{\infty} S(x_n) = \{\bar{x}\}.$$

This implies that $\bar{x} \in S(x_0) = S(\hat{x})$, that is,

$$f(\bar{x}) - f(\hat{x}) + d_e(\hat{x}, \bar{x}) \leq 0$$

and so (aa) holds.

Moreover, $\bar{x}$ also belongs to all $S(x_n)$ and, since $S(\bar{x}) \subseteq S(x_n)$ for all n, we have

$$S(\bar{x}) = \{\bar{x}\}.$$

It follows that $x \notin S(\bar{x})$ whenever $x \neq \bar{x}$ implying that

$$f(x) - f(\bar{x}) + d_e(\bar{x}, x) > 0,$$

that is, (bb) holds. □

Remark 3.4. Theorems 3.1 and 3.2 are equivalent.

Proof. Assume that Theorem 3.2 and the hypothesis of Theorem 3.1 hold. By (aa), we have $f(\bar{x}) \leq f(\hat{x}) - ed(\hat{x}, \bar{x})$. Since $e > 0$, we obtain $f(\bar{x}) \leq f(\hat{x})$ and so (a) holds.

By hypothesis of Theorem 3.1, $f(\hat{x}) \leq \inf_X f + \varepsilon$, that is, $f(\hat{x}) - \inf_X f \leq \varepsilon$. Therefore, in particular, $f(\hat{x}) - f(\bar{x}) \leq \varepsilon$. Set $e = \frac{\varepsilon}{\lambda}$. Then by (aa) of Theorem 3.2, we have

$$d(\hat{x}, \bar{x}) \leq \frac{\lambda}{\varepsilon}\left[f(\hat{x}) - f(\bar{x})\right] \leq \frac{\lambda}{\varepsilon}\varepsilon = \lambda,$$

and (b) of Theorem 3.1 follows.

Conversely, suppose that Theorem 3.1 and the hypothesis of Theorem 3.2 hold. Set $\varepsilon = e\lambda$ and let $\hat{x} \in dom(f)$ and $e > 0$ be given. Take $f(\hat{x}) - \inf_X f \leq \varepsilon$, and consider

$$S_1 := \{x \in X : f(x) - f(\hat{x}) \leq -\varepsilon\},$$

and

$$S_2 := \{x \in X : f(x) - f(\hat{x}) + ed(\hat{x}, x) \leq 0\}.$$

The lower semicontinuity of f implies that S_1 and S_2 are closed. Furthermore, $S_2 \neq \emptyset$ as $\hat{x} \in S_2$. By applying Theorem 3.1 for ε, λ and $S_1 \cup S_2$ instead of X, we obtain $\bar{x} \in S_1 \cup S_2$ such that (a), (b) and

$$(cc) \qquad f(x) - f(\bar{x}) + \frac{\varepsilon}{\lambda} d(x, \bar{x}) > 0 \quad \text{for all } x \in S_1 \cup S_2 \setminus \{\bar{x}\}$$

hold.

If $\bar{x} \in S_2$, then (aa) holds. If $\bar{x} \in S_1$, then we have

$$f(\bar{x}) \leq f(\hat{x}) - \varepsilon.$$

Adding this inequality and (b) multiplied by e, we obtain

$$f(\bar{x}) - f(\hat{x}) + ed(\hat{x}, \bar{x}) \leq 0,$$

that is, $\bar{x} \in S_2$. So (aa) holds.

To show (bb), it is sufficient to check (cc) is satisfied also for $x \notin S_1 \cup S_2$. By the definition of S_2, $x \notin S_2$ means $f(x) - f(\hat{x}) + ed(\hat{x}, x) > 0$. Adding this inequality and (aa), we get

$$f(x) - f(\bar{x}) + ed(x, \bar{x}) \geq 0,$$

that is, (bb) holds. □

The following result is called the weak formulation of Ekeland's variational principle.

Corollary 3.1 (Weak Form of Ekeland's Variational Principle). *Let (X, d) be a complete metric space and $f : X \to \mathbb{R} \cup \{+\infty\}$ be a proper, bounded below and lower semicontinuous functional. Then, for any given $\varepsilon > 0$, there exists $\bar{x} \in X$ such that*

$$f(\bar{x}) \leq \inf_X f + \varepsilon,$$

and

$$f(\bar{x}) < f(x) + \varepsilon d(x, \bar{x}), \quad \textit{for all } x \in X \textit{ with } x \neq \bar{x}. \tag{3.5}$$

Proof. It follows from the fact that there always exists some point $\hat{x} \in X$ such that

$$f(\hat{x}) \leq \inf_X f + \varepsilon.$$

Then by Theorem 3.1, there exists $\bar{x} \in X$ (by taking $\lambda = 1$) such that

$$f(\bar{x}) \leq f(\hat{x}) \leq \inf_X f + \varepsilon,$$

and

$$f(\bar{x}) < f(x) + \varepsilon d(x, \bar{x}), \quad \text{for all } x \in X \text{ with } x \neq \bar{x}.$$

□

Chen et al. [28] introduced the following concept of lower semicontinuity from above.

Definition 3.3. Let X be a metric space. A function $f : X \to \mathbb{R} \cup \{+\infty\}$ is said to be *lower semicontinuous from above* at a point $x \in X$ if $x_n \to x$ as $n \to \infty$ and $f(x_1) \geq f(x_2) \geq \cdots f(x_n) \geq \cdots$ imply that $f(x) \leq \lim_{n \to \infty} f(x_n)$.

Obviously, lower semicontinuity implies lower semicontinuity from above, but the converse does not hold.

Example 3.3. Let $f : \mathbb{R} \to \mathbb{R}$ be a function defined by

$$f(x) = \begin{cases} x + \frac{1}{2}, & \text{if } x < 0, \\ x^2 + 1, & \text{if } x \geq 0. \end{cases}$$

Then, f is lower semicontinuous from above at $x = 0$, but not lower semicontinuous at that point.

Chen et al. [28] showed that the Weierstrass's theorem, Ekeland's variational principle, and Caristi's fixed point theorem hold for lower semicontinuity from above.

The property of Ekeland's variational principle for proper but extended real-valued lower semicontinuous and bounded below functions on a metric space characterizes compactness of the metric space.

Theorem 3.3 (Converse of Ekeland's Variational Principle). *A metric space (X, d) is complete if for every functional $f : X \to \mathbb{R} \cup \{+\infty\}$ which is proper, bounded below and lower semicontinuous on X and for every given $\varepsilon > 0$, there exists $\bar{x} \in X$ such that*

$$f(\bar{x}) \leq \inf_X f + \varepsilon,$$

and

$$f(\bar{x}) \leq f(x) + \varepsilon d(x, \bar{x}), \quad \textit{for all } x \in X.$$

Proof. Consider a Cauchy sequence $\{x_n\}$ in X and define a functional $f : X \to [0,+\infty)$ by

$$f(x) = \lim_{n\to\infty} d(x_n,x), \quad \text{for all } x \in X.$$

Then, f is well-defined, nonnegative, and continuous. Since $\{x_n\}$ is Cauchy, we have

$$f(x_m) = \lim_{n\to\infty} d(x_n,x_m) \to 0, \quad \text{as } m \to \infty.$$

Therefore, $\inf_X f = 0$.

For given $0 < \varepsilon < 1$, choose $\bar{x} \in X$ such that $f(\bar{x}) \le \varepsilon$ and

$$f(\bar{x}) \le f(x) + \varepsilon d(x,\bar{x}), \quad \text{for all } x \in X. \tag{3.6}$$

Letting $x = x_n$ in (3.6) and taking limit as $n \to \infty$, we obtain $f(\bar{x}) \le \varepsilon f(\bar{x})$, and so that $f(\bar{x}) = 0$. This implies that $\{x_n\}$ converges to $\bar{x}$. □

3.3 Applications to Fixed Point Theorems

As a first application of Ekeland's variational principle, we prove the well-known Banach contraction theorem.

Theorem 3.4 (Banach Contraction Theorem). *Let X be a complete metric space and $T : X \to X$ be a contraction mapping. Then T has a unique fixed point in X.*

Proof. Consider the functional $f : X \to [0,\infty)$ defined by

$$f(x) = d(x,T(x)), \quad \text{for all } x \in X.$$

Then, f is bounded below and continuous on X. Choose ε such that $0 < \varepsilon < 1-\alpha$, where α is the Lipschitz constant. By Corollary 3.1, there exists $\bar{x} \in X$ (depending on ε) such that

$$f(\bar{x}) \le f(x) + \varepsilon d(x,\bar{x}), \quad \text{for all } x \in X.$$

Putting $x = T(\bar{x})$, we have

$$\begin{aligned} d(\bar{x},T(\bar{x})) &\le d\left(T(\bar{x}),T(T(\bar{x}))\right) + \varepsilon d(\bar{x},T(\bar{x})) \\ &\le \alpha d(\bar{x},T(\bar{x})) + \varepsilon d(\bar{x},T(\bar{x})) \\ &= (\alpha+\varepsilon) d(\bar{x},T(\bar{x})). \end{aligned}$$

If $\bar{x} \neq T(\bar{x})$, then we obtain $1 \leq \alpha + \varepsilon$, which contradicts to our assumption that $\alpha + \varepsilon < 1$. Therefore, we have $\bar{x} = T(\bar{x})$. The uniqueness of $\bar{x}$ can be proved as in the original proof of the Banach contraction theorem. □

As a second application of Ekeland's variational principle, we prove Caristi's fixed point theorem.

Theorem 3.5 (Caristi's Fixed Point Theorem [24]). *Let X be a complete metric space and $T : X \to X$ be a mapping such that*

$$d(x, T(x)) + \varphi(T(x)) \leq \varphi(x), \quad \textit{for all } x \in X, \tag{3.7}$$

where $\varphi : X \to \mathbb{R} \cup \{+\infty\}$ is a proper, lower semicontinuous and bounded below functional. Then, there exists $\bar{x} \in X$ such that $T(\bar{x}) = \bar{x}$ and $\varphi(\bar{x}) < \infty$.

Proof. By using Corollary 3.1 with $\varepsilon = 1$, we obtain $\bar{x} \in X$ such that

$$\varphi(\bar{x}) < \varphi(x) + d(x, \bar{x}), \quad \text{for all } x \in X \setminus \{\bar{x}\}. \tag{3.8}$$

We claim that $\bar{x} = T(\bar{x})$. Otherwise all $y = T(\bar{x}) \in X$ are such that $y \neq \bar{x}$. Then from (3.7) and (3.8), we have

$$d(y, \bar{x}) + \varphi(y) \leq \varphi(\bar{x}) \quad \text{and} \quad \varphi(\bar{x}) < \varphi(y) + d(y, \bar{x})$$

which cannot hold simultaneously. □

Daneš [33] proved that the Daneš drop theorem [32], Krasnoselskii–Zabreiko renorming theorem [89], Brower's generalization of the Bishop–Phelps theorem [17, 22], Caristi's fixed point theorem, and Ekeland's variational principle are all equivalent.

By using Ekeland's variational principle, Mizoguchi and Takahashi [61] derived the following Caristi and Kirk's theorem [25], which is the set-valued version of Theorem 3.5.

Theorem 3.6 (Caristi–Kirk's Fixed Point Theorem). *Let (X, d) be a complete metric space and $T : X \rightrightarrows X$ be a set-valued map with nonempty values such that for each $x \in X$, there exists $y \in T(x)$ satisfying*

$$d(x, y) + \varphi(y) \leq \varphi(x), \tag{3.9}$$

where $\varphi : X \to \mathbb{R} \cup \{+\infty\}$ is a proper, lower semicontinuous and bounded below functional. Then, T has a fixed point, that is, there exists $\bar{x} \in X$ such that $\bar{x} \in T(\bar{x})$.

Proof. For each $x \in X$, we put $f(x) = y$, where $y \in X$ such that $y \in T(x)$ and $d(x, y) + \varphi(y) \leq \varphi(x)$. Then, f defines a mapping from X into itself such that

$$d(x, f(x)) + \varphi(f(x)) \leq \varphi(x), \quad \text{for all } x \in X.$$

Since φ is proper, there exists $u \in X$ with $\varphi(u) < +\infty$. So, let

$$X' = \{x \in X : \varphi(x) \leq \varphi(u) - d(u,x)\}.$$

Then, X' is nonempty and closed. We can also see that X' is invariant under the mapping f. In fact, for each $x \in X'$, we have

$$d(x, f(x)) + \varphi(f(x)) \leq \varphi(x) \leq \varphi(u) - d(u,x),$$

and hence,

$$\begin{aligned}\varphi(f(x)) &\leq \varphi(u) - \{d(u,x) + d(x, f(x))\} \\ &\leq \varphi(u) - d(u, f(x)).\end{aligned}$$

This implies that $f(x) \in X'$. Now, from Caristi's fixed point theorem, we obtain $z \in X'$ such that $z = f(z) \in T(z)$. □

Now we prove Ekeland's variational principle (Theorem 3.1) by using Caristi and Kirk's Theorem 3.6.

Proof (Proof of Ekeland's Variational Principle [61]). Without loss of generality, we may assume that $\lambda = 1$. Let $\varepsilon > 0$ be given and choose $\hat{x} \in X$ such that

$$f(\hat{x}) \leq \inf_X f + \varepsilon.$$

Let $X' = \{x \in X : f(x) \leq f(\hat{x}) - \varepsilon d(\hat{x}, x)\}$. Then, X' is nonempty. By lower semicontinuity of f, X' is closed subset of a complete metric space, and hence, a complete metric space. For each $x \in X'$, let

$$S(x) = \{y \in X : y \neq x,\ f(y) \leq f(x) - \varepsilon d(x,y)\},$$

and then define

$$T(x) = \begin{cases} x, & \text{if } S(x) = \emptyset, \\ S(x), & \text{if } S(x) \neq \emptyset. \end{cases}$$

Then, T is a set-valued map from X' to itself with nonempty values. Indeed, $T(x) = x \in X'$ if $S(x) = \emptyset$. Since $T(x) = S(x)$ if not, we have for all $y \in T(x)$,

$$\begin{aligned}\varepsilon d(\hat{x}, y) &\leq \varepsilon d(\hat{x}, x) + \varepsilon d(x, y) \\ &\leq f(\hat{x}) - f(x) + f(x) - f(y) \\ &= f(\hat{x}) - f(y),\end{aligned}$$

and hence, $y \in X'$. For all $x \in X'$ and $y \in T(x)$, we have

$$\frac{1}{\varepsilon} f(y) + d(x,y) \le \frac{1}{\varepsilon} f(x).$$

From Theorem 3.6, T has a fixed point $\bar{x} \in X'$. Consequently, $S(\bar{x}) = \emptyset$, that is, $f(x) > f(\bar{x}) - \varepsilon d(\bar{x},x)$ for all $x \in X$ with $x \neq \bar{x}$. Since $\bar{x} \in X'$, we have

$$f(\bar{x}) \le f(\hat{x}) - \varepsilon d(\hat{x},\bar{x}) \le f(\hat{x}).$$

Further, we have

$$\begin{aligned} \varepsilon d(\hat{x},\bar{x}) &\le f(\hat{x}) - f(\bar{x}) \\ &\le f(\hat{x}) - \inf_{x \in X} f(x) \\ &\le \varepsilon, \end{aligned}$$

and hence, $d(\hat{x},\bar{x}) \le 1$. □

By using Ekeland's variational principle, Mizoguchi and Takahashi [61] also derived several fixed point theorems.

Definition 3.4. Let (X,d) be a metric space. For any $x,y \in X$, the *segment* between x and y is defined by

$$[x,y] = \{z \in X : d(x,z) + d(z,y) = d(x,y)\}. \tag{3.10}$$

Definition 3.5. Let (X,d) be a metric space. A mapping $T : X \to X$ is said to be a *directional contraction* if

(i) T is continuous, and
(ii) there exists $\alpha \in (0,1)$ such that for any $x \in X$ with $T(x) \neq x$, there exists $z \in [x,T(x)] \setminus \{x\}$ such that

$$d(T(x),T(z)) \le \alpha d(x,z).$$

Theorem 3.7 (Clarke's Fixed Point Theorem). *Let (X,d) be a complete metric space and $T : X \to X$ be a directional contraction mapping. Then, T has a fixed point.*

Proof. Define the functional $f : X \to [0,+\infty)$ by

$$f(x) = d(x,T(x)), \quad \text{for all } x \in X.$$

Then, T is continuous and bounded below. Applying Corollary 3.1 to f with $\varepsilon \in (0,1-\alpha)$, we conclude that there exists $\bar{x} \in X$ such that

$$f(\bar{x}) \leq f(x) + \varepsilon d(x,\bar{x}), \quad \text{for all } x \in X. \tag{3.11}$$

If $T(\bar{x}) = \bar{x}$, then we are done. Otherwise, by the definition of a directional contraction mapping, there exists a point $z \neq \bar{x}$ with $z \in [\bar{x}, T(\bar{x})]$, that is,

$$d(\bar{x},z) + d(z,T(\bar{x})) = d(\bar{x},T(\bar{x})) = f(\bar{x}), \tag{3.12}$$

satisfying

$$d(T(\bar{x}),T(\bar{z})) \leq \alpha d(\bar{x},z). \tag{3.13}$$

Setting $x = z$ in (3.11), we get

$$f(\bar{x}) \leq f(z) + \varepsilon d(z,\bar{x}).$$

By using the above inequality and (3.12), we obtain

$$d(\bar{x},z) + d(z,T(\bar{x})) \leq d(z,T(z)) + \varepsilon d(z,\bar{x}),$$

or

$$d(\bar{x},z) \leq d(z,T(z)) - d(z,T(\bar{x})) + \varepsilon d(z,\bar{x}). \tag{3.14}$$

By the triangle inequality and (3.13), we obtain

$$d(z,T(z)) - d(z,T(\bar{x})) \leq d(T(\bar{x}),T(z)) \leq \alpha d(\bar{x},z). \tag{3.15}$$

Combining (3.14) and (3.15), we have

$$d(\bar{x},z) \leq (\alpha + \varepsilon) d(\bar{x},z),$$

a contradiction. □

Remark 3.5. It is clear that every contraction mapping is a directional contraction. Therefore, Clarke's fixed point theorem is more general than Banach contraction theorem.

In the following example, we show that Theorem 3.7 is applicable; however, Theorem 3.4 is not.

Example 3.4. Let $X = \mathbb{R}^2$ be a metric space with the metric

$$d(x,y) = |x_1 - y_1| + |x_2 - y_2|, \quad \text{for all } x = (x_1,x_2),\ y = (y_1,y_2) \in \mathbb{R}^2.$$

A segment between two points (a_1,a_2) and (b_1,b_2) consists of the closed rectangle having the two points as diagonally opposite corners. Define

$$T(x) = \left(\frac{3x_1}{2} - \frac{x_2}{3}, x_1 + \frac{x_2}{3}\right), \quad \text{for all } x = (x_1, x_2) \in \mathbb{R}^2.$$

Then, T is a directional contraction mapping. Indeed, if $y = (y_1, y_2) = T(x) \neq x = (x_1, x_2)$, then, $y_2 \neq x_2$ (for otherwise we will also have $y_1 = x_1$). Now the set $[x, y]$ contains points of the form (x_1, t) with t arbitrarily close to x_2 but not equal to x_2. For such points, we have

$$d(T(x_1, t), T(x_1, x_2)) = \frac{2}{3} d((x_1, t), (x_1, x_2)).$$

Hence, T is a directional contraction mapping. We can easily check that the fixed points of T are all points of the form $(x, 3x/2)$. Since T has more than one fixed point clearly then Banach contraction theorem does not apply to this mapping.

Remark 3.6. The fixed point theorems for set-valued directional contraction have been studied widely in the literature, see, for example, [68,69,74] and the references therein.

3.4 Applications to Optimization

We derive the following existence result for a solution of an optimization problem without compactness and convexity assumptions on the underlying space.

Theorem 3.8 (Takahashi's Minimization Theorem). *Let X be a complete metric space and $f : X \to \mathbb{R} \cup \{+\infty\}$ be a proper, bounded below and lower semicontinuous functional. Suppose that, for each $\hat{x} \in X$ with $\inf_X f < f(\hat{x})$, there exists $z \in X$ such that $z \neq \hat{x}$ and*

$$f(z) + d(\hat{x}, z) \leq f(\hat{x}).$$

Then, there exists $\bar{x} \in X$ such that $f(\bar{x}) = \inf_{x \in X} f(x)$, that is, $\bar{x}$ is a solution of OP.

Proof. Assume to the contrary that $\inf_{x \in X} f(x) < f(y)$ for all $y \in X$ and let $\hat{x} \in X$ with $f(\hat{x}) < +\infty$. We define inductively a sequence $\{x_n\}$ in X, starting with $x_1 = \hat{x}$. Suppose that $x_n \in X$ is known. Then, choose $x_{n+1} \in S_{n+1}$ such that

$$S_{n+1} = \{x \in X : f(x) \leq f(x_n) - d(x_n, x)\},$$

and

$$f(x_{n+1}) \leq \frac{1}{2} \left\{ \inf_{x \in S_{n+1}} f(x) + f(x_n) \right\}. \tag{3.16}$$

We claim that $\{x_n\}$ is a Cauchy sequence. Indeed, if $m > n$, then

$$\begin{aligned} d(x_n, x_m) &\leq \sum_{j=n}^{m-1} d(x_j, x_{j+1}) \\ &\leq \sum_{j=n}^{m-1} \{f(x_j) - f(x_{j+1})\} \\ &= f(x_n) - f(x_m). \end{aligned} \tag{3.17}$$

Since $\{f(x_n)\}$ is a decreasing sequence and the functional f is bounded below, there exists $\varepsilon > 0$ such that

$$f(x_n) - f(x_m) < \varepsilon, \quad \text{for all } m > n.$$

Therefore, from (3.17), we have

$$d(x_n, x_m) < \varepsilon, \quad \text{for all } m > n,$$

and hence, $\{x_n\}$ is a Cauchy sequence in X. Since X is a complete metric space, there exists $\tilde{x} \in X$ such that $x_n \to \tilde{x}$. Then, if $m \to \infty$ in (3.17), the lower semicontinuity of f and continuity of d imply that

$$d(x_n, \tilde{x}) \leq f(x_n) - \lim_{m \to \infty} f(x_m) \leq f(x_n) - f(\tilde{x}). \tag{3.18}$$

On the other hand, by hypothesis, there exists a $z \in X$ such that $z \neq \tilde{x}$ and

$$f(z) + d(\tilde{x}, z) \leq f(\tilde{x}). \tag{3.19}$$

By using (3.18) and (3.19), we have

$$\begin{aligned} f(z) &\leq f(\tilde{x}) - d(\tilde{x}, z) \\ &\leq f(\tilde{x}) - d(\tilde{x}, z) + f(x_n) - f(\tilde{x}) - d(x_n, \tilde{x}) \\ &= f(x_n) - \{d(x_n, \tilde{x}) + d(\tilde{x}, z)\} \\ &\leq f(x_n) - d(x_n, z). \end{aligned}$$

Consequently, $z \in S_{n+1}$ for all $n \in \mathbb{N}$. Using (3.16), we obtain

$$2f(x_{n+1}) - f(x_n) \leq \inf_{x \in S_{n+1}} f(x) \leq f(z).$$

Hence,

$$f(\tilde{x}) \leq \lim_{n \to \infty} f(x_n) \leq f(z) \leq f(\tilde{x}) - d(\tilde{x}, z) < f(\tilde{x}),$$

which is a contradiction. Therefore, there exists $\bar{x} \in X$ such that $f(\bar{x}) = \inf_{x \in X} f(x)$. □

Remark 3.7. Takahashi's Minimization Theorem 3.8 and Ekeland's Variational Principle Theorem 3.1 are equivalent.

Proof. We first prove Theorem 3.1 by using Theorem 3.8. Let

$$X_0 = \{x \in X : f(x) \leq f(\hat{x}) - \varepsilon d_\lambda(\hat{x},x)\}.$$

Since $\hat{x} \in X_0$, X_0 is nonempty. By lower semicontinuity of f and continuity of d_λ, X_0 is closed. Further, for each $x \in X_0$,

$$\varepsilon d_\lambda(\hat{x},x) \leq f(\hat{x}) - f(x) \leq f(\hat{x}) - \inf_{y \in X} f(y) \leq \varepsilon,$$

and hence, $d_\lambda(\hat{x},x) \leq 1$, and thus, $d(\hat{x},x) \leq \lambda$. We also have $f(x) \leq f(\hat{x})$.

Assume to the contrary of conclusion (c) in Theorem 3.1 that for every $x \in X_0$, there exists $y \in X$ such that $y \neq x$ and $f(y) \leq f(x) - \varepsilon d_\lambda(y,x)$. Then,

$$\begin{aligned} \varepsilon d_\lambda(\hat{x},y) &\leq \varepsilon d_\lambda(\hat{x},x) + \varepsilon d_\lambda(x,y) \\ &\leq f(\hat{x}) - f(x) + f(x) - f(y) \\ &= f(\hat{x}) - f(y), \end{aligned}$$

and hence, $y \in X_0$. Then, by Theorem 3.8, there exists $\bar{x} \in X$ such that $f(\bar{x}) = \inf_{x \in X_0} f(x)$. This is a contradiction of the hypothesis that there exists $y_0 \in X_0$ with $f(y_0) < f(\bar{x})$.

Now we prove Theorem 3.8 by using Theorem 3.1. By Theorem 3.1, for any given $\varepsilon > 0$, there exists $\bar{x} \in X$ such that

$$f(\bar{x}) < f(x) + \varepsilon d(x,\bar{x}), \quad \text{for all } x \in X \text{ with } x \neq \bar{x}. \tag{3.20}$$

We claim that $f(\bar{x}) = \inf_{x \in X} f(x)$.

Assume to the contrary that there exists $w \in X$ such that $f(w) > \inf_{x \in X} f(x)$. By hypothesis, there exists $z \in X$ such that $z \neq w$ and

$$f(z) + d(w,z) \leq f(w)$$

contradicting (3.20). Hence, $f(\bar{x}) = \inf_{x \in X} f(x)$. □

The following theorem due to Takahashi [84] characterizes the completeness of the underlying metric space X.

Theorem 3.9 (84, Theorem 3]). *A metric space X is complete if for every uniformly continuous function $f : X \to \mathbb{R} \cup \{+\infty\}$ and every $\hat{x} \in X$ with $\inf_X f < f(\hat{x})$, there exists $z \in X$ such that $z \neq \hat{x}$ and*

$$f(z)+d(\hat{x},z) \le f(\hat{x}),$$

then, there exists $\bar{x} \in X$ *such that* $f(\bar{x}) = \inf_{x\in X} f(x)$.

Proof. Let $\{x_n\}$ be a Cauchy sequence in X. Consider the function $f : X \to [0,+\infty)$ defined by

$$f(x) = \lim_{n\to\infty} d(x_n,x), \quad \text{for all } x \in X.$$

Then, f is uniformly continuous and $\inf_{x\in X} f(x) = 0$. Let $f(\hat{x}) > 0$. Then, there exists an $x_m \in X$ such that

$$x_m \neq \hat{x}, \quad f(x_m) < \frac{1}{3} f(\hat{x}) \quad \text{and} \quad d(x_m,\hat{x}) - f(\hat{x}) < f(\hat{x}).$$

Thus, we have

$$3f(x_m) + d(x_m,\hat{x}) < f(\hat{x}) + 2f(\hat{x}) = 3f(\hat{x}).$$

Therefore, there exists an $\bar{x} \in X$ such that $f(\bar{x}) = \inf_{x\in X} f(x) = 0$, and so, $0 = f(\bar{x}) = \lim_{n\to\infty} d(x_n,\bar{x})$. Thus, $\{x_n\}$ converges to $\bar{x}$, and hence, X is complete. □

We give an application of Takahashi's Minimization Theorem 3.8 to prove the Nadler's fixed point theorem for set-valued maps [62]. Let us present some definitions and notations before deriving the Nadler's fixed point theorem.

Let (X,d) be a metric space, $x \in X$ and A be a subset of X. We define

$$d(x,A) = \inf\{d(x,y) : y \in A\}.$$

$CB(X)$ denotes the class of all nonempty bounded closed subsets of X. For any $A, B \in CB(X)$, the *Hausdorff metric* is defined as

$$\mathscr{H}(A,B) = \max\left\{\sup_{x\in A} d(x,B), \sup_{y\in B} d(y,A)\right\}.$$

A set-valued mapping $T : X \to CB(X)$ is said to be *contraction* if there exists $\alpha \in [0,1)$ such that

$$\mathscr{H}(T(x),T(y)) \le \alpha d(x,y), \quad \text{for all } x,y \in X.$$

For further details on set-valued mappings and Hausdorff metric, we refer to [3] and the references therein.

Theorem 3.10 (Nadler's Fixed Point Theorem). *Let (X,d) be a complete metric space and $T : X \to CB(X)$ be a contraction set-valued mapping. Then, T has a fixed point, that is, there exists $\bar{x} \in X$ such that $\bar{x} \in T(\bar{x})$.*

Proof. Assume that T does not have any fixed point. Then, $d(x,T(x)) > 0$ for all $x \in X$. Choose a number $\varepsilon > 0$ with $\varepsilon < \frac{1}{\alpha} - 1$. Then, for every $x \in X$, there exists $y \in X$ such that $y \in T(x)$ and

$$d(x,y) \leq (1+\varepsilon)d(x,T(x)).$$

Since

$$\begin{aligned} d(y,T(y)) &\leq \mathscr{H}(T(x),T(y)) \leq \alpha d(x,y) \\ &\leq \alpha(1+\varepsilon)d(x,T(x)), \end{aligned}$$

we have $\inf_{x \in X} d(x,T(x)) = 0$. Further, since

$$\begin{aligned} d(x,T(x)) - d(y,T(y)) &\geq \frac{1}{1+\varepsilon} d(x,y) - \alpha d(x,y) \\ &= \left(\frac{1}{1+\varepsilon} - \alpha \right) d(x,y), \end{aligned}$$

we obtain,

$$f(y) + d(x,y) \leq f(x),$$

where f is a continuous function defined by

$$f(x) = \left(\frac{1}{1+\varepsilon} - \alpha \right)^{-1} d(x,T(x)), \quad \text{for all } x \in X.$$

By Theorem 3.8, there exists $\bar{x} \in X$ such that $f(\bar{x}) = 0$, that is, $d(\bar{x},T(\bar{x})) = 0$, a contradiction. Hence, T has a fixed point. □

Remark 3.8. Takahashi [84] also derived Caristi's fixed point theorem [24] and Fan's minimization theorem [39] by using Theorem 3.8.

3.5 Applications to Weak Sharp Minima

Daffer et al. [30] and Hamel [48] gave an application of Takahashi's minimization Theorem 3.8 to prove the existence of a weak sharp minima for a class of lower semicontinuous functions.

Let (X,d) be a complete metric space and $f : X \to \mathbb{R} \cup \{+\infty\}$ be a lower semicontinuous function. We define

$$m = \inf\{f(x) : x \in X\}$$

and

$$M = \{y \in X : f(y) = m\}. \tag{3.21}$$

We say that f has *weak sharp minima* if

$$d(x,M) \leq f(x) - m, \qquad \text{for all } x \in X.$$

Theorem 3.11 ([30]). *Let (X,d) be a complete metric space and $f : X \to \mathbb{R} \cup \{+\infty\}$ be a proper, bounded below and lower semicontinuous functional. Suppose that, for each $\hat{x} \in X$ with $\inf_X f < f(\hat{x})$, there exists $z \in X$ such that $z \neq \hat{x}$ and*

$$d(\hat{x},z) \leq f(\hat{x}) - f(z).$$

Then, M defined by (3.21) *is nonempty and f has weak sharp minima.*

Proof. For $\hat{x} \in X$, define

$$S(\hat{x}) = \{z \in X : d(\hat{x},z) \leq f(\hat{x}) - f(z)\}.$$

By the lower semicontinuity of f, $S(\hat{x})$ is closed. By Takahashi's Minimization Theorem 3.8, $M \neq \emptyset$.

We note that for all $z \in S(\hat{x})$, $f(z) \leq f(\hat{x})$. Assume to the contrary that there exists $x_0 \in X$ such that

$$d(x_0,M) > f(x_0) - m. \tag{3.22}$$

Then, $x_0 \notin M$ and this is true for all $z \in S(x_0)$.

Indeed, if there were $z \in S(x_0)$ with $f(z) = m$, then we have

$$d(x_0,M) \leq d(x_0,z) \leq f(x_0) - m,$$

which contradicts (3.22).

We also note that (3.22) holds for all $z \in S(x_0)$. Indeed, take $z \in S(x_0)$, $y \in M$, so that

$$d(x_0,y) \leq d(x_0,z) + d(z,y) \leq f(x_0) - f(z) + d(z,y),$$

and this yields,

$$d(x_0,M) \leq f(x_0) - f(z) + d(z,M).$$

But $d(x_0,M) > f(x_0) - m$. Then, from (3.22), we obtain

$$f(x_0) - m < f(x_0) - f(z) + d(z,M),$$

that is, $f(z) - m < d(z,M)$. This gives (3.22) with y is in place of x_0.

Since $f(x_0) > m$, by hypothesis there is $x_1 \in S(x_0)$ with $x_1 \neq x_0$. Since $f(x_1) - m < d(x_1,M)$, again it is clear that $x_1 \notin M$, $f(x_1) < f(x_0)$ and we can again show as above that

$$f(z) - m < d(z,M), \quad \text{for all } z \in S(x_1) \quad \text{and} \quad S(x_1) \cap M = \emptyset.$$

In addition, we select x_1 such that $f(x_1) = \inf\{f(z) : z \in S(x_1)\}$. This is possible since X is complete, f is lower semicontinuous, and $S(x_1)$ is nonempty and closed. Continuing in this way, we generate a sequence $\{x_n\}$ with the above properties. Namely, if $x_0, x_1, \ldots, x_n$ have been chosen so that at $x_i \in S(x_{i-1})$, $f(x_i) < f(x_{i-1})$, $f(x_i) = \inf\{f(z) : z \in S(x_i)\}$, $S(x_{i-1}) \cap M = \emptyset$, for $i = 1,2,\ldots,n$ and $f(z) - m < d(z,M)$ for all $z \in \bigcup_{i=1}^{n} S(x_{i-1})$, then, since $x_n \notin M$, we can choose $x_{n+1} \in S(x_n)$, $x_{n+1} \neq x_n$ with $f(x_{n+1}) = \inf\{f(z) : z \in S(x_n)\}$, and as above we will have

$$f(x_{n+1}) > m, \;\; f(x_{n+1}) < f(x_n) \text{ and } f(z) - m < d(z,M), \text{ for all } z \in S(x_{n+1}).$$

To see the latter, we write again, just as above, $f(x_n) - m < d(x_n,M) \leq f(x_n) - f(x_{n+1}) + d(x_{n+1},M)$, giving $f(x_{n+1}) - m < d(x_{n+1},M)$. Hence, $A(x_{n+1}) \cap M = \emptyset$.

We now have our sequence $\{x_n\}$ consisting of all different elements and $f(x_{n+1}) < f(x_n)$. Since

$$d(x_{n+k},x_n) \leq \sum_{i=1}^{k} d(x_{n+i},x_{n+i-1}) \leq \sum_{i=1}^{k} (f(x_{n+i-1}) - f(x_{n+i})) = f(x_n) - f(x_{n+k}),$$

and noting that $f(x_n)$ monotonically decreases to some c, $\{x_n\}$ must be Cauchy. Since X is complete, we can assume that x_n converges to $x \in X$. We now show that $x \in \bigcap_{i=0}^{\infty} S(x_i)$. We first show that, for every n, $x_n \in \bigcap_{i=0}^{n-1} S(x_i)$. This follows from the following:

$$\begin{aligned} d(x_{n-k},x_n) &\leq \sum_{j=0}^{k-1} d(x_{n-k+j},x_{n-k+j+1}) \\ &\leq \sum_{j=0}^{k-1} \big(f(x_{n-k+j}) - f(x_{n-k+j+1})\big) \\ &= f(x_{n-k}) - f(x_n), \end{aligned}$$

which shows (recall that all the x_i are outside M) that $x_n \in S(x_{n-k})$, $k = 1,2,\ldots,n$, hence $x_n \in \bigcap_{i=0}^{n-1} S(x_i)$. It follows immediately from this that $x_k \in \bigcap_{i=0}^{n-1} S(x_i)$ for all $k \geq n$. Since $\bigcap_{i=0}^{n-1} S(x_i)$ is a closed set, $x \in \bigcap_{i=0}^{\infty} S(x_i)$. Thus, $x \in S(x_n)$, and $x \neq x_n$; hence, $f(x) < f(x_n)$, and this is a contradiction, since $f(z) \geq f(x_n)$ for all $z \in S(x_n)$.

□

3.6 Equilibrium Problems and Extended Ekeland's Variational Principle

3.6.1 Equilibrium Problems

The mathematical equilibrium problem (in short, EP) is to find an element $\bar{x}$ of a set K such that

$$F(\bar{x},y) \geq 0 \quad \text{for all } y \in K, \tag{3.23}$$

where $F : K \times K \to \mathbb{R}$ is a bifunction such that $F(x,x) = 0$ for all $x \in K$. It seems the most general problem and unified model of several fundamental mathematical problems, namely, optimization problem, saddle point problem, fixed point problem, minimax inequality problems, Nash equilibrium problem, complementarity problem, variational inequality problems, etc. In 1955, Nikaido and Isoda [66] first considered EP as an auxiliary problem to establish existence results for Nash's equilibrium points in non-cooperative games. In the theory of EPs, the key contribution was made by Ky Fan [40], whose new existence results contained the original techniques which became a basis for most further existence theorems in topological spaces. Within the context of calculus of variations, motivated mainly by the works of Stampacchia [75], there arises the work of Brézis, Niremberg, and Stampacchia [23] establishing a more general result than that in [40]. In the last two decades, it emerges as a new direction of research in nonlinear analysis, optimization, optimal control, game theory, mathematical economics, etc. Most of the results on the existence of solutions of equilibrium problems are studied in the setting of topological vector spaces by using some kind of fixed point (Fan–Browder type fixed point) theorem or KKM-type theorem. After the work of Blum and Otteli [18], many mathematicians have started to study the EP again. For further details on equilibrium problems, we refer to [6, 12–14, 18, 23, 26, 27, 43, 44, 47] and the references therein.

Example 3.5. (a) *Minimization Problem.* Let K be a nonempty set and $f : K \to \mathbb{R}$ be a real-valued function. The *minimization problem* (in short, MP) is to find $\bar{x} \in K$ such that

$$f(\bar{x}) \leq f(y), \quad \text{for all } y \in K. \tag{3.24}$$

If we set $F(x,y) = f(y) - f(x)$ for all $x, y \in K$, then MP is equivalent to EP.

(b) *Saddle Point Problem.* Let K_1 and K_2 be nonempty sets and $L : K_1 \times K_2 \to \mathbb{R}$ be a real-valued bifunction. The *saddle point problem* (in short, SPP) is to find $(\bar{x}_1, \bar{x}_2) \in K_1 \times K_2$ such that

$$L(\bar{x}_1, y_2) \leq L(\bar{x}_1, \bar{x}_2) \leq L(y_1, \bar{x}_2), \quad \text{for all } (y_1, y_2) \in K_1 \times K_2. \tag{3.25}$$

Set $K := K_1 \times K_2$ and define $F : K \times K \to \mathbb{R}$ by

$$F((x_1,x_2),(y_1,y_2)) = L(y_1,x_2) - L(x_1,y_2), \tag{3.26}$$

for all $(x_1,x_2),(y_1,y_2) \in K_1 \times K_2$. Then, SPP coincides with EP.

(c) *Nash Equilibrium Problem.* Let I be a finite set of players. For each $i \in I$, let K_i be the strategy set of the ith player. Let $K := \prod_{i\in I} K_i$. For each player $i \in I$, let $\varphi_i : K \to \mathbb{R}$ be the loss function of the ith player, depending on the strategies of all players. For $x = (x_1,x_2,\dots,x_n) \in K$, we define $x^i = (x_1,\dots,x_{i-1},x_{i+1},\dots,x_n)$. The *Nash equilibrium problem* (in short, NEP) [63,64] is to find $\bar{x} \in K$ such that for each $i \in I$,

$$\varphi_i(\bar{x}) \le \varphi_i(\bar{x}^i, y_i), \quad \text{for all } y_i \in K_i. \tag{3.27}$$

Define

$$F(x,y) = \sum_{i=1}^{n} \left(\varphi_i(x^i,y_i) - \varphi_i(x)\right).$$

Then, NEP is same as EP.

(d) *Fixed Point Problem.* Let K be a nonempty subset of $\mathbb{R}^n$ and $f : K \to K$ be a given mapping. The *fixed point problem* (in short, FPP) is to find $\bar{x} \in K$ such that $f(\bar{x}) = \bar{x}$.

Setting $F(x,y) = \langle x - f(x), y - x\rangle$, where $\langle .,.\rangle$ denotes the inner product on $\mathbb{R}^n$. Then, $\bar{x}$ is a solution of FPP if and only if it is a solution of EP.

(e) *Variational Inequality Problem.* Let K be a nonempty subset of $\mathbb{R}^n$. Let $\Phi : K \to \mathbb{R}^n$ be a mapping. The *variational inequality problem* (in short, VIP) is to find $\bar{x} \in K$ such that

$$\langle \Phi(\bar{x}), y - \bar{x}\rangle \ge 0, \quad \text{for all } y \in K, \tag{3.28}$$

where $\langle .,.\rangle$ denotes the inner product on $\mathbb{R}^n$. We set $F(x,y) = \langle \Phi(x), y - x\rangle$ for all $x,y \in K$. Then, VIP is equivalent to EP. For further details on variational inequality problems in the setting of finite dimensional spaces, we refer to [4,38].

3.6.2 Extended Ekeland's Variational Principle

In the recent past, Ekeland's variational principle has been extended for equilibrium-type bifunctions. Such extended forms of EVP is known as extended Ekeland's variational principle (in short, EEVP). It has been used to study the existence of a solution of EP without any convexity assumption on the underlying set; see, for example, [1–3,5,14,18,56,58,67,71] and the references therein.

We present the following equilibrium version of Ekeland's variational principle.

Theorem 3.12 (Extended Ekeland's Variational Principle). *Let K be a nonempty closed subset of a complete metric space (X,d) and $F : K \times K \to \mathbb{R}$ be a bifunction. Assume that $\varepsilon > 0$ and the following assumptions are satisfied:*

(i) For all $x \in K$, $L := \{y \in K : F(x,y) + \varepsilon d(x,y) \leq 0\}$ is closed,;
(ii) $F(x,x) = 0$, for all $x \in K$;
(iii) $F(x,y) \leq F(x,z) + F(z,y)$, for all $x,y,z \in K$.

If $\inf\limits_{y \in K} F(x_0,y) > -\infty$ for some $x_0 \in K$, then there exists $\bar{x} \in K$ such that

(aaa) $F(x_0,\bar{x}) + \varepsilon d(x_0,\bar{x}) \leq 0,$
(bbb) $F(\bar{x},x) + \varepsilon d(\bar{x},x) > 0,$ *for all $x \in K, x \neq \bar{x}$.*

Proof. For the sake of convenience, we set $d_\varepsilon(u,v) = \varepsilon d(u,v)$. Then, d_ε is equivalent to d and (X, d_ε) is complete. For all $x \in X$, define

$$S(x) = \{y \in K : F(x,y) + d_\varepsilon(x,y) \leq 0\}.$$

By condition (i), $S(x)$ is closed for every $x \in K$. From condition (ii), $x \in S(x)$ for all $x \in X$, and therefore, $S(x)$ is nonempty for all $x \in X$. So, we can assume that $y \in S(x)$, that is,

$$F(x,y) + d_\varepsilon(x,y) \leq 0, \tag{3.29}$$

and, also let $z \in S(y)$. Then,

$$F(y,z) + d_\varepsilon(z,y) \leq 0. \tag{3.30}$$

Adding inequalities (3.29) and (3.29), and using condition (iii), we obtain

$$0 \geq F(x,y) + d_\varepsilon(x,y) + F(y,z) + d_\varepsilon(z,y) \geq F(x,z) + d_\varepsilon(x,z).$$

Therefore, $z \in S(x)$, which implies that $S(y) \subseteq S(x)$. Now we show that there is a sequence $\{x_n\}$ of K such that

$$x_{n+1} \in S(x_n), \quad F(x_n, x_{n+1}) < \inf_{z \in S(x_n)} F(x_n, z) + \frac{1}{n+1}, \quad \text{for all } n \geq 0.$$

Define

$$\mathscr{V}(x_0) := \inf_{z \in S(x_0)} F(x_0, z) > -\infty,$$

and construct a sequence in the following manner: There exists $x_1 \in S(x_0)$ such that

$$F(x_0, x_1) < \mathscr{V}(x_0) + \frac{1}{1}.$$

Since $x_1 \in S(x_0)$, we have $S(x_1) \subseteq S(x_0)$. By condition (iii), we have

$$\begin{aligned}\mathscr{V}(x_1) := \inf_{z\in S(x_1)} F(x_1,z) &\geq \inf_{z\in S(x_1)} F(x_0,z) - F(x_0,x_1)\\ &\geq \mathscr{V}(x_0) - F(x_0,x_1) > -\infty.\end{aligned}$$

Then, there exists $x_2 \in S(x_1)$ such that

$$F(x_1,x_2) \leq \mathscr{V}(x_1) + \frac{1}{2}.$$

Again, since $x_2 \in S(x_1)$, we have $S(x_2) \subseteq S(x_1)$. By condition (iii), we have

$$\begin{aligned}\mathscr{V}(x_2) := \inf_{z\in S(x_2)} F(x_2,z) &\geq \inf_{z\in S(x_2)} F(x_1,z) - F(x_1,x_2)\\ &\geq \mathscr{V}(x_1) - F(x_1,x_2) > -\infty.\end{aligned}$$

Continuing in this way, we obtain a sequence $\{x_n\}$ such that

$$x_{n+1} \in S(x_n), \quad F(x_n,x_{n+1}) < \mathscr{V}(x_n) + \frac{1}{n+1},$$

$$\mathscr{V}(x_{n+1}) \geq \mathscr{V}(x_n) - F(x_n,x_{n+1}), \quad \text{for all } n \geq 0,$$

which imply that

$$-\mathscr{V}(x_n) \leq -F(x_n,x_{n+1}) + \frac{1}{n+1} \leq \mathscr{V}(x_{n+1}) - \mathscr{V}(x_n) + \frac{1}{n+1}.$$

Consequently, we obtain

$$\mathscr{V}(x_{n+1}) + \frac{1}{n+1} \geq 0, \quad \text{for all } n \geq 0. \tag{3.31}$$

If $z_1, z_2 \in S(x_n)$, then

$$d(z_1,z_2) \leq d(x_n,z_1) + d(x_n,z_2) \leq -F(x_n,z_1) - F(x_n,z_2) \leq -2\mathscr{V}(x_n).$$

Hence, the diameter of $S(x_n)$, $\delta(S(x_n)) \leq -2\mathscr{V}(x_n)$. Thus, by inequality (3.31), $\delta(S(x_n)) \to 0$ as $n \to \infty$. Since $\{S(x_n)\}$ is a family of closed sets such that $S(x_{n+1}) \subseteq S(x_n)$ for every $n \geq 0$ and $\delta(S(x_n)) \to 0$ as $n \to \infty$, by Cantor's Intersection Theorem 3.18, there exists exactly one point $\bar{x} \in X$ such that

$$\bigcap_{n=0}^{\infty} S(x_n) = \{\bar{x}\}.$$

This implies that $\bar{x} \in S(x_0)$, that is,

$$F(x_0, \bar{x}) + d_\varepsilon(x_0, \bar{x}) \leq 0,$$

and so, (aaa) holds.

Moreover, $\bar{x}$ also belongs to all $S(x_n)$ and, since $S(\bar{x}) \subseteq S(x_n)$ for all n, we have $S(\bar{x}) = \{\bar{x}\}$. It follows that $x \notin S(\bar{x})$ whenever $x \neq \bar{x}$ implying that

$$F(\bar{x}, x) + d_e(\bar{x}, x) > 0,$$

that is, (bbb) holds. □

Remark 3.9. If for each fixed $x \in K$, the function $F(x, .)$ as a mapping on K, is lower semicontinuous, then condition (i) of Theorem 3.12 holds for all $\varepsilon > 0$.

The following example shows that the converse of Remark 3.9 need not be true in general.

Example 3.6. Let $K = [0, \infty)$ and $F : K \times K \to \mathbb{R}$ be defined by

$$F(x,y) = \begin{cases} 0, & \text{if}(x,y) \in \{0\} \times K or x = y, \\ 1, & \text{otherwise.} \end{cases}$$

It is easy to see that F satisfies all of the assumptions of Theorem 3.12 for $\varepsilon > 0$, but $F(x, .)$ is not lower semicontinuous at 0.

The following example shows that condition (iii) of Theorem 3.12 cannot be omitted.

Example 3.7. Let $K = [0, 1]$ and $F : K \times K \to \mathbb{R}$ be defined by $F(x,y) = -\frac{1}{3}\sqrt{|x-y|}$. If in Theorem 3.12, we let $\varepsilon = \frac{1}{2}$, then F satisfies all the assumptions of Theorem 3.12 except condition (iii), but the conclusion of Theorem 3.12 does not hold.

We mention some equivalences among extended Ekelend's variational principle, extended Takahashi's minimization theorem, Caristi–Kirk fixed point theorem for set-valued maps, and Oettli–Théra theorem.

Theorem 3.13. *Let (X, d) be a complete metric space. Let $F : X \times X \to \mathbb{R} \cup \{+\infty\}$ be a bifunction such that it is lower semicontinuous in the second argument and the following conditions hold:*

(i) $F(x, x) = 0$ for all $x \in X$;
(ii) $F(x, y) \leq F(x, z) + F(z, y)$ for all $x, y, z \in X$.

Assume that there exists $\hat{x} \in X$ such that $\inf_{x \in X} F(\hat{x}, x) > -\infty$. Let

$$\hat{S} := \{x \in X : F(\hat{x}, x) + d(\hat{x}, x) \leq 0\} \tag{3.32}$$

(From (i) it follows that $\hat{x} \in \hat{S} \neq \emptyset$*). Then, the following statements are equivalent:*

(a) (Extended Ekeland's Variational Principle). There exists $\bar{x} \in \hat{S}$ *such that*

$$F(\bar{x},x) + d(\bar{x},x) > 0, \quad \text{for all } x \neq \bar{x}. \tag{3.33}$$

(b) (Extended Takahashi's Minimization Theorem). Assume that

$$\begin{cases} \text{for every } \hat{x} \in \hat{S} \text{ with } \inf_{x \in X} F(\hat{x},x) < 0 \text{ there exists} \\ x \in X \text{ such that } F(\hat{x},x) + d(\hat{x},x) \leq 0, \quad \text{for all } x \neq \hat{x}. \end{cases} \tag{3.34}$$

Then, there exists $\bar{x} \in \hat{S}$ *such that* $F(\bar{x},x) \geq 0$ *for all* $x \in X$*.*

(c) (Caristi–Kirk Fixed Point Theorem). Let $T : X \rightrightarrows X$ *be a set-valued map such that*

$$\begin{cases} \text{for every } \hat{x} \in \hat{S} \text{ there exists} \\ x \in T(\hat{x}) \text{ satisfying } F(\hat{x},x) + d(\hat{x},x) \leq 0. \end{cases} \tag{3.35}$$

Then, there exists $\bar{x} \in \hat{S}$ *such that* $\bar{x} \in T(\bar{x})$*.*

(d) (Oettli–Théra Theorem). Let $D \subset X$ *have the property that*

$$\begin{cases} \text{for every } \hat{x} \in \hat{S} \setminus D \text{ there exists} \\ x \in X \text{ such that } F(\hat{x},x) + d(\hat{x},x) \leq 0, \quad \text{for all } x \neq \hat{x}. \end{cases} \tag{3.36}$$

Then, there exists $\bar{x} \in \hat{S} \cap D$*.*

Proof. (a) $\Rightarrow$ (d): Let (a) and the hypothesis of (d) hold. Then, (a) gives $\bar{x} \in \hat{S}$ such that $F(\bar{x},x) + d(\bar{x},x) > 0$, for all $x \neq \bar{x}$. From (3.36), we have $\bar{x} \in D$. Hence, $\bar{x} \in \hat{S} \cap D$ and (d) holds.

(d) $\Rightarrow$ (a): Let (d) hold. For all $\hat{x} \in X$, define

$$\Gamma(\hat{x}) = \{x \in X : F(\hat{x},x) + d(\hat{x},x) \leq 0, \; x \neq \hat{x}\}.$$

Choose $D := \{\hat{x} \in X : \Gamma(\hat{x}) = \emptyset\}$. If $\hat{x} \notin D$, then from the definition of D, there exists $x \in \Gamma(\hat{x})$. Hence, (3.36) is satisfied, and by (d), there exists $\bar{x} \in \hat{S} \cap D$. Then, $\Gamma(\bar{x}) = \emptyset$, that is, $F(\hat{x},x) + d(\hat{x},x) > 0$ for all $x \neq \bar{x}$. Hence, (a) holds.

(b) $\Rightarrow$ (d): Suppose that both (b) and the hypothesis of (d) hold. Assume to the contrary that $\hat{x} \notin D$ for all $\hat{x} \in \hat{S}$. Then, by (3.36) for all $\hat{x} \in \hat{S}$

$$\text{there exists } x \neq \hat{x} \text{ with } F(\hat{x},x) + d(\hat{x},x) \leq 0. \tag{3.37}$$

Hence, (3.35) is satisfied. By (b), there exists $\bar{x} \in \hat{S}$ such that $F(\bar{x},x) \geq 0$, for all $x \in X$. This implies that $F(\bar{x},x) + d(\bar{x},x) > 0$, for all $x \in X$, $x \neq \bar{x}$, a contradiction with (3.37). Hence, $\hat{x} \in D$ for some $\hat{x} \in \hat{S}$ and (d) holds.

(d) $\Rightarrow$ (b): Suppose that both (d) and the hypothesis of (b) hold. Choose $D := \{\hat{x} \in X : \inf_{x \in X} F(\hat{x},x) \geq 0\}$. Then, (3.36) follows from (3.34), and (d) furnishes some $\bar{x} \in \hat{S} \cap D$. It follows from the definition of D that $\inf_{x \in X} F(\bar{x},x) \geq 0$. Hence, (b) holds.

(c) $\Rightarrow$ (d): Let (c) and the hypothesis of (d) hold. Define a set-valued map $T : X \rightrightarrows X$ by

$$T(\hat{x}) = \{x \in X : x \neq \hat{x}\}.$$

Assume to the contrary that $\hat{x} \notin D$ for all $\hat{x} \in \hat{S}$. Then, (3.35) follows from (3.36), and by (c) there exists $\bar{x} \in T(\bar{x})$. But this is clearly impossible from the definition of T. Hence, $\hat{x} \in D$ for some $\hat{x} \in \hat{S}$ and (d) holds.
(d) $\Rightarrow$ (c): Suppose that both (d) and the hypothesis of (c) hold. Choose $D := \{\hat{x} \in X : \hat{x} \in T(\hat{x})\}$. Then, (3.36) follows from (3.35), and (d) furnishes some $\bar{x} \in \hat{S} \cap D$ which, from the definition of D, necessarily belongs to $T(\bar{x})$. Hence, (c) holds. □

Remark 3.10. Theorem 3.13 (b) guarantees the existence of a solution of EP without any compactness or convexity assumption on the underlying set.

As applications of Theorem 3.12, we derive the following existence results for a solution of EP.

Theorem 3.14. *If the following condition*

(iv) for every $x \in K$ with $\inf_{y \in K} F(x,y) < 0$, there exists $z \in K$ such that $z \neq x$ and $F(x,z) + \varepsilon d(x,z) \leq 0$,

together with the assumptions of Theorem 3.12 are satisfied, then the solution set of EP is nonempty, that is, there exists $\bar{x} \in K$ such that $F(\bar{x},y) \geq 0$, for all $y \in K$.

Proof. By Theorem 3.12, there exists $\bar{x} \in K$ such that

$$F(\bar{x},y) + \varepsilon d(\bar{x},y) > 0, \quad \text{for all } y \in K,\ y \neq \bar{x}. \tag{3.38}$$

We claim that $\bar{x}$ is a solution of EP. Otherwise, there exists $y \in K$ such that $F(\bar{x},y) < 0$. From assumption (iv), we obtain $z \in K$ with $z \neq \bar{x}$ and $F(\bar{x},z) + \varepsilon d(\bar{x},z) \leq 0$ which contradicts inequality (3.38). □

The following result guarantees the existence of a solution of the minimization problem for a lower bounded function on a closed set.

Theorem 3.15. *Let K be a nonempty closed subset of a complete metric space (X,d), $f : K \to \mathbb{R}$ be a lower bounded function. If for every $x \in K$ with $\inf_{y \in K} f(y) < f(x)$, there exists $z \in K$ such that $z \neq x$ and $f(z) + \varepsilon d(x,z) \leq f(x)$, then there exists $\bar{x} \in K$ such that $f(\bar{x}) \leq f(y)$ for all $y \in K$.*

Proof. Define $F : K \times K \to \mathbb{R}$ by

$$F(x,y) = f(y) - f(x), \quad \text{for all } x,y \in K.$$

Then, F satisfies all of the assumptions of Theorem 3.14. So, there exists $\bar{x} \in K$ such that $F(\bar{x},y) \geq 0$, for all $y \in K$. □

Remark 3.11. Let $\{t_n\}$ be a sequence of positive real numbers which is bounded below by some positive numbers. If conditions of Theorem 3.12 hold for $\varepsilon = t_n$, for all n, then the sequence of approximate solutions $\{x_n\}$ obtained by Theorem 3.12 corresponding t_n is bounded. Indeed, putting $\alpha = \inf_{y\in K} F(x_0,y)$ and $\beta = \inf_{n\in\mathbb{N}} t_n$ and using Theorem 3.12 with $\varepsilon = t_n$ for every n, we have

$$d(x_0,x_n) \le \frac{1}{t_n} - F(x_0,x_n) \le \frac{1}{t_n} - \inf_{y\in K} F(x_0,y) \le \frac{1}{\beta} - \alpha.$$

Definition 3.6. Let K be a nonempty set, $F : K \times K \to \mathbb{R}$ be a bifunction and $\varepsilon > 0$ be given. The element $\bar{x} \in K$ is said to be an *ε-solution* of EP if

$$F(\bar{x},y) \ge -\varepsilon d(\bar{x},y) \quad \text{for all } y \in K. \tag{3.39}$$

It is called *strictly ε-solution* of EP if the inequality (3.39) is strict for all $x \ne y$.

Theorem 3.16. *Let K be a nonempty compact subset of a metric space (X,d), $F : K \times K \to \mathbb{R}$ be a bifunction and $\{t_n\}$ be a decreasing sequence of positive real numbers such that $t_n \to 0$. Assume that*

(i) $L := \{y \in K : F(x,y) + t_n d(x,y) \le 0\}$ is closed for every $x \in K$ and for all $n \in \mathbb{N}$,
(ii) $F(x,x) = 0$, for all $x \in K$,
(iii) $F(x,y) \le F(x,z) + F(z,y)$, for all $x,y,z \in K$,
(iv) $U := \{y \in K : F(y,x) + t_n d(y,x)\}$ is closed for every $x \in K$ and for all $n \in \mathbb{N}$.

If $\inf_{y\in K} F(x_0,y) > -\infty$ for some $x_0 \in K$, then the set of solutions of EP is nonempty.

Proof. By Theorem 3.12, for each $n \in \mathbb{N}$, there exists t_n-equilibrium point of f, that is, there exists $x_n \in K$ such that

$$F(x_n,y) \ge -t_n d(x_n,y), \quad \text{for all } y \in K. \tag{3.40}$$

Since K is compact, we can choose a subsequence $\{x_{n_k}\}$ of $\{x_n\}$ such that $x_{n_k} \to \bar{x}$ as $k \to \infty$. By using inequality (3.40) and since $\{t_n\}$ is decreasing sequence, for every fixed positive integer k_0, we have

$$x_{n_k} \in \{y \in K : F(x,y) + t_{n_{k_0}} d(x,y)\}, \quad \text{for all } k \ge k_0.$$

By condition (iv) and $x_{n_k} \to \bar{x}$, we deduce that

$$F(\bar{x},y) + t_{n_{k_0}} d(\bar{x},y) \ge 0, \quad \text{for all } y \in K. \tag{3.41}$$

Since n_{k_0} is arbitrary and t_{n_k} approaches to zero as $k \to +\infty$, inequality (3.41) implies that $\bar{x}$ is a solution of EP. □

Corollary 3.2 ([15, Proposition 3.2]). *Let K be a compact subset of a complete metric space (X,d) and $F : K \times K \to \mathbb{R}$ be a function satisfying the following conditions.*

(i) $F(x,.)$ *is lower semicontinuous, for every* $x \in K$;
(ii) $F(x,x) = 0$, *for every* $x \in K$;
(iii) $F(x,y) \leq F(y,z) + F(z,x)$ *for every* $x,y,z \in K$;
(iv) $F(.,y)$ *is upper semicontinuous, for every* $y \in K$.

Then, the set of solutions of EP is nonempty and compact.

We now consider the noncompact case. Rest of this section, we assume that (X,d) is a metric space with Heine–Borel property, that is, each closed bounded subset of X is compact. For instance, if X is a reflexive Banach space with sparable dual, then X is equipped with metrizable weak topology has Heine–Borel property. Let K be a closed subset of X and $F : K \times K \to \mathbb{R}$ be a given function.

Consider the following coercivity condition:

$$\exists\, \mathbb{B}(c,r) : \forall x \in K \backslash K_r,\ \exists\, y \in K \text{ satisfying } d(y,c) < d(x,c) \text{ and } F(x,y) \leq 0, \quad (3.42)$$

where $K_r = K \cap \mathbb{B}(c,r)$ and $\mathbb{B}(c,r) = \{y \in X : d(c,y) \leq r\}$.

If $X = \mathbb{R}^n$, then Euclidean space and $c = 0$, then the above coercivity condition reduces to the coercivity introduced in [15].

Theorem 3.17. *Let* K *be a nonempty closed subset of* (X,d) *and* $\{t_n\}$ *be a decreasing sequence of positive real numbers such that* $t_n \to 0$. *Suppose that* $F : K \times K \to \mathbb{R}$ *satisfies conditions (i)–(iv) of Theorem 3.16. If* $\inf_{y \in K} f(x_0,y) > -\infty$ *for some* $x_0 \in K$ *and coercivity condition* (3.42) *hold, then the set of solutions of EP is nonempty.*

Proof. For each $x \in K$, consider the nonempty set

$$S(x) = \{y \in K : d(y,c) \leq d(x,c),\ F(x,y) \leq 0\}.$$

Observe that for every $x,y \in K_r$, $y \in S(x)$ implies that $S(y) \subseteq S(x)$. Indeed, for $z \in S(y)$ we have $d(z,c) \leq d(y,c) \leq d(x,c)$ and by condition (iii) of Theorem 3.16 $F(x,z) \leq F(x,y) + F(y,z) \leq 0$. Since K_r is nonempty and compact, by Theorem 3.16 there exists $x_r \in K_r$ such that

$$F(x_r,y) \geq 0, \quad \text{for all } y \in K_r. \quad (3.43)$$

Suppose that there exists $x \in K$ with $F(x_r,x) < 0$ and put

$$a = \min_{y \in S(x)} d(y,c).$$

We distinguish two cases.

case 1: $a \leq r$. Let $y_0 \in S(x)$ such that $d(y_0,c) = a \leq r$. Then, we have $F(x,y_0) \leq 0$. Since $F(x_r,x) < 0$, it follows by condition (iii) of Theorem 3.16 that

$$F(x_r, y_0) \leq F(x_r, x) + F(x, y_0) < 0,$$

which contradicted by (3.43).

case 1: $a > r$. Let again $y_0 \in S(x)$ such that $d(y_0, c) = a > r$. Then by condition (3.42), we can choose an element $y_1 \in K$ with $d(y_1, c) < d(y_0, c) = a$ such that $F(y_0, y_1) \leq 0$. Thus, $y_1 \in S(y_0) \subseteq S(x)$. Hence

$$d(y_1, c) < a = \min_{y \in S(x)} d(y, c),$$

which is a contradiction. Therefore, there is no $x \in K$ such that $F(x_r, x) < 0$, that is, x_r is a solution of EP on K. □

Appendix A

Theorem 3.18 (Cantor's Intersection Theorem). *Let (X, d) be a complete metric space and let $\{D_n\}$ be a decreasing sequence (that is, $D_{n+1} \subseteq D_n$) of nonempty closed subsets of X such that the diameter of D_n, $\delta(D_n) \to 0$ as $n \to \infty$. Then, the intersection $\bigcap_{n=1}^{\infty}$ contains exactly one point.*

References

1. Al-Homidan, S., Ansari, Q.H., Yao, J.-C.: Some generalizations of Ekeland-type variational principle with applications to equilibrium problems and fixed point theory. Nonlinear Anal. **69**, 126–139 (2008)
2. Amini-Harandi, A., Ansari, Q.H., Farajzadeh, A.P.: Existence of equilibria in complete metric spaces. Taiwanese J. Math. **16**, 777–785 (2012)
3. Ansari, Q.H.: Metric Spaces: Including Fixed Point Theory and Set-Valued Maps. Narosa Publishing House, New Delhi (2010)
4. Ansari, Q.H., Lalitha, C.S., Mehta, M.: Generalized Convexity, Nonsmooth Variational Inequalities, and Nonsmooth Optimization. CRC Press, Taylor and Francis, London, New York (2013)
5. Ansari, Q.H., Lin, L.-J.: Ekeland-type variational principle and equilibrium problems. In: Mishra, S.K. (ed.) Topics in Nonconvex Optimization. Springer Optimization and Its Applications, vol. 59, pp. 147–174. Springer, New York (2011)
6. Ansari, Q.H., Wong, N.C., Yao, J.-C.: The existence of nonlinear inequalities. Appl. Math. Lett. **12**, 89–92 (1999)
7. Aubin, J.-P.: Mathematical Methods of Game and Economic Theory. North-Holland Publishing, Amsterdam, New York, Oxford (1979)
8. Aubin, J.-P., Ekeland, I.: Applied Nonlinear Analysis. Wiley, New York, Chichester, Brisbane, Toronto, Singapore (1984)
9. Aubin, J.-P., Frankowska, H.: Set-Valued Analysis. Birkhäuser, Boston, Basel, Berlin (1990)

10. Bae, J.S., Park, S.: Remarks on the Caristi-Kirk fixed point theorem. Bull. Kor. Math. Soc. **19**, 57–60 (1983)
11. Bao, T.Q., Khanh, P.Q.: Are several recent generalizations of Ekeland's variational principle more general than the original principle? Acta Math Vietnam. **28**, 345–350 (2003)
12. Bianchi, M., Schaible, S.: Generalized monotone bifunctions and equilibrium problems. J. Optim. Theor. Appl. **90**, 31–43 (1996)
13. Bianchi, M., Schaible, S.: Equilibrium problems under generalized convexity and generalized monotonicity. J. Global Optim. **30**, 121–134 (2004)
14. Bianchi, M., Kassay, G., Pini, R.: Existence of equilibria via Ekeland's principle. J. Math. Anal. Appl. **305**, 502–512 (2005)
15. Bianchi, M., Pini, R.: A note on equilibrium problems with properly quasimonotone functions. J. Global Optim. **20**, 67–76 (2001)
16. Bianchi, M., Pini, R.: Coercivity conditions for equilibrium problems. J. Optim. Theor. Appl. **124**, 79–92 (2005)
17. Bishop, E., Phelps, R.R.: The support function of a convex set. In: Klee, P. (ed.) Convexity, Proc. Sympo. Pure Math., vol. 7, pp. 27–35. American Mathematical Society, Providence, RI (1963)
18. Blum, E., Oettli, W.: From optimization and variational inequalities to equilibrium problems. Math. Student **63**, 123–145 (1994)
19. Borwein, J.M., Preiss, D.: A smooth variational principle with applications to subdiffferentiability and to differentiability of convex functions. Trans. Am. Math. Soc. **303**, 517–527 (1987)
20. Borwein, J.M., Zhu, Q.J.: Techniques of Variational Analysis. Springer, Berlin, Heidelberg, New York (2005)
21. Bosch, C., Garcia, A., Garcia, C.L.: An extension of Ekeland's variational principle to locally complete spaces. J. Math. Anal. Appl. **328**, 106–108 (2207)
22. Brézis, H., Browder, F.: A general principle on ordered sets in nonlinear functional analysis. Adv. Math. **21**, 355–364 (1976)
23. Brézis, H., Nirenberg, L., Stampacchia, G.: A remark on Ky Fan's minimax principle. Boll. Unione Mat. Ital. **6**, 293–300 (1972)
24. Caristi, J.: Fixed point theorems for mappings satisfying inwardness conditions. Trans. Am. Math. Soc. **215**, 241–251 (1976)
25. Caristi, J., Kirk, W.A.: Geometric fixed point theory and inwardness conditions. In The Geometry of Metric and Linear Spaces. Lecture Notes in Mathematics, vol. 490, pp. 74–83. Springer, Berlin, Heidelberg, New York (1975)
26. Chadli, O., Chbani, Z., Riahi, H.: Equilibrium problems with generalized monotone bifunctions and applications to variational inequalities. J. Optim. Theor Appl. **105**, 299–323 (2000)
27. Chadli, O., Wong, N.C., Yao, J.-C.: Equilibrium problems with applications to eigenvalue problems. J. Optim. Theor. Appl. **117**, 245–266 (2003)
28. Chen, Y., Cho, Y.J., Yang, L.: Note on the results with lower semicontinuity. Bull. Kor. Math. Soc. **39**, 535–541 (2002)
29. Clarke, F.: Pointwise contraction criteria for the existence of fixed points. MRC Technical Report **1658**, July 1976, University of Wisconsin, Madison, Wisconsin, 1976
30. Daffer, P.Z., Kaneko, H., Li, W.: Variational principle and fixed points. SIMAA **4**, 129–136 (2002)
31. Dancs, S., Hegedũs, M., Medvegyev, P.: A general ordering and fixed-point principle in complete metric space. Acta Sci. Math. **46**, 381–388 (1983)
32. Daneš, J.: A geometric theorem useful in nonlinear analysis. Boll. Unione Mat. Ital. **6**, 369–372 (1972)
33. Daneš, J.: Equivalence of some geometric and related results of nonlinear functional analysis. Comment. Math. Univ. Carolinae **26**, 443–454 (1985)
34. De Figueiredo, D.G.: The Ekeland Variational Principle with Applications and Detours. Tata Institute of Fundamental Research, Bobmay (1989)
35. Ekeland, I.: Sur les prolèms variationnels. C. R. Acad. des Sci. Paris **275**, 1057–1059 (1972)

36. Ekeland, I.: On the variational principle. J. Math. Anal. Appl. **47**, 324–353 (1974)
37. Ekeland, I.: On convex minimization problems. Bull. Am. Math. Soc. **1**, 445–474 (1979)
38. Facchinei, F., Pang, J.-S.: Finite Dimensional Variational Inequalities and Complementarity Problems, I and II. Springer, New York, Berlin, Heidelberg (2003)
39. Fan, K.: Extensions of two point theorems of F.E. Browder. Math. Z. **112** 234–240 (1969)
40. Fan, K.: A minimax inequlity and applications. In: Shisha, O. (ed.) Inequality III, pp. 103–113. Academic Press, New York (1972)
41. Farkas, C.: A generalizated form of Ekeland's variational principle. An. Şt. Univ. Ovidius Constanţa **20**, 101–112 (2012)
42. Feng, Y.Q., Liu, S.Y.: Fixed point theorems for multivalued contractive mappings and multivalued Caristi type mappings. J. Math. Anal. Appl. **317**, 103–112 (2006)
43. Flores-Bazán, F.: Existence theorems for generalized noncoercive equilibrium problems: The quasi-convex case. SIAM J. Optim. **11**, 675–690 (2000)
44. Flores-Bazán, F.: Existence theory for finite-dimensional pseudomonotone equilibrium problems. Acta Appl. Math. **77**, 249–297 (2003)
45. Georgiev, P.G.: The strong Ekeland variational principle, the strong drop theorem and applications. J. Math. Anal. Appl. **131**, 1–21 (1988)
46. Göfert, A., Tammer, Chr., Riahi, H., Zălinescu, C.: Variational Methods in Partially Ordered Spaces. Springer, New York, Berlin, Heidelberg (2003)
47. Hadjisavvas, N., Komlósi, S., Schaible, S.: Handbook of Generalized Convexity and Generalized Monotonicity. Springer, Berlin, Heidelberg, New York (2005)
48. Hamel, A.: Remarks to an equivalent formulation of Ekeland's variational principle. Optimization **31**, 233–238 (1994)
49. Hamel, A.: Phelps' lemma, Daneš' drop theorem and Ekeland principle in locally convex spaces. Proc. Am. Math. Soc. **131**, 3025–3038 (2003)
50. Hamel, A.: Equivalents to Ekeland's variational principle in uniform spaces. Nonlinear Anal. **62**, 913–924 (2005)
51. Isac, G.: Ekeland principle and nuclear cones: A geometrical aspect. Math. Comput. Model. **26**, 111–116 (1997)
52. Jing-Hui, Q.: Ekeland's variational principle in locally complete spaces. Math. Nachr. **257**, 55–58 (2003)
53. Kada, O., Suzuki, T., Takahashi, W.: Nonconvex minimization theorems and fixed point theorems in complete metric spaces. Math. Japonica **44**, 381–391 (1996)
54. Kas, P., Kassay, G., Sensoy, Z.B.: On generalized equilibrium points. J. Math. Anal. Appl. **296**, 619–633 (2004)
55. Khamsi, M.A., Kirk, W.A.: An Introdunction to Metric Spaces and Fixed Point Theory. Wiley, New York, Chichester, Brisbane, Singapore, Toronto (2001)
56. Lin, L.-J., Du, W.S.: Ekeland's variational principle, minimax theorems and existence of nonconvex equilibria in complete metric spaces. J. Math. Anal. Appl. **323**, 360–370 (2006)
57. Lin, L.-J., Du, W.S.: Some equivalent formulations of generalized Ekeland's variational principle and their applications. Nonlinear Anal. **67**, 187–199 (2007)
58. Lin, L.-J., Du, W.S.: On maximal element theorems, varinats of Ekeland's variational principle and their applications. Nonlinear Anal. **68**, 1246–1262 (2008)
59. McLinden, L.: An application of Ekeland's theorem to minimax problems. Nonlinear Anal. Theor. Meth. Appl. **6**, 189–196 (1982)
60. Meghea, I.: Ekeland Variational Principle with Generalizations and Variants. Old City Publishing, Philadelphia (2009)
61. Mizoguchi, N., Takahashi, W.: Fixed point theorems for multivalued mappings on complete metric spaces. J. Math. Anal. Appl. **141**, 177–188 (1989)
62. Nadler, S.B. Jr.: Multivalued contraction mappings. Pac. J. Math. **30**, 475–488 (1969)
63. Nash, J.F.: Equilibrium points in n-person games. Proc. Natl. Acad. Sci. USA **36**, 48–49 (1950)
64. Nash, J.F.: Noncooperative games. Ann. Math. **54**, 286–295 (1951)

65. Nemeth, A.B.: A nonconvex vector minimization problem. Nonlinear Anal. Theor. Meth. Appl. **10**, 669–678 (1986)
66. Nikaido, H., Isoda, K.: Note on noncooperative convex games. Pac. J. Math. **5**, 807–815 (1955)
67. Oettli, W., Théra, M.: Equivalents of Ekeland's principle. Bull. Aust. Math. Soc. **48**, 385–392 (1993)
68. Park, S.: On fixed points of set-valued directional contractions. Int. J. Math. Math. Sci. **8**, 663–667 (1985)
69. Park, S.: Some applications of Ekeland's variational principle to fixed point theory. In: Singh, S.P. (ed.) Approximation Theory and Applications. Reseach Notes in Mathematics, vol. 133, pp. 159–172. Pitman, Boston (1985)
70. Park, S.: Equivalent formulations of Ekeland's variational principle for approximate soltuions of minimization problems and their applications. In: Singh, S.P., Sehgal, V.M., Burry, J.H.W. (eds.) Operator Equations and Fixed Point Theorems, pp. 55–68. MSRI-Korea Publishers 1, Soul (1986)
71. Park, S.: On generalizations of the Ekeland-type variational principles. Nonlinear Anal. **39**, 881–889 (2000)
72. Penot, J.-P.: The drop theorem, the petal theorem and Ekeland's variational principle. Nonlinear Anal. Theor. Meth. Appl. **10**, 813–822 (1986)
73. Rockafellar, R.T.: Directionally Lipschitz functions and subdifferential calculus. Proc. Lond. Math. Soc. **39**, 331–355 (1979)
74. Sehgal, V.M.: Some fixed point theorems for set-valued directional contraction mappings. Int. J. Math. Math. Sci. **3**, 455–460 (1980)
75. Stampacchia, G.: Formes bilinéaires coercitives sur les ensembles convexes. C. R. Acad. Sci. Paris **258**, 4413–4416 (1964)
76. Sullivan, F.: A characterization of complete metric spaces. Proc. Am. Math. Soc. **83**, 345–346 (1981)
77. Shioji,N., Suzuki, T., Takahashi, W.: Contractive mappings, Kannan mappings and metric completeness. Proc. Am. Math. Soc. **126**, 3117–3124 (1998)
78. Squassina, M.: On Ekeland's variational principle. J. Fixed Point Theor. Appl. **10**, 191–195 (2011)
79. Suzuki, T.: Generalized distnace and existence theorems in complete metric spaces. J. Math. Anal. Appl. **253**, 440–458 (2001)
80. Suzuki, T.: On Downing-Kirk's theorem. J. Math. Anal. Appl. **286**, 453–458 (2003)
81. Suzuki, T.: Generalized Caristi's fixed point theorems by Bae and others. J. Math. Anal. Appl. **302**, 502–508 (2005)
82. Suzuki, T.: The strong Ekeland variational principle. J. Math. Anal. Appl. **320**, 787–794 (2006)
83. Suzuki, T., Takahashi, W.: Fixed point theorems and characterizations of metric completeness. Top. Meth. Nonlinear Anal. **8**, 371–382 (1996)
84. Takahashi, W.: Existence theorems generalizing fixed point theorems for multivalued mappings. In: Baillon, J.-B., Théra, M. (eds.) Fixed Point Theory and Applications. Pitman Research Notes in Mathematics, vol. 252, pp. 397–406. Longman, Harlow (1991)
85. Takahashi, W.: Nonlinear Functional Analysis. Yokohama Publishers, Yokohama (2000)
86. Tammer, Chr.: A generalization of Ekeland's variational principle. Optimization **25**, 129–141 (1992)
87. Tataru, D.: Viscosity solutions of Hamilton-Jacobi equations with unbounded nonlinear terms. J. Math. Anal. Appl. **163**, 345–392 (1992)
88. Yongsin, L., Shuzhong, S.: A generalization of Ekeland's ε-variational principle and its Borwein-Preiss variant. J. Math. Anal. Appl. **246**, 308–319 (2000)

89. Zabreiko, P.P., Krasnoselskii, M.A.: On the sovability of nonlinear operator equations. Funct. Anal. Priložž **5**, 42–44 (1971) (In Russian)
90. Zhong, C.-K.: On Ekeland's variational principle and a minimax theorem. J. Math. Anal. Appl. **205**, 239–250 (1997)
91. Zhong, C.-K.: A generalization of Ekeland's variational principle and application to the study of the relation between the P.S. condition and coercivity. Nonlinear Anal. Theor. Meth. Appl. **29**, 1421–1431 (1997)

Chapter 4
Fixed Point Theory in Hyperconvex Metric Spaces

Rafael Espínola and Aurora Fernández-León

4.1 Introduction and Basic Definitions

Hyperconvexity finds its roots in the theory of Banach spaces as *injective spaces* or $\mathscr{P}_1$*-spaces*. A Banach space X is said to be a $\mathscr{P}_1$-space if for every Banach space Z containing X, there is a linear projection P from Z onto X with $\|P\| \leq 1$. This property can be easily re-written in terms of extension of operators in the following way [17, p. 123]: For every Banach space Y, for every linear subspace Z of Y and for every bounded linear operator $T\colon Z \to X$, there exists a liner extension $\tilde{T}\colon Y \to X$ of T such that $\|T\| = \|\tilde{T}\|$. The best well-known $\mathscr{P}_1$-spaces are $\ell_\infty(\Gamma)$ for some set $\Gamma \neq \emptyset$. This can be shown as an easy application of Hahn–Banach theorem coordinatewise.

A very large collection of works was devoted to the better understanding and characterization of $\mathscr{P}_1$-spaces in the first decades of the second half of the twentieth century for both the real and the complex cases. The interested reader may check the works by Nachbin [65], Goodner [32], Kelley [43], and Grothendieck [33] for real Banach spaces and by Hasumi [34] and Sakai [74] for complex Banach spaces.

In 1950, Nachbin [65] proved that a real Banach space is a $\mathscr{P}_1$-space if and only if its closed balls have the *binary intersection property* (that is, if every family of closed balls, which are pairwise intersecting, has nonempty intersection). We will see later in this section that this is actually the property that lives at the heart of the notion of *hyperconvexity* introduced by Aronszajn and Panitchpakdi [4] in 1956. A counterpart intersection of ball properties was also determined for complex $\mathscr{P}_1$-spaces. In 1973, Hustad [39, 40] proved that a complex Banach space is a $\mathscr{P}_1$-space if and only if every family of closed balls with the so-called *weak intersection property* has a nonempty intersection. A family of closed balls is said to have the

R. Espínola (✉) • A. Fernández-León
Universidad de Sevilla, Dpto. de Análisis Matemático-IMUS,
P.O.Box 1160, 41080-Sevilla, Spain
e-mail: espinola@us.es; aurorafl@us.es

S. Almezel et al. (eds.), *Topics in Fixed Point Theory*, DOI 10.1007/978-3-319-01586-6_4, 101

weak intersection property if for every linear functional the images of the balls by this functional have nonempty intersection.

It is our intention in the chapter to focus on the nonlinear theory of injective spaces and, in particular, on their properties in relation to the existence of fixed points for metric and topological operators. This nonlinear theory finds its origin in the work by Aronszain and Panitchpakdi [4] from 1956. In this work the authors studied injectivity from a purely metric point of view and called it *hyperconvexity*. However, before going directly to the definition of hyperconvex metric spaces we will look at the Hahn–Banach theorem for real Banach spaces, which is actually the idea behind injectivity. In fact, injective linear spaces can be regarded as those linear spaces which can replace the role of the real line in Hahn–Banach theorem.

We begin with a particular case of Helly's theorem [7, pp. 90–91], which is nothing else but the binary intersection property for the real line.

Theorem 4.1. *Let $\{I_\alpha\}_{\alpha\in\Gamma}$ be a collection of closed and bounded intervals in the real line $\mathbb{R}$ such that*

$$I_\alpha \cap I_\beta \neq \emptyset,$$

for any $\alpha,\beta\in\Gamma$, then

$$\bigcap_{\alpha\in\Gamma} I_\alpha \neq \emptyset.$$

Proof. Make $I_\alpha = [a_\alpha, b_\alpha]$ and let $A = \sup\{a_\alpha\colon \alpha\in\Gamma\}$ and $B = \inf\{b_\alpha\colon \alpha\in\Gamma\}$. If we take $\alpha,\beta\in\Gamma$, then

$$[a_\alpha, b_\alpha]\cap[a_\beta, b_\beta]\neq\emptyset$$

and so $a_\alpha \leq b_\beta$. Therefore, since α and β are arbitrary, a_α is a lower bound for the set $\{b_\gamma\colon \gamma\in\Gamma\}$, from where $a_\alpha \leq B$ for each α and A is a lower bound of $\{b_\gamma\colon \gamma\in\Gamma\}$. In the same manner, $b_\alpha \geq A$ for each u_α and B is an upper bound of $\{a_\gamma\colon \gamma\in\Gamma\}$. Consequently, since $A\leq B$, the interval $[A,B]\neq\emptyset$ and $[A,B]\subseteq[a_\alpha,b_\alpha]$ for each α, which completes the proof. □

Next we show Hahn–Banach theorem. This version of Hahn–Banach theorem refers to a *convex functional* on a real vector space X, which is a function $q\colon X\to\mathbb{R}$ such that $q(x+y)\leq q(x)+q(y)$ and $q(\alpha x)=\alpha q(x)$ for $x,y\in X$ and $\alpha\geq 0$. The proof we provide is based on the original proof of S. Banach (1929) [15, p. 106].

Theorem 4.2. *Let X be a real vector space, Y a vector subspace of X, q a convex functional on X, and u a linear form on Y, which is dominated by q:*

$$u(y)\leq q(y) \text{ for } y\in Y.$$

Then, u can be extended to all X as a linear form v dominated by q:

$$v(x) \leq q(x) \text{ for } x \in X.$$

Proof. If $X \neq Y$ we begin with a simple extension of u to the subspace

$$Y \oplus [x] = \{z + tx\colon z \in Y,\, t \in \mathbb{R}\},$$

the linear span of Y and $x \notin Y$.

Then, if $s \in \mathbb{R}$, we consider the linear extension of u to $Y \oplus [x]$ given by $v(z+tx) = u(z) + ts$. Our aim is to show that

$$v(z+tx) = u(z) + ts \leq q(z+tx)$$

by choosing an adequate value for s to make equal to $v(x)$. Since q is a convex functional, if this estimate holds for $t = \pm 1$, then

$$v(z+tx) = |t| v\left(\frac{1}{|t|} z \pm x\right) \leq |t| q\left(\frac{1}{|t|} z \pm x\right) = q(z+tx).$$

Hence all we need are the inequalities

$$u(z) + s \leq q(z+x) \text{ and } u(z') - s \leq q(z'-x) \text{ for } z, z' \in Y;$$

that is, $u(z') - q(z'-x) \leq s \leq q(z+x) - u(z)$. Therefore we only need to show that the collection of intervals $\{[u(z) - q(z-x), q(z+x) - u(z)]\}_{z \in Y}$ is in the conditions of Theorem 4.1; however, this trivially follows from the properties of u and q, and the fact that u is dominated by q. It is then possible to choose s so that

$$\sup_{z' \in Y}(u(z') - q(z'-x)) \leq s \leq \inf_{z \in Y}(q(z+x) - u(z))$$

and so v is dominated by q.

Once we know that a one-dimensional extension is always possible, we can continue with a standard application of Zorn's lemma as follows.

Consider the family Φ of all extensions ℓ of u to vector subspaces L of X that are dominated by q, and we order Φ by $(L_1, \ell_1) \leq (L_2, \ell_2)$ meaning that $L_1 \subseteq L_2$ and ℓ_2 coincides with ℓ_1 on L_1.

Every totally ordered subset $\{(L_\alpha, \ell_\alpha)\}$ of Φ has the upper bound (L, ℓ) obtained by defining $L = \cup_\alpha L_\alpha$ and $\ell(x) = \ell_\alpha(x)$ if $x \in L_\alpha$. If also $x \in L_{\alpha'}$, then $\ell_\alpha(x) = \ell_{\alpha'}(x)$ by the total ordering of the set $\{(L_\alpha, \ell_\alpha)\}$. For the same reason, L is a vector subspace of X and ℓ is a linear extension of all the linear forms ℓ_α.

By Zorn's lemma, there is a maximal element $(\tilde{F}, \tilde{\ell})$ in Φ. But according to the first part of this proof, the domain of this extension must be the whole space X since otherwise we could always extend our maximal operator to $\tilde{F} \oplus [x]$ for any $x \in X \setminus \tilde{F}$. □

Remark 4.1. A reasoning by transfinite induction, as Banach actually did in his original proof, or, alternatively, by Zorn's lemma is very recurrent in proofs about extension of operators once we have been able to extend our operator to one extra point. We will see this same idea in several proofs along this chapter.

By a direct examination of the proof of Hahn–Banach theorem, it is immediate to realize that the key fact for it to hold is precisely Theorem 4.1. This intersecting ball property would be isolated by Aronszajn and Panitchpakdi in [4] to study the extension of uniformly continuous operators between metric spaces. In fact, Aronszajn and Panitchpakdi proved that for a uniformly continuous mapping with a subadditive modulus of continuity from a metric space N into another metric space M to be extended with same modulus of continuity to any metric space Z metrically containing N is sufficient and necessary the target space M to be *hyperconvex*.

Definition 4.1. A metric space M is said to be *hyperconvex* if given any family $\{x_\alpha\}_{\alpha\in\mathscr{A}}$ of points of M and any family $\{r_\alpha\}_{\alpha\in\mathscr{A}}$ of nonnegative real numbers satisfying

$$d(x_\alpha, x_\beta) \leq r_\alpha + r_\beta$$

for any α and β in $\mathscr{A}$, then

$$\bigcap_{\alpha\in\mathscr{A}} B(x_\alpha, r_\alpha) \neq \emptyset,$$

where $B(x,r)$ stands for the closed ball of center x and radius r.

This definition admits many graduations attending the cardinality of the index set $\mathscr{A}$ leading to the notion of m-hyperconvexity [4]. The weakest of all them is the metric convexity.

Definition 4.2. Let M be a metric space. We say that M is *metrically convex* if for any points x, y in M and nonnegative numbers α and β such that $d(x,y) \leq \alpha + \beta$, then

$$B(x,\alpha) \cap B(y,\beta) \neq \emptyset.$$

An example of a hyperconvex space that we already know is the real line and, after the work of Nachbin [65], it is clear that a real Banach space is hyperconvex if and only if it is a $\mathscr{P}_1$-space. Some other examples are listed next:

Example 4.1. (a) Any closed ball of a hyperconvex metric space is hyperconvex itself with the induced metric, this will be shown in Sect. 4.2. In particular, any closed real interval is hyperconvex.

(b) It is immediate to see that hyperconvexity is a purely metric property and so it is invariant by isometries; therefore, any isometric set to a closed real interval is hyperconvex.

(c) $\mathbb{R}$-*trees*, or *real trees*, are also well known to be hyperconvex whenever they are complete [53]. An $\mathbb{R}$-tree is a metric space T such that:

(i) there is a unique geodesic segment (denoted by $[x,y]$) joining each pair of points $x,y \in T$;
(ii) if $[y,x] \cap [x,z] = \{x\}$, then $[y,x] \cup [x,z] = [y,z]$.

(d) $\mathbb{R}^2$ with the following metric, known as the *river metric*, is a complete $\mathbb{R}$-tree:

$$d(v_1,v_2) = \begin{cases} |\, y_1 - y_2 \,|, & \text{if} \quad x_1 = x_2, \\ |\, y_1 \,| + |\, y_2 \,| + |\, x_1 - x_2 \,|, & \text{if} \quad x_1 \neq x_2, \end{cases}$$

where $v_1 = (x_1,y_1)$, $v_2 = (x_2,y_2) \in \mathbb{R}^2$.

As a particular case we show next that any ℓ_∞-space over any index set Γ is hyperconvex.

Proposition 4.1. *Let $\{B_\alpha\}_{\alpha\in\mathscr{A}}$ be a family of closed balls in $\ell_\infty(\Gamma)$ such that $B_\alpha \cap B_\beta \neq \emptyset$ for each $\alpha,\beta \in \mathscr{A}$, then*

$$\bigcap_{\alpha\in\mathscr{A}} B_\alpha \neq \emptyset.$$

Proof. Take $\gamma \in \Gamma$ and fix the set L_γ of all points in $\ell_\infty(\Gamma)$ with coordinates equal to zero but, at most, the one in position γ. Obviously L_γ is isometric to $\mathbb{R}$ and $\{B_\alpha \cap L_\gamma\}_\alpha$ is a family of closed and bounded intervals in L_γ each two of them with nonempty intersection. Thus, for each $\gamma \in \Gamma$,

$$\bigcap_{\alpha\in\mathscr{A}} (L_\gamma \cap B_\alpha) \neq \emptyset.$$

Take, for each γ, x_γ in the above intersection. Then, it follows that the element $(x_\gamma)_{\gamma\in\Gamma}$ is in $\ell_\infty(\Gamma)$ and belongs to $\cap_\alpha B_\alpha$ which is nonempty. □

Proposition 4.2. *Let $\ell_\infty(\Gamma)$ as above. If $B(x,r)$ and $B(y,s)$ are two closed balls in $\ell_\infty(\Gamma)$, then they have nonempty intersection if and only if $\|x-y\|_\infty \leq r+s$.*

Proof. The first implication is a direct consequence of the triangle inequality. For the converse we have from hypothesis that

$$|x_\gamma - y_\gamma| \leq r+s$$

for each $\gamma \in \Gamma$. For each γ fix $z_\gamma = \dfrac{rx_\gamma + sy_\gamma}{r+s}$ and consider the element $z = (z_\gamma)$ in $\ell_\infty(\Gamma)$. Then, it follows that $|x_\gamma - z_\gamma| \leq r$ and $|y_\gamma - z_\gamma| \leq s$ for each γ, and so $z \in B(x,r) \cap B(y,s)$. □

The next theorem follows as an immediate consequence of the two previous propositions.

Theorem 4.3. *Let Γ be an index set and consider $\ell_\infty(\Gamma)$ the real ℓ_∞ space over the index set Γ. Then, it is hyperconvex.*

We will look closer at the structure of hyperconvex metric spaces later on, now we recall Aronszajn and Panitchpakdi result on extension of uniformly continuous mappings. We begin by recalling the definition of modulus of uniform continuity.

Definition 4.3. An extended valued nonnegative real function $\delta\colon (0,\infty) \to [0,\infty]$ is a *modulus of (uniform) continuity* if it is nondecreasing and converges to 0 as ε goes to 0.

A modulus of continuity is said to be *subadditive* if

$$\delta(\varepsilon_1+\varepsilon_2) \leq \delta(\varepsilon_1)+\delta(\varepsilon_2)$$

for any $\varepsilon_1 > 0$ and $\varepsilon_2 > 0$.

Given a mapping T from a metric space Y into another metric space X, we say that $\delta(\varepsilon)$ is *modulus of continuity* of T if it is a modulus of continuity and, for $x,y \in Y$ such that $d_Y(x,y) \leq \varepsilon$,

$$d_X(T(x),T(y)) \leq \delta(\varepsilon),$$

where d_X and d_Y are the respective metrics on X and Y.

Now we give the main result of [4]. Recall that a metric space Y is *isometrically embedded* into another metric space Z if there exists a distance preserving mapping $j\colon Y \to Z$, that is, such a mapping j that $d_Z(j(x),j(y)) = d_Y(x,y)$ for all $x,y \in Y$.

Theorem 4.4. *Let T be a uniformly continuous mapping, with a subadditive modulus of continuity $\delta(\varepsilon)$, from a metric space Y into another metric space X. Then, T admits an extension with same modulus of continuity to any metric space Z containing Y isometrically embedded if and only if X is hyperconvex.*

Proof. We prove *necessity* first. Let $\{x_\alpha\} \subseteq X$ and $\{r_\alpha\} \subseteq [0,\infty)$ given as in the definition of hyperconvexity. Make $Y = \{x_\alpha\}_{\alpha\in\mathscr{A}}$ with the induced metric and T the identity map from Y into X. Then, $\delta(\varepsilon) = \varepsilon$ is a subadditive modulus of continuity of T. Add a new point $\{z\}$ to Y and consider the set $Z = Y \cup \{z\}$. Define the following metric on Z:

$$d_Z(x,y) = \begin{cases} d_X(x_\alpha,x_\beta), & \text{if} \quad x = x_\alpha,\ y = x_\beta \text{ for some } \alpha,\beta \in \mathscr{A}, \\ r_\alpha, & \text{if} \qquad\qquad x = x_\alpha,\ y = z. \end{cases}$$

It is immediate to check that (Z,d) is a metric space from the conditions on $\{x_\alpha\}$ and $\{r_\alpha\}$. From our hypothesis we can extend T to any metric space Z containing X isometrically with the same modulus of continuity $\delta(\varepsilon) = \varepsilon$, and so, it must be the case that

$$\tilde{T}(z) \in \bigcap_{\alpha\in\mathscr{A}} B(x_\alpha,r_\alpha).$$

For sufficiency, we will only prove how to extend to one extra point, then the proof can be completed by following a Zorn's lemma or transfinite induction reasoning as in Hahn–Banach theorem (Theorem 4.2). Let T, Y, and X be as in the statement with X hyperconvex. Consider $Z = Y \cup \{z\}$ as a new metric space adding one point to Y and containing Y isometrically. We want to extend T to Z with equal modulus of convexity and still with target space X. For y in Y take $r_y = d_Z(y,z)$, then

$$d_X(T(y_1),T(y_2)) \leq \delta(d_Y(y_1,y_2)) \leq \delta(r_{y_1}+r_{y_2}) \leq \delta(r_{y_1})+\delta(r_{y_2}).$$

Now, from the hyperconvexity of X, there exists $z' \in X$ such that

$$d_X(z',T(y)) \leq \delta(r_y),$$

for each $y \in Y$. The proof is finished by defining $\tilde{T}(z) = z'$. □

We will prove more elementary properties of hyperconvex metric spaces in Sect. 4.2. To finish this section we recall some basic definitions that will be needed all along this chapter. We begin with the definition of some distinguished subsets of hyperconvex metric spaces. Remember that the distance from a point x to a subset A of a metric space M is given by

$$\operatorname{dist}(x,A) = \inf\{d(x,y) \colon y \in A\}.$$

Definition 4.4. Let M be a metric space.

- A subset A of M is said to be an *admissible* subset of M if A coincides with the intersection of all closed balls containing it. The class of admissible subsets of M will be denoted by $\mathscr{A}(M)$.
- A subset E of M is said to be *externally hyperconvex* if given any family $\{x_\alpha\}$ of points in M and any family $\{r_\alpha\}$ of nonnegative real numbers satisfying

  $$d(x_\alpha,x_\beta) \leq r_\alpha + r_\beta \text{ and } \operatorname{dist}(x_\alpha,E) \leq r_\alpha,$$

  it follows that

  $$\bigcap_\alpha B(x_\alpha,r_\alpha) \cap E \neq \emptyset.$$

  The class of externally hyperconvex subsets of M will be denoted by $\mathscr{E}(M)$;
- A subset H of M is said to be hyperconvex if it is hyperconvex as a metric space with the induced metric. The class of hyperconvex subsets of M will be denoted by $\mathscr{H}(M)$.

We end this section with a glossary of terms. Let M be a metric space, $x \in M$ and A and B two subsets of M, then:

- $r_x(A) = \sup\{d(x,y) \colon y \in A\}$ is the *radius of A with respect to x.*
- $r(A) = \inf\{r_x(A) \colon x \in M\}$ is the *(Chebyshev) radius of A relative to M.*
- $r_B(A) = \inf\{r_x(A) \colon x \in B\}$ is the *(Chebyshev) radius of A relative to B.*
- $\operatorname{diam}(A) = \sup\{d(x,y) \colon x,y \in A\}$ is the *diameter of A.*
- $C(A) = \{x \in M \colon r_x(A) = r(A)\}$ is the *(Chebyshev) center of A.*

4.2 Some Basic Properties of Hyperconvex Metric Spaces

Hyperconvexity provides us with a very rich geometric structure which does not depend on any linear underlying setting. We begin this section with the fact that even when we are dealing with a linear hyperconvex space, that is, a $\mathscr{P}_1$-space, there is no direct relation between linear convexity and metric hyperconvexity.

Example 4.2. Let X be $\mathbb{R}^2$ with the maximum norm. This space is a hyperconvex as we have seen in Sect. 4.1. Consider

$$A = \{(x,y) : x = y \text{ if } 0 \le x \le 1 \text{ and } x = 2 - y \text{ if } 1 \le x \le 2\},$$

then $T : A \to [0,2]$ given by $T(x,y) = x$ is an isometry and so A is hyperconvex itself while it is a nonconvex set of $\mathbb{R}^2$.

The term *hyperconvexity* comes actually from the idea of *metric convexity* or *Menger convexity*.

Definition 4.5. A metric space M is said to be *Menger convex* if given any two points x, y in M there exists a third point $z \in M$ such that $d(x,z) + d(z,y) = d(x,y)$.

Obviously, any hyperconvex metric space is metrically convex. This fact brings another characterization of hyperconvexity.

Theorem 4.5. *A metric space M is hyperconvex if and only if it is metrically convex and has the binary intersection property.*

In Definition 4.4 we introduced three classes of distinguished subsets of a metric space M: admissible sets $\mathscr{A}(M)$, externally hyperconvex sets $\mathscr{E}(M)$ and hyperconvex sets $\mathscr{H}(M)$. Next we study some relations among them.

Theorem 4.6. *Let M be a hyperconvex metric space. Then:*

(a) Any $A \in \mathscr{A}(M)$, $E \in \mathscr{E}(M)$ or $H \in \mathscr{H}(M)$ is hyperconvex with the induced metric.
(b) The set of contentions

$$\mathscr{A}(M) \subseteq \mathscr{E}(M) \subseteq \mathscr{H}(M)$$

holds and, in general, are strict contentions.
(c) If E is externally hyperconvex and A is admissible, then $E \cap A$ is externally hyperconvex.

Proof. (a) is a consequence of (b), so it suffices to prove (b). From the definition of externally hyperconvex subset it is immediate that such a set is also hyperconvex. Let us see that $\mathscr{A}(M) \subseteq \mathscr{E}(M)$. Let A be an admissible subset of M and let $\{x_\alpha\} \subseteq M$ and $\{r_\alpha\}$ corresponding nonnegative numbers such that

$$d(x_\alpha, x_\beta) \le r_\alpha + r_\beta \text{ and } \operatorname{dist}(x_\alpha, A) \le r_\alpha,$$

for each α and β. Since A is an intersection of closed balls and M is metrically convex, then $B(x_\alpha, r_\alpha) \cap B$ is nonempty for each closed ball determining the set A and so, from hyperconvexity of M,

$$B(x_\alpha, r_\alpha) \cap A \neq \emptyset,$$

for each α. Again, from the hyperconvexity of M and the fact that pairs of balls in $\{B(x_\alpha, r_\alpha)\}$ intersect each of the balls determining A, it follows that

$$A \bigcap \left(\bigcap_\alpha B(x_\alpha, r_\alpha) \right) \neq \emptyset,$$

which proves the inclusion.

For the strict contentions consider M as the right half real plane endowed with the maximum metric, that is, $M = \{(x,y) \in \mathbb{R}^2 \colon x \ge 0\}$. Let $E = B((0,-1),2) \cap M$, where $B((0,-1),2)$ stands for the ball in $\mathbb{R}^2$ with the infinity norm, and $H = \{(x, x/2) \colon 0 \le x \le 1\}$. Then, E is externally hyperconvex but not admissible and H is hyperconvex but not externally hyperconvex.

(c) Let $\{x_\alpha\} \subseteq M$ and $\{r_\alpha\}$ as usual with $\operatorname{dist}(x_\alpha, E \cap A) \le r_\alpha$ for each $\alpha \in \mathscr{A}$. Since A is admissible, $A = \cap_{i \in I} B(x_i, r_i)$ and since $\operatorname{dist}(x_\alpha, A) \le r_\alpha$, it follows that $d(x_\alpha, x_i) \le r_\alpha + r_i$ for each $\alpha \in \mathscr{A}$ and $i \in I$. It is also clear (A nonempty) that $d(x_i, x_j) \le r_i + r_j$ for each $i, j \in I$. Therefore, the external hyperconvexity of E implies that

$$\left(\bigcap_{i \in I} B(x_i, r_i) \right) \cup \left(\bigcap_{\alpha \in \mathscr{A}} B(x_\alpha, r_\alpha) \right) \cap E \neq \emptyset,$$

which completes the proof. □

A very interesting property of subsets of a metric spaces when dealing with projections, minimal distance points, and fixed points under boundary conditions is that of being *proximinal*.[1]

Definition 4.6. A subset A of a metric space M is said to be *proximinal* if for every point $x \in M$ there exists at least a point $y \in A$ such that $d(x,y) = \operatorname{dist}(x, A)$, that is, if

$$B(x, \operatorname{dist}(x, A)) \cap A \neq \emptyset$$

for all $x \in M$.

[1] As long as the authors know, the word proximinal was proposed by Robert Phelps as a mixture of the words *proximinty* and *minimal* to embrace in one word the meaning of both words separately. It is also usual to find the word *proximal* in the literature meaning the same as proximinal.

To determine that a given set is proximinal some compactness conditions are usually needed. This will not be the case for externally hyperconvex subsets of hyperconvex spaces.

Lemma 4.1. *If A is a externally hyperconvex subset of a hyperconvex metric space M, then A is proximinal in M.*

Proof. For an easier exposition we will only show the admissible case. The general case follows in the same way. Let $A = \bigcap_{\alpha \in \mathscr{A}} B_\alpha$. Then, for any $\varepsilon > 0$ there exists $a_\varepsilon \in A$ such that

$$d(x, a_\varepsilon) \leq \operatorname{dist}(x, A) + \varepsilon.$$

Then, $B(x, \operatorname{dist}(x,A) + \varepsilon) \cap B_\alpha \neq \emptyset$ for every $\alpha \in \mathscr{A}$, and so, from the hyperconvexity of M,

$$A \cap B(x, \operatorname{dist}(x,A) + \varepsilon) = \bigcap_{\alpha \in \mathscr{A}} B_\alpha \bigcap \left(\bigcap_{\varepsilon > 0} B(x, \operatorname{dist}(x,A) + \varepsilon) \right) \neq \emptyset,$$

which completes the proof. □

Remark 4.2. Another class of subsets of a hyperconvex metric space has been proved to be a class of proximinal subsets. This is the class of *weakly externally hyperconvex* subsets introduced for the first time in [29] (see also [23, 25, 28] for further readings). This class is shown to be larger than that of externally hyperconvex subsets [25, p. 398] and is still relevant to some fixed point questions. We have decided to leave it out of reach of this chapter for expository reasons.

A very important tool in linear spaces is to associate with any given set the minimal convex set containing it, that is, its convex hull. As we have shown we cannot count on linear convexity when dealing with hyperconvex spaces. This role will be played, however, most of the times by admissible subsets.

Definition 4.7. Let M be a metric space and $A \subseteq M$ nonempty. Then, the *admissible cover*, or *admissible hull*, of A, denoted by $\operatorname{cov}(A)$, is the minimal, with respect to the set inclusion, admissible set containing A.

Not surprisingly, admissible subsets have played a major role in the nonlinear theory on hyperconvex spaces. R. Sine proved many relevant results on this class of sets in works as [79–81] and some others (see also reviews [25, 28]).

Admissible hulls can be determined in a precise way as the following proposition shows.

Proposition 4.3. *Let A be a subset of a metric space M. Then,*

$$cov(A) = \bigcap_{x \in M} B(x, r_x(A)).$$

Proof. Since $A \subseteq B(x, r_x(A))$ then it is clear that $\cap_{x \in M} B(x, r_x(A))$ is an admissible set containing A. Let $B = \cap_{\alpha \in \mathscr{A}} B(x_\alpha, r_\alpha)$ another admissible set containing A. Then, since $A \subseteq B(x_\alpha, r_\alpha)$, it is clear that $r_{x_\alpha}(A) \leq r_\alpha$, and so, $B(x_\alpha, r_{x_\alpha}(A)) \subseteq B(x_\alpha, r_\alpha)$, thus

$$\bigcap_{x \in M} B(x, r_x(A)) \subseteq \bigcap_{\alpha \in \mathscr{A}} B(x_\alpha, r_\alpha).$$

□

Proposition 4.4. *If M is a hyperconvex metric space, then it is complete.*

Proof. Let $\{x_n\}$ be a Cauchy sequence in M and for each $n \in \mathbb{N}$, set

$$r_n = \sup\{d(x_n, x_m) : m \geq n\}.$$

Then, for $m \geq n$, $d(x_n, x_m) \leq r_n \leq r_n + r_m$. Thus, the hyperconvexity of M implies that there exists $z \in M$ such that

$$z \in \bigcap_{n=1}^{\infty} B(x_n, r_n).$$

Now, since $r_n \to 0$ then it must be the case that $x_n \to z$. □

Next we show some metric properties of admissible hulls in hyperconvex metric spaces which have been extensively applied in metric fixed point theory.

Lemma 4.2. *Let A be a bounded subset of a hyperconvex metric space M. Then:*

(a) $r_x(cov(A)) = r_x(A)$, *for any* $x \in M$.
(b) $r(cov(A)) = r(A)$.
(c) $r(A) = \frac{1}{2} diam(A)$.
(d) $diam(cov(A)) = diam(A)$.
(e) If A is hyperconvex then $r(A) = r_A(A)$. In particular, $r_A(A) = 1/2\, diam(A)$.

Proof. (a) From Proposition 4.3, we have that

$$r_x(\mathrm{cov}(A)) = \sup\left\{ d(x,y) : y \in \bigcap_{z \in M} B(z, r_z(A)) \right\}.$$

Therefore, if $y \in \mathrm{cov}(A)$, then $y \in B(x, r_x(A))$ for each $x \in M$. Hence, $d(x,y) \leq r_x(A)$, and so, $r_x(\mathrm{cov}(A)) \leq r_x(A)$. The reverse inequality is trivial since $A \subseteq \mathrm{cov}(A)$.

(b) This is immediate from (a).

(c) Let $\delta = \mathrm{diam}(A)$ and consider the family of balls $\{B(x, \delta/2)\}_{x \in A}$. If $x, y \in A$, then $d(x,y) \leq \delta/2 + \delta/2$, and so, by hyperconvexity,

$$C = \bigcap_{a \in A} B\left(a, \frac{\delta}{2}\right) \neq \emptyset.$$

Let $x \in C$, then $d(x,a) \leq \delta/2$ for each $a \in A$ and so $r_x(A) \leq \delta/2$ which gives one inequality.

For the reverse inequality, we have that $d(a,b) \leq d(a,x) + d(x,b)$ for any $a, b \in A$ and $x \in M$ so $\delta \leq 2r_x(A)$. Since x is arbitrary then the statement is proved.

(d) Using (b) and (c), we get that $\mathrm{diam}(A) = 2r(A) = 2r(\mathrm{cov}(A)) = \mathrm{diam}(\mathrm{cov}(A))$.

(e) Obviously, it must be the case that $r_A(A) \geq 1/2\ \mathrm{diam}(A)$. For the reverse inequality, it suffices to recall that A is hyperconvex, and so

$$C = A \bigcap \left(\bigcap_{a \in A} B(a, \frac{\delta}{2}) \right) \neq \emptyset.$$

It is enough to take $a \in C$ to prove (e). □

Notice that in general one only has that $r(A) \leq r_A(A)$, however equality is guaranteed for hyperconvex subsets of metric spaces.

Corollary 4.1. *Let A be a hyperconvex subset of a metric space, then*

$$r(A) = r_A(A) = 1/2 diam(A).$$

Definition 4.8. Given a subset A of a metric space M, the *Chebyshev center* of A, denoted by $C(A)$, is given by

$$C(A) = \{x \in M \colon r_x(A) = r(A)\}.$$

The previous corollary is extraordinarily important in metric fixed point theory as it is providing hyperconvex metric spaces with normal structure. We recall that a Banach space X has *normal structure* if any bounded and convex subset C of X with more than one point contains a *nondiametral point*, that is, a point $x_0 \in C$ such that

$$r_{x_0}(C) < \mathrm{diam}(C).$$

Normal structure was then taken into a more abstract setting for metric spaces [48, Chap. 5]. The reader may find more on this in Sect. 1.10 of Chap. 1.

Definition 4.9. A class $\mathscr{A}$ of subsets of a metric space is said to be *uniformly normal* if there exists $c \in (0,1)$ such that $r_D(D) \leq c\ \mathrm{diam}(D)$ for each $D \in \mathscr{A}$.

Attending to this abstract formulation we have the following corollary.

Corollary 4.2. *Let M be a hyperconvex metric space, then $\mathscr{A}(M)$ is uniformly normal with the smallest possible constant $c = 1/2$. Moreover, if $A \in \mathscr{A}(M)$, then $C(A) \in \mathscr{A}(M)$, $C(A) \cap A \neq \emptyset$ and $diam(C(A) \cap A) = (1/2)diam(A)$.*

Proof. It only needs to be proved that $C(A) \cap A \neq \emptyset$ and that $\mathrm{diam}(C(A) \cap A) = (1/2)\mathrm{diam}(A)$. The first fact directly follows from hyperconvexity. The second one is consequence from the fact that $C(A) \cap A \subseteq B(x, \mathrm{diam}(A)/2)$ for any $x \in A$. Therefore, if $x, y \in C(A) \cap A$ then $y \in B(x, \mathrm{diam}(A)/2)$. □

When dealing with abstract structures, a very interesting property for a directed chain of sets is that of being compact. That is, that the intersection of all sets in the directed chain is nonempty and an element of the same structure. To prove that the family of admissible subsets of a hyperconvex metric space is a rather simple fact that we show next.

Theorem 4.7. *Let M be a hyperconvex metric space, $\mathscr{A}(M)$ its class of admissible subsets and Γ a totally ordered index set. Then if $\{A_\alpha\}_{\alpha\in\Gamma}$ is a decreasing family of nonempty elements of $\mathscr{A}(M)$, then $\bigcap_{\alpha\in\Gamma} A_\alpha$ is nonempty and admissible.*

Proof. Call $A = \bigcap_{\alpha\in\Gamma} A_\alpha$, then, since A_α is an intersection of balls for each α, it is clear that A is an intersection of balls too, in fact, from Proposition 4.3

$$A = \bigcap_{\alpha\in\Gamma}\left(\bigcap_{x\in M} B(x, r_x(A_\alpha))\right).$$

We only need to prove that A is nonempty but, since M is hyperconvex, it is enough to show that each two balls conforming A have nonempty intersection. Let $B(x, r_x(A_\alpha))$ and $B(y, r_y(A_\beta))$ be two such balls. Suppose that $\alpha \le \beta$, then, since

$$A_\beta \subseteq B(y, r_y(A_\beta)),\ A_\alpha \subseteq B(x, r_x(A_\alpha)) \text{ and } A_\beta \subseteq A_\alpha,$$

it must be the case that $B(x, r_x(A_\alpha)) \cap B(y, r_y(A_\beta)) \neq \emptyset$, and so the conclusion follows. □

Remark 4.3. Notice that the above theorem is true for general families of admissible subsets given that they have nonempty intersection or assuming that the empty set is an admissible set too.

Early investigators in metric hyperconvexity wondered whether this compactness property holds also for the family of hyperconvex sets. Notice that the family of hyperconvex sets is not closed under arbitrary intersections. Actually the intersection of two hyperconvex subsets of a hyperconvex metric space need not be hyperconvex, so it was asked whether any descending chain of nonempty hyperconvex sets has a nonempty intersection. This question was answered by Baillon in the affirmative in [6] (see also [25, p. 406] for a more elementary proof). We do not include the proof for this result.

Theorem 4.8. *Let M be a bounded metric space and Γ a totally ordered index set. Let $(H_\alpha)_{\alpha\in\Gamma}$ be a decreasing family of nonempty hyperconvex subsets of M, then*

$$\bigcap_{\alpha\in\Gamma} H_\alpha$$

is nonempty and hyperconvex.

After this theorem, we can prove a counterpart of Theorem 4.7 for externally hyperconvex sets.

Corollary 4.3. *Let M be a bounded hyperconvex metric space and Γ a totally ordered index set. Let $(E_i)_{i\in\Gamma}$ be a decreasing family of nonempty externally hyperconvex subsets of M, then*

$$\bigcap_{i\in\Gamma} E_i$$

is nonempty and externally hyperconvex.

Proof. Theorem 4.8 implies that $E = \bigcap_{i\in\Gamma} E_i$ is nonempty. Let us see that it is externally hyperconvex too. Let $\{x_\alpha\} \subseteq M$ and $\{r_\alpha\}$ as usual with $\operatorname{dist}(x_\alpha, E) \leq r_\alpha$. Since M is hyperconvex we have that

$$A = \bigcap_{\alpha} B(x_\alpha, r_\alpha) \neq \emptyset.$$

Also, since $\operatorname{dist}(x_\alpha, E) \leq r_\alpha$ we have that $\operatorname{dist}(x_\alpha, E_i) \leq r_\alpha$ for each i, so, by external hyperconvexity, we have that $A \cap E_i \neq \emptyset$ for each i. By (c) of Theorem 4.6, we know that $(A \cap E_i)_{i\in\Gamma}$ is a descending chain of hyperconvex subsets of M, now a direct application of Theorem 4.8 concludes the proof. □

4.3 Hyperconvexity, Injectivity, and Retractions

As it was explained in Sect. 4.1, metric hyperconvexity reduces to injectivity in the case of linear spaces. We will look closer at this relation in this section as well as at the also related problem of existence of *nice* retractions in hyperconvex metric spaces. First, we recall the definition of nonexpansive mapping.

Definition 4.10. Let (X, d_X) and (Y, d_Y) be two given metric spaces, then a mapping $T\colon X \to Y$ is *nonexpansive* if it is Lipschitzian with Lipschitz constant 1, that is, if

$$d_Y(T(x), T(y)) \leq d_X(x, y), \quad \text{for any } x, y \in X.$$

Now we can give the definition of injectivity in the metric setting.

Definition 4.11. A metric space M is said to be *injective* if it has the extension property that whenever two metric spaces X and Y are given with Y metrically embedded into X and $T\colon Y \to M$ is nonexpansive, then T has a nonexpansive extension $\tilde{T}\colon X \to M$.

As a direct consequence of Theorem 4.4 we obtain the equivalence between hyperconvexity and metric injectivity.

Theorem 4.9. *Let M be a metric space. Then, M is hyperconvex if and only if it is injective.*

Proof. Necessity directly follows from Theorem 4.4 since nonexpansive mappings are uniformly continuous mappings with subadditive modulus of continuity $\delta(\varepsilon) = \varepsilon$.

For sufficiency, we find here a formally weaker condition than the one given in Theorem 4.4; however, if we look at the proof given for this theorem we can see that we were working with nonexpansive mappings and so the same proof works for our present case. □

The previous theorem is a very useful one. Particularly it finds many applications in metric fixed point theory when we reformulate it in terms of *projections* since it distinguishes hyperconvex spaces with a very strong property, that of being *absolute nonexpansive retracts*, *ANR* for short.

Definition 4.12. A subset A of a metric space M is said to be a *nonexpansive retract* (of M) if there exists a nonexpansive retraction from M onto A, that is, a nonexpansive mapping $R\colon M \to A$ such that $Rx = x$ for each $x \in A$.

A metric space M is said to be an *absolute nonexpansive retract*, *ANR* for short, if it is a nonexpansive retract of any metric space where it may be metrically embedded.

Hyperconvex metric spaces are nonexpansive retracts of any metric space where are isometrically embedded. Even more, this property characterizes hyperconvex subsets of a hyperconvex metric space.

Theorem 4.10. *Let M be a metric space. Then, M is hyperconvex if and only if it is an ANR.*

Proof. This theorem follows after small modifications in the proof of Theorem 4.9. Indeed, let N be another metric space metrically containing the hyperconvex space M. Consider $I\colon M \to M$ as the identity mapping which, obviously, is nonexpansive. After Theorem 4.9 we can find a nonexpansive extension $\tilde{I}\colon N \to M$ which is the requested nonexpansive retraction. *Sufficiency* also follows in the same way as in Theorem 4.4. □

This theorem leads to the following characterization of hyperconvex subsets of a hyperconvex space as a corollary.

Corollary 4.4. *Let M be a hyperconvex metric space, then $A \subseteq M$ is hyperconvex if and only if it is a nonexpansive retract of M.*

Proof. If A is hyperconvex, then it is an ANR and so, in particular, a nonexpansive retract of M. Let us see that if it is a nonexpansive retract of M, then it is hyperconvex. For that let us see that it is an ANR.

To see that A is hyperconvex given that it is a nonexpansive retract of M, take as usual $\{x_\alpha\} \subseteq A$ and $\{r_\alpha\}$ with $d(x_\alpha, x_\beta) \leq r_\alpha + r_\beta$. Since M is hyperconvex and $A \subseteq M$, then there exists

$$x_0 \in \bigcap_{\alpha \in \mathscr{A}} B(x_\alpha, r_\alpha).$$

Let now $R\colon M \to A$ be the nonexpansive retraction, then

$$d(x_\alpha, R(x_0)) = d(R(x_\alpha), R(x_0)) \le d(x_\alpha, x_0) \le r_\alpha,$$

which, recalling that $R(x_0) \in A$, completes our proof. □

Khamsi gave a new reformulation of the above theorem after introducing the concept of 1-*local retract* in [45].

Definition 4.13. Let M be a metric space. A subset N of M is called a 1-*local retract* of M if for any point $x \in M \setminus N$, there exists a nonexpansive retraction from $N \cup \{x\}$ onto N.

Corollary 4.5. *A metric space M is hyperconvex if and only if it is a* 1*-local retract of any other metric space where it may be metrically embedded.*

As we have just seen, hyperconvexity exhibits a very good behavior with respect to nonexpansive retractions. However, much more can be said and, in fact, what we have shown so far is just the starting point for a much deeper study of nonexpansive retractions on hyperconvex spaces. This study can be set to have begun in the nonlinear setting with seminal works of Sine, see, for instance, [79]. His results are crucial in investigating nonexpansive mappings defined on hyperconvex metric spaces. Sine work mainly focused on admissible subsets of hyperconvex spaces. We state some of his pioneering results next to prove them later as consequences of some more recent ones.

Definition 4.14. Let A be a subset of a metric space M, then for any positive number r the r-*parallel* set of A is the set given by

$$A + r = \bigcup_{a \in A} B(a, r).$$

The following result is about the structure of admissible subsets of hyperconvex spaces. The attentive reader may easily check that the next lemma very seldom holds in general metric spaces or even normed spaces.

Lemma 4.3. *Let M be a hyperconvex metric space and A an admissible subset of M. Then, for every $r \ge 0$, we have*

$$A + r = \bigcap_{x \in M} B(x, r_x(A) + r).$$

In other words, the r-parallel sets of an admissible subset of a hyperconvex metric space are also admissible sets.

Proof. Let $y \in A + r$. Then, there exists $a \in A$ such that $d(y, a) \le r$. Hence,

$$d(x, y) \le d(x, a) + d(a, y) \le r_x(A) + r, \quad \text{for all } x \in M.$$

Thus,

$$A+r \subseteq \bigcap_{x\in M} B(x, r_x(A)+r).$$

For the reverse inclusion, let $y \in \cap_{x\in M} B(x, r_x(A)+r)$. So we have that $d(x,y) \leq r_x(A)+r$ for all $x \in M$. Now, since M is hyperconvex,

$$A \cap B(y,r) = \left(\bigcap_{x\in M} B(x, r_x(A))\right) \bigcap B(y,r) \neq \emptyset$$

which, recalling that $B(a,r) \subseteq A+r$, completes our proof. □

The corresponding result for externally hyperconvex subsets is also true.

Lemma 4.4. *Let M be a hyperconvex metric space and $A \subseteq M$ externally hyperconvex, then $A+r$ is externally hyperconvex for any $r \geq 0$.*

Proof. Let $\{x_\alpha\} \subseteq M$ and $\{r_\alpha\} \subseteq [0,+\infty)$ given as usual. Suppose that $\operatorname{dist}(x_\alpha, A+r) \leq r_\alpha$, then $\operatorname{dist}(x_\alpha, A) \leq r_\alpha + r$, and so

$$\left(\bigcap_{\alpha\in\mathscr{A}} B(x_\alpha, r_\alpha + r)\right) \bigcap A \neq \emptyset,$$

consequently, there exists $a \in A$ such that $d(a, x_\alpha) \leq r_\alpha + r$ for every α. Finally, the hyperconvexity of M implies that

$$\left(\bigcap_{\alpha\in\mathscr{A}} B(x_\alpha, r_\alpha)\right) \bigcap B(a,r) \neq \emptyset,$$

which proves the lemma. □

As a corollary of this lemma we obtain the following.

Corollary 4.6. *Let M be a hyperconvex metric space and $A \subseteq M$ externally hyperconvex, then $A+r$ is proximinal for any $r \geq 0$.*

Now we recall the concept of ε-*constant* mapping.

Definition 4.15. A mapping T from a metric space M into itself is said to be ε-*constant* if $d(x, T(x)) \leq \varepsilon$ for every $x \in M$.

Then, Sine's result [79] reads as follows.

Theorem 4.11. *Let M be a hyperconvex metric space and A an admissible subset of M. Then, for any $\varepsilon > 0$, there exists an ε-constant nonexpansive retraction from $A+\varepsilon$ onto A.*

To prove this theorem we will look at another kind of retracts, the so-called *proximinal nonexpansive retracts*. We need first to recall the definition of *metric projection*.

Definition 4.16. Let M be a metric space and $A \subseteq M$ nonempty. The *metric projection* onto A is then defined by

$$P_A(x) = B(x, \text{dist}(x,A)) \cap A.$$

Metric projections play a major role in approximation theory. Notice that if $P_A(x)$ is nonempty for each $x \in M$, then A is a proximinal subset of M (Definition 4.6). In general, the values $P_A(x)$ need not be singletons so this mapping is usually regarded as a multivalued mapping if the set A is proximinal.

Definition 4.17. An application T between two sets M and N is said to be a *multivalued mapping* if it is defined on M and takes values in the set of nonempty subsets of N, that is, if $T\colon M \to \mathscr{P}(N)$. For simplicity we will always write $T\colon M \to N$ too for multivalued mappings.

Definition 4.18. A (single-valued) mapping $f\colon M \to N$ will be said to be a *selection* of a multivalued mapping T if $f(x) \in T(x)$ for every $x \in M$.

The problem of finding nice selections of the metric projection is a very deep and interesting one. It is of special relevance in questions related to best approximation theory and nearest and farthest points between sets. Many authors working in the linear setting have studied the problem of finding selections of metric projections with some degree of continuity. There exists a very large and interesting literature on this topic, see, for instance, [11] and references therein. However, when working under hyperconvex conditions, much more can be said. In this case the natural question is to determine those subsets A of a hyperconvex space for which the metric projection P_A admits a nonexpansive selection. That is, those sets being *nonexpansive proximinal retracts*.

Definition 4.19. Let A be a subset of a metric space M, then, if A is proximinal in M and P_A is nonexpansive or admits a nonexpansive selection, A is said to be a *proximinal nonexpansive retract* of M.

Remark 4.4. It is a very well-known fact, recorded in basically every book in functional analysis, that a subset of a Hilbert space is a proximinal nonexpansive retract if and only if it is nonempty, closed, and convex [15, p. 47].

To study the nature of proximinal nonexpansive retracts of subsets of a hyperconvex space is a much more complicated task than that of characterizing nonexpansive retracts, see Corollary 4.4. The following result was given in [80] for admissible subsets, we offer here a proof for externally hyperconvex subsets which is very close to the one given by Sine in [80] for admissible subsets and which can also be found in [25, p. 426]. This proof requires of a result on existence of nonexpansive selections of nonexpansive multivalued mappings that will be given later on in

Sect. 4.7. We recall some definitions about multivalued mappings first. We begin with the Pompeiu–Hausdorff distance (for more on this, see [48, p. 24], [31, p. 19] or [2, p. 72]).

Definition 4.20. Let M be a metric space and let $\mathscr{M}$ denote the family of all nonempty bounded closed subsets of M. For $A \in \mathscr{M}$ and $\varepsilon > 0$ define the ε-neighborhood of A to be the set

$$N_\varepsilon(A) = \{\varepsilon > 0 \colon \operatorname{dist}(x, A) < \varepsilon\}.$$

Now for $A, B \in \mathscr{M}$, set

$$H(A,B) = \inf\{\varepsilon > 0 \colon A \subseteq N_\varepsilon(B) \text{ and } B \subseteq N_\varepsilon(A)\}.$$

Then, $(\mathscr{M}, H)$ is a metric space, and H is called the *Pompeiu–Hausdorff distance* on $\mathscr{M}$.

A nonexpansive multivalued mapping is then defined as follows.

Definition 4.21. Let M be a metric space and $T \colon M \to M$ a multivalued mapping with closed values. T is said to be *nonexpansive* if

$$H(T(x), T(y)) \leq d(x,y), \quad \text{for all } x, y \in M.$$

We will get back on these concepts in Sect. 4.7. Now we state and prove the result on proximinal nonexpansive retracts.

Theorem 4.12. *Let A be an admissible subset of a hyperconvex metric space M. Then, A is a proximinal nonexpansive retract of M.*

Proof. For each $x \in M$ consider its metric projection onto A, that is, $P_A(x)$ which is nonempty because of the hyperconvexity of M. Consider $x \in M$ fixed, then for each $y \in M$ consider the $d(x,y)$-parallel set of $P_A(y)$, that is, the set $P_A(y) + d(x,y)$. Then call

$$C(x) = \bigcap_{y \in M} \left(P_A(y) + d(x,y)\right).$$

Obviously $C(x)$ is nonempty and, from Lemma 4.3, it is admissible for every $x \in M$. Furthermore, $C(x) \subseteq P_A(x)$ because $x \in M$. We show next that $C \colon M \to A$ is a nonexpansive multivalued mapping. Pick $u, v \in M$, then it suffices to show that $C(u) \subseteq C(v) + d(u,v)$, or equivalently,

$$\bigcap_{z \in M} (P_A(z) + d(u,z)) \subseteq \bigcap_{z \in M} (P_A(z) + d(z,v) + d(u,v)),$$

which is clear if one recalls the triangle inequality. Now the theorem follows after applying the selection theorem for nonexpansive multivalued mappings, Theorem 4.26, to C. □

Theorem 4.11 follows now as a corollary of Theorem 4.12.

Proof of Theorem 4.11. Let A be an admissible subset of a hyperconvex space M, then A is a proximinal nonexpansive retract and so there exists $R\colon M \to A$ such that $\operatorname{dist}(x,A) = d(x,R(x))$. In particular, for $x \in A + \varepsilon$ we have that $d(x,R(x)) \leq \varepsilon$. □

Remark 4.5. Stronger versions of Theorem 4.12 can be found in the literature (see [24] and [25, Sect. 5]). In fact, the work begun in [29] about the study of proximinal nonexpansive retracts of hyperconvex spaces was completed with a characterization of such retracts in [24]. The main result of [24] asserts that *a subset of a hyperconvex metric space is a proximinal nonexpansive retract if and only if it is weakly externally hyperconvex*. Notice that this result also applies to externally hyperconvex subsets. Weakly externally hyperconvex subsets were introduced in [29] to study this problem. We are not covering these sets and their associated results in this chapter to keep our proofs as simple as possible.

Properties exhibit by proximinal nonexpansive retracts may be of great importance in several situations as, for instance, in Ky-Fan best approximation results (see Sect. 4.6). This is mainly due to the following result.

Proposition 4.5. *Let A be a proximinal nonexpansive retract of a metric space M, then there exists a nonexpansive retraction $R\colon M \to A$ such that $R(M \setminus A) \subseteq \partial A$, where ∂A stands for the boundary of A in M.*

Proof. Consider $R\colon M \to A$ the nonexpansive proximinal retraction from M onto A, then $R(x) \in P_A(x)$ and $P_A(x) \subseteq \partial A$ for any $x \in M \setminus A$. □

4.4 Isbell's Hyperconvex Hull

Embeddings are a very powerful tool in basically all branches of mathematics. The idea is always to associate a given object in a certain category to a larger one where somehow we can find our original object embedded under some structural conditions. Usually we look for a larger element with nice properties which ultimately may allow us to work in a better way with our original element. One of the most illustrative examples of this is the convex hull of a subset of a linear space. *Isbell's hyperconvex hull* (also *injective hull* or *tight span* [20]) looks for the same idea but without leaving the class of metric spaces. Before explaining the ideas behind Isbell's hyperconvex hull, we are going to describe one of the most well-known and extensively used embeddings, the so-called *Kuratowski embedding* of a metric space [35].

Theorem 4.13. *Every metric space M embeds isometrically in the Banach space $\ell_\infty(M)$ of bounded functions on M with the supremum norm:*

$$\|s\|_\infty = \sup_{x\in M} |s(x)|, \ s\colon M \to \mathbb{R}.$$

This embedding is known as the Kuratowski embedding *of M.*

Proof. Fix $x_0 \in M$ and define $s\colon M \to \ell_\infty(M)$ by $x \mapsto s^x$, where

$$s^x(a) = d(x,a) - d(a,x_0).$$

Then by the triangle inequality

$$|s^x(a)| \le d(x,x_0),$$

which proves that $s^x \in \ell_\infty(M)$, and

$$|s^x(a) - s^y(a)| = |d(x,a) - d(y,a)| \le d(x,y),$$

with equality if $a = x$ or $a = y$, which proves that s is an isometry. □

The importance of good embeddings is enormous. For a nice survey on this topic for metric spaces the reader may consult the nice and deep exposition by Heinonen [35].

We will see applications of Kuratowski embedding to fixed point theory in Sects. 4.5 and 4.8. Notice that Kuratowski embedding looks for a linear space where to embed our metric space. The idea of Isbell [41], given in 1965, was, however, to embed our given metric space into another metric space (not necessarily linear) with good properties, which in this case means into a hyperconvex metric space. The same idea would be re-discovered about three decades later by Dress [20] who called it *tight span* of a metric space. Dress would show it to be a very powerful tool in *phylogenetic theory*, the interested reader may consult [22] for more information on this topic.

Isbell showed that any metric space has an *injective envelope* or *hyperconvex hull*. The hyperconvex hull of a metric space M is another metric space $\tilde{M}$ that contains an isometric copy of M and which is minimal in the sense that it is isometric with a subspace of any hyperconvex metric space which also contains M metrically. We show in this section how this hyperconvex hull is constructed.

Let M be a metric space. For any $x \in M$ define the positive real-valued function $f_x\colon M \to [0,\infty)$ by

$$f_x(y) = d(x,y).$$

Property 4.1. Let f_x be as above, then:

(a) For every $x,y,a \in M$ we have that $d(x,y) \le f_a(x) + f_a(y)$ and $f_a(x) - f_a(y) \le d(x,y)$.
(b) Functions f_a are pointwise minimal, that is, if $f\colon M \to [0,\infty)$ is such that $d(x,y) \le f(x) + f(y)$ for any $x,y \in M$ and for some $a \in M$ we have that $f(x) \le f_a(x)$ for any $x \in M$, then $f = f_a$.

Proof. (a) Using the triangle inequality, we have that

$$d(x,y) \le f_a(x) + f_a(y)$$

and, by the reverse triangle inequality as in Kuratowski embedding theorem,

$$f_a(x) \le d(x,y) + f_a(y).$$

(b) Let f be as in 2, then in particular we have that $f(a) \le f_a(a) = 0$ and so $f(a) = 0$. Now, from the property of f,

$$f_a(x) = d(a,x) \le f(x) + f(a) = f(x),$$

and so, $f = f_a$. □

We need the notion of *extremal function*.

Definition 4.22. Let M be a metric space, then $f\colon M \to [0,\infty)$ is said to be an *extremal function* if f satisfies

$$d(x,y) \le f(x) + f(y) \tag{4.1}$$

for all $x,y \in M$ and, moreover, f is pointwise minimal with respect to this property, that is, if $g\colon M \to [0,\infty)$ satisfies $d(x,y) \le g(x) + g(y)$ for all $x,y \in M$ and $g(x) \le f(x)$, then $f = g$.

Remark 4.6. Notice that after Property 4.1, functions f_a are extremal.

Now we can define the *hyperconvex hull* of a metric space.

Definition 4.23. Let M be a metric space, then the *hyperconvex hull* of M, denoted by εM, is the set of all extremal functions defined on M.

Remark 4.7. Although the hyperconvex hull of a metric space is univocally determined by the above definition, all its interesting properties are kept invariant under isometries. Therefore, it is usual to refer to any isometric copy of εM as a hyperconvex hull of M.

In a very similar way as in the Kuratowski embedding case, we can prove that any metric space is metrically embedded in its hyperconvex hull.

Theorem 4.14. *Let M be a metric space and εM its hyperconvex hull, then there is a metric embedding from M into εM.*

Proof. Consider the mapping $e\colon M \to \varepsilon M$ given by $x \mapsto f_x$, that is, $e(x) = f_x$. Then, the mapping e is an isometry over its image since, for $a,b \in M$,

$$d(e(a),e(b)) = \sup_{x\in M} |f_a(x) - f_b(x)| = \sup_{x\in M} |d(a,x) - d(b,x)| = d(a,b).$$

□

Next we will study the properties of εM and its elements, that is, the extremal functions of M. We begin with the following lemma.

Lemma 4.5. *Let A be a subset of a metric space M. Let $r\colon A \to [0,\infty)$ be such that $d(x,y) \leq r(x)+r(y)$ for every $x,y \in A$. Then, there exists $R\colon M \to [0,\infty)$ an extension of r such that $d(x,y) \leq R(x)+R(y)$ for every $x,y \in M$. Moreover, there exists an extremal function f on M such that $f(x) \leq R(x)$ for $x \in M$.*

Proof. To define the extension of r, fix $x_0 \in A$ and define

$$R(x) = \begin{cases} r(x_0)+d(x,x_0), & \text{if } \; x \notin A, \\ r(x), & \text{if } \; x \in A. \end{cases}$$

It is immediate to check that R is the required extension. The second part of the lemma is trivial since if there is no such f then we can take $f = R$. □

Now we give some properties of extremal functions.

Proposition 4.6. *Let M be a metric space, then the following statements are true:*

(a) If $f \in \varepsilon M$, then $f(x) \leq d(x,y)+f(y)$ for $x,y \in M$. Moreover, we have

$$f(x) = \sup_{y\in M} \mid f(y) - f_x(y) \mid = d(f,e(x)).$$

(b) For any $f \in \varepsilon M$, $\delta > 0$ and $x \in M$ there exists $y \in M$ such that

$$f(x)+f(y) < d(x,y)+\delta.$$

(c) If M is compact, then εM is compact.
(d) If s is an extremal function on the metric space εM, then $s \circ e$ is extremal on M where e stands for the embedding mapping given in Theorem 4.14.

Proof. (a) Assume not. Then, there exist $x_0, y_0 \in M$ such that $d(x_0,y_0)+f(y_0) < f(x_0)$. Set

$$g(x) = \begin{cases} f(x), & \text{if } \; x \neq x_0, \\ d(x_0,y_0)+f(y_0), & \text{if } \; x = x_0. \end{cases}$$

It is clear that we have $g(x) \leq f(x_0)$ for $x \in M$ and, in particular, $g(x_0) < f(x_0)$. Let us show that for any $x,y \in M$ it must be the case that

$$d(x,y) \leq g(x)+g(y).$$

Indeed, if both x and y are different from x_0, we use the properties of f. So we can assume that $x = x_0$ and $y \neq x_0$. We have then

$$d(x,y) = d(x_0,y) \leq d(x_0,y_0)+d(y_0,y) \leq d(x_0,y_0)+f(y_0)+f(y),$$

which implies that

$$d(x_0,y) \leq g(x_0)+g(y).$$

The minimality of f gives that $f = g$ which is a contradiction.

Combining this inequality with (4.1), we get

$$| f(y) - f_x(y) | \leq f(x), \quad \text{for any } y \in M.$$

The equality holds for $y = x$ which clearly implies that

$$f(x) = \sup_{y \in M} | f(y) - f_x(y) | = d(f, e(x)).$$

(b) Assume not. Then, there exist $x \in M$ and $\delta > 0$ (and smaller than $f(x)$ if needed) such that for any $y \in M$, we have

$$d(x,y) + \delta \leq f(x) + f(y).$$

Set

$$h(z) = \begin{cases} f(z), & \text{if} \quad z \neq x, \\ f(x) - \delta, & \text{if} \quad z = x. \end{cases}$$

It is easy to check that $g(y,z) \leq h(y) + h(z)$ for $y, z \in M$. Since $h \leq f$ and $h(x) < f(x)$, we get a contradiction which completes the proof.

(c) From (a), we have

$$| f(x) - f(y) | \leq d(x,y),$$

which basically means that the family of extremal functions is equicontinuous. Therefore, this statement is a direct consequence of Ascoli–Arzelà theorem [15, p. 115].

(d) Let s be a extremal function on the metric space εM. Note that for any $x, y \in M$ we have

$$d(x,y) = d(f_x, f_y) \leq s(f_x) + s(f_y) = s \circ e(x) + s \circ e(y).$$

Assume that $s \circ e$ is not an extremal function on M. Then, there exists $h \in \varepsilon M$ such that $h(x) \leq s \circ e(x)$ for $x \in M$ and the inequality is strict at some $x_0 \in M$. Define the function $t \colon \varepsilon M \to \mathbb{R}$ as:

$$t(f) = \begin{cases} s(f), & \text{if} \quad f \neq e(x_0), \\ h(x_0), & \text{if} \quad f = e(x_0). \end{cases}$$

Let us show that t satisfies the inequality $d(f,g) \leq t(f) + t(g)$ for $f, g \in \varepsilon M$. We only need to prove the above inequality for $g = e(x_0)$, that is,

$$d(f, e(x_0)) \leq t(f) + t(e(x_0)).$$

For $\delta > 0$, there exists $y \in M$ such that $f(x_0) + f(y) < d(x_0, y) + \delta$. If $y = x_0$, then we must have $f(x_0) < (1/2)\delta$. Hence,

$$d(f, e(x_0)) + f(x_0) \le \frac{1}{2}\delta + t(f) + t(e(x_0)).$$

On the other hand, if $y \neq x_0$ and $f \neq e(x_0)$, then

$$d(f, e(x_0)) + f(y) - \delta = f(x_0) + f(y) - \delta < d(x_0, y)$$

and

$$d(x_0, y) \le h(x_0) + h(y) \le h(x_0) + s \circ e(y) = t(e(x_0)) + t(e(y)).$$

Since s is an extremal function, then

$$t(e(y)) = s(e(y)) \le s(f) + d(f, e(y)) = t(f) + f(y),$$

where we are applying that $d(f, e(y)) = f(y)$ because f is extremal. So we have

$$d(f, e(x_0)) + f(y) - \delta < t(e(x_0)) + t(e(y)),$$

and

$$t(e(y)) \le t(f) + f(y).$$

Adding the above two inequalities, we get

$$d(f, e(x_0)) + f(y) - \delta + t(e(y)) < t(e(x_0)) + t(e(y)) + t(f) + f(y),$$

which leads to

$$d(f, e(x_0)) - \delta < t(e(x_0)) + t(f).$$

Since δ is arbitrary, our proof is completed. □

Remark 4.8. (c) in the above proposition is a particular case of Lemma 4.6 in Sect. 4.5.

Next we prove some properties of the hyperconvex hull.

Theorem 4.15. *Let M be a metric space. Then the following statements hold:*

(a) εM is hyperconvex.
(b) εM is minimal in the sense that no proper subset of εM which contains M metrically is hyperconvex. Moreover, any hyperconvex metric space H which contains M metrically and is minimal (that is, any proper subset of H which contains M is not hyperconvex) is isometric to εM.

Proof. (a) To prove that εM is hyperconvex, consider $\{f_\alpha\}_{\alpha\in\mathscr{A}} \subseteq \varepsilon M$ and $\{r_\alpha\}_{\alpha\in\mathscr{A}} \subseteq [0,\infty)$ with $d(f_\alpha, f_\beta) \le r_\alpha + r_\beta$, and define $r\colon \{f_\alpha\}_{\alpha\in\mathscr{A}} \to [0,\infty)$ by $r(f_\alpha) = r_\alpha$. By Lemma 4.5, we extend r to the entire εM in such a way that

$$d(f,g) \le r(f) + r(g), \quad \text{for all } f,g \in \varepsilon M.$$

Again, from Lemma 4.5, there exists h an extremal function on εM such that $h \le r$. Using (d) in Proposition 4.6, we know that $h\circ e$ is an extremal function on M, that is, $h\circ e \in \varepsilon M$. Then,

$$h\circ e \in \bigcap_{f\in\varepsilon M} B(f, r(f)) \subseteq \bigcap_{\alpha\in\mathscr{A}} B(f_\alpha, r_\alpha).$$

Indeed, we have

$$h\circ e(x) - f(x) = h\circ e(x) - d(f, e(x)) \le h(f) \le r(f),$$

and

$$f(x) - h\circ e(x) = d(f, e(x)) - h\circ e(x) \le h(f) \le r(f), \quad \text{for any } x \in M.$$

Hence, $d(f, h\circ e) \le r(f)$ for $f \in \varepsilon M$ and the proof is completed.

(b) Recall that M and $e(M)$ are isometric, and let H be a subset of εM such that $e(M) \subseteq H$. Assume that H is hyperconvex. Since εM is hyperconvex, there exists a nonexpansive retraction $R\colon \varepsilon M \to H$. Let $f \in \varepsilon M$. We have

$$d(R(f), e(x)) = R(f)(x) \le d(f, e(x)) = f(x), \quad \text{for all } x \in M.$$

Since f is an extremal function, we must have that $R(f) = f$. This implies that $H = \varepsilon M$ and completes the first part of the proof.

Let N be now a hyperconvex metric space containing M metrically. Let $i\colon M \to N$ be the metric embedding. Then, since N is hyperconvex, we can apply Theorem 4.4 to extend i as a nonexpansive mapping $R_1\colon \varepsilon M \to N$. Since

$$e\circ i^{-1}\colon i(M) \to e(M) \subseteq \varepsilon M,$$

there exists $R_2\colon N \to \varepsilon M$ a nonexpansive mapping extending $e\circ i^{-1}$. The composition $R_2\circ R_1\colon \varepsilon M \to \varepsilon M$ is nonexpansive and extends the identity map on $e(M)$. Therefore it must be the case that $R_2\circ R_1$ is the identity on $e(M)$ which completes the proof. □

The next very useful corollary follows through Zorn's lemma and Theorem 4.8.

Corollary 4.7. *Let A be a subset of a hyperconvex bounded metric space M. Then, there exists $h(A) \subseteq M$ such that $A \subseteq h(A)$ and $h(A)$ is isometric to εA, that is, $h(A)$ is a realization for the hyperconvex hull of A within M.*

The next example shows that $h(A)$ needs not be unique.

Example 4.3. Make $A = \{(0,0),(2,0)\} \subseteq \mathbb{R}^2$ where $\mathbb{R}^2$ is given with the maximum norm. Let $h_1(A)$ be the linear segment connecting the points of A, and $h_2(A)$ the set from Example 4.2. Then $h_1(A)$ and $h_2(A)$ are two different, although necessarily isometric, realizations of the hyperconvex hull of A in $\mathbb{R}^2$.

Unicity of hyperconvex hulls in the sense of Corollary 4.7 has recently been studied in [8] where this problem is considered under *strict convexity* conditions and with a special emphasis on $\mathbb{R}$-trees (see Example 4.1).

4.5 Topological Fixed Point Property and Hyperconvexity

It is in the Topological Theory of Fixed Point where the hyperconvex hull of Isbell has found its most natural applications to the existence of fixed points for operators defined on hyperconvex metric spaces. The goal of this section is to study some of the most basic Topological Fixed Point results in hyperconvex spaces, starting from *Schauder fixed point theorem* and ending with some non-compact versions of Schauder theorem. In the way, we will need to recall the notion of *measure of noncompactness* and its main properties.

We state next Schauder fixed point theorem in its Banach space version. As it is well known, Schauder fixed point theorem has found several extensions to broader classes of spaces but we will not cover that subject in this chapter. The interested reader may found a lot of information about this in [5, 18].

Theorem 4.16. *If C is a convex subset of a Banach space X and $T: C \to C$ is continuous with $T(C)$ contained in a compact subset of C, then T has a* fixed point, *that is, there exists a point $x \in C$ such that $T(x) = x$.*

This theorem has found many and important applications in the theory of differential and integral equations, especially in its version better known as Leray–Schauder theorem for non-self operators—that is, $T: C \to X$—which we will not cover in this chapter.

As an example of the power of metric embeddings, we will show how easily Schauder theorem for hyperconvex metric spaces follows from Theorem 4.16.

Corollary 4.8. *Let M be a hyperconvex metric space and $T: M \to M$ a continuous mapping such that $\overline{T(M)}$, its topological closed closure, is compact, then T has a fixed point.*

Proof. To prove this just consider the Kuratowski embedding $s(M)$ of M into $\ell_\infty(M)$. Then, we can define

$$T_1 : s(M) \to s(M) \text{ as } T_1(y) = s(T(x)), \text{ where } y = s(x).$$

It is clear that T has a fixed point if and only if T_1 has it too. Now consider N as the linear convex hull of $s(M)$ in $\ell_\infty(M)$. Since $s(M)$ is hyperconvex, there exists a nonexpansive retraction $R\colon N \to s(M)$. Then $T_1 \circ R$ is a continuous mapping from the convex subset N in $\ell_\infty(M)$ into a set contained in a compact subset of N. Therefore, we can apply Theorem 4.16 to deduce that $T_1 \circ R$ has a fixed point which, necessarily, must be a fixed point of T_1. □

One of the most natural extensions of Schauder theorem is the well-known *Darbo–Sadovskii theorem* which requires the notion of *measure of noncompactness*. There are several notions of measure of noncompactness being three of them the most well-understood ones; *Kuratowski measure of noncompactness*, *Haurdorff (or ball) measure of noncompactness*, and *separation (or* β*) measure of noncompactness*. The two first have been proved to behave well for hyperconvex metric spaces where they actually coincide. For an in-depth treatment of these notions, the reader may consult [5].

Definition 4.24. Let M be a metric space and let $\mathscr{B}(M)$ denote the class of nonempty and bounded subsets of M. The *Kuratowski measure of noncompactness*, $\alpha\colon \mathscr{B}(M) \to [0,\infty)$, is defined by

$$\alpha(A) = \inf\left\{\varepsilon > 0\colon A \subset \bigcup_{i=1}^{n} A_i \text{ with diam } (A_i) \leq \varepsilon\right\}.$$

The *Hausdorff measure of noncompactness*, $\chi\colon \mathscr{B}(M) \to [0,\infty)$, is defined by

$$\chi(A) = \inf\left\{r > 0\colon A \subset \bigcup_{i=1}^{n} B(x_i, r) \text{ with } x_i \in M\right\}.$$

These two measures are closely related to each other and to compactness. Indeed, the following classical properties are well known and their proofs can be found in [5, p. 18].

Property 4.2. Let M be a metric space and $\mathscr{B}(M)$ be as above. The following statements are true for $A, B \in \mathscr{B}(M)$:

(a) $0 \leq \alpha(A) \leq \text{diam}(A)$.
(b) $\alpha(A) = 0$ if and only if A is precompact.
(c) $\alpha(A \cup B) = \max\{\alpha(A), \alpha(B)\}$.
(d) $\chi(A) \leq \alpha(A) \leq 2\chi(A)$.
(e) If $(A_i)_{i\in I}$ is a decreasing chain of closed and bounded sets such that $\inf_{i\in I} \alpha(A_i) = 0$, then $\cap_{i\in I} A_i$ is nonempty and compact.

Remark 4.9. A very important property when measures of noncompactness are studied in linear spaces is that they remain invariant under the action of taking convex hull. That is, if A is a bounded subset of a Banach space and γ is a measure of noncompactness, then $\gamma(\overline{\text{conv}}(A)) = \gamma(A)$ where $\overline{\text{conv}}(A)$ stands for the closed and convex hull of A.

A very famous extension of Schauder theorem is *Darbo–Sadovskii theorem*, which we describe next for linear spaces. For a proof, consult [5, p. 38].

Definition 4.25. Let M be a metric space and γ a measure of noncompactness. Then, a mapping $T\colon X \to X$ is said to be *γ-condensing* if T is continuous and $\gamma(T(A)) < \gamma(A)$ for every bounded and nonprecompact subset A of X.

Theorem 4.17. *Let M be a Banach space and γ a measure of noncompactness which is invariant under passage to the convex hull. Let M be a nonempty bounded closed and convex subset of X and $T\colon M \to M$ a γ-condensing mapping. Then, T has a fixed point.*

It was noticed in [59] that Darbo-Sadowski theorem can be deduced from its linear version through Kuratowski embedding for hyperconvex spaces.

Corollary 4.9. *Let M be a nonempty and bounded hyperconvex metric space and γ a measure of noncompactness which is invariant under passage to the convex hull. If $T\colon M \to M$ is a γ-condensing mapping, then it has a fixed point.*

Proof. It follows the same pattern as Corollary 4.8 after composing with the retraction R. □

As it has been announced earlier, these Kuratowski and Hausdorff measures reduce to the same measure in hyperconvex spaces.

Proposition 4.7. *Let M be a hyperconvex metric space and A a bounded subset of M, then $\alpha(A) = 2\chi(A)$.*

Proof. From (d) in Property 4.2, we only need to prove that $2\chi(A) \leq \alpha(A)$. Let $\varepsilon > \alpha(A)$. Then, there exist subsets $A_1,\dots,A_n$ of A such that $A = \cup_{i=1}^n A_i$ with $\mathrm{diam}(A_i) \leq \varepsilon$ for $i = 1,\dots,n$. From the hyperconvexity of M ((c) in Lemma 4.2) for each i, there exists $h_i \in M$ such that $A_i \subseteq B(h_i, \varepsilon/2)$. Hence,

$$A \subseteq \bigcup_{1\leq i\leq n} B(h_i, \varepsilon/2),$$

which gives $\chi(A) \leq \varepsilon/2$. Therefore, $\chi(A) \leq \alpha(A)/2$, and the proof is completed. □

When we say that the Kuratowski and Hausdorff measures have been proved to behave well in hyperconvex spaces, it is because they also exhibit an invariance property with respect to the hyperconvex hull. The following technical result will be needed for this invariance.

Lemma 4.6. *Let M be a metric space and consider $\lambda_{[a,b]}(M)$ the space of Lipschitzian real-valued functions defined on M with Lipschitz constant less than λ and taking values in the interval $[a,b]$, endowed with the supremum norm. Then we have*

$$\alpha(\lambda_{[a,b]}(M)) \leq 2\lambda\, \chi(M).$$

Proof. Let $\varepsilon_0 > \chi(M)$. Then we can assume that there exist $x_1,\dots,x_n$ in M such that for any $x \in M$ there exists $i \in \{1,\dots,n\}$ such that $d(x,x_i) \leq \varepsilon_0$. Since $[a,b]$ is compact, for any $\varepsilon > 0$ there exist $c_1, \dots, c_m$ in $[a,b]$ such that for any $c \in [a,b]$ there exists $i \in \{1, \dots, m\}$ with $|c - c_i| \le \varepsilon$. Let $\phi \colon \{1, \dots, n\} \to \{1, \dots, m\}$ be an application. Define

$$\lambda_\phi = \left\{ f \in \lambda_{[a,b]}(M) \colon \sup_{1 \le i \le n} |f(x_i) - c_{\phi(i)}| \le \varepsilon \right\}.$$

Then, we have

$$\lambda_{[a,b]}(M) = \bigcup_{\phi \in \{1,\dots,n\}^{\{1,\dots,m\}}} \lambda_\phi .$$

Let $f, g \in \lambda_\phi$, for any $x \in M$, there exists $i \in \{1, \dots, n\}$ such that $d(x, x_i) \le \varepsilon_0$. Then,

$$|f(x) - g(x)| \le |f(x) - f(x_i)| + |f(x_i) - g(x_i)| + |g(x_i) - g(x)|,$$

which implies that $|f(x) - g(x)| \le \lambda \varepsilon_0 + 2\varepsilon + \lambda \varepsilon_0$. Hence,

$$\sup_{x \in M} |f(x) - g(x)| \le 2\lambda \varepsilon_0 + 2\varepsilon .$$

Since the set $\{1, \dots, n\}^{\{1,\dots,m\}}$ is finite, we get

$$\alpha(\lambda_{[a,b]}(M)) \le 2\lambda \varepsilon_0 + 2\varepsilon .$$

So, since ε is arbitrary, we obtain that $\alpha(\lambda_{[a,b]}(M)) \le 2\varepsilon_0$ and the lemma follows. □

Remark 4.10. Notice that in the construction of Isbell hyperconvex hull, $h(M)$ is included in $\lambda_{[0,\delta]}$ where $\lambda = 1$ and δ is the diameter of M.

After the previous lemma we have the announced invariance result.

Corollary 4.10. *Let M be a bounded metric space and $h(M)$ a realization of the hyperconvex hull of M. Then, $\gamma(h(M)) = \gamma(M)$ for $\gamma = \chi, \alpha$.*

Proof. Since $h(M)$ may be chosen so that $h(M) \subseteq \lambda_{[0,\delta]}$ as explained in Remark 4.10, it follows that $\alpha(h(M)) \le 2\chi(M)$. Therefore, from Proposition 4.7, we have then that $\chi(h(M)) \le \chi(M)$. For the reverse inequality, remember that $h(M)$ contains M isometrically, and α and χ are invariant under isometries and monotone by set inclusion, so we have that $\chi(M) \le \chi(h(M))$. Now, Proposition 4.7 gives the same equality for α. □

The above result was first proved in [23] where a proof for Darbo–Sadovskii theorem in hyperconvex spaces was given without the need of Kuratowski embedding. We can regard this proof, which follows the same patterns as the one for Banach spaces, as an intrinsic proof for this theorem. After that, the same idea was used by different authors to give more general versions of Darbo–Sadovskii theorem which have not been proved to follow from Kuratowski embedding technique. We will reproduce next the study on ultimately compact mappings given in [27]. After that, we will close this section with two more extensions of Darbo–Sadovskii theorem in hyperconvex spaces by means of the hyperconvex hull.

Definition 4.26. Let A be a subset of a hyperconvex metric space M. Given an operator $T\colon A \to M$, we will say that (T_α) is a *transfinite sequence* associated with T on A if

(i) $T_0 = h\big(T(A)\big)$
(ii) $T_\alpha = h\big(f(A \cap T_{\alpha-1})\big)$ if $\alpha - 1$ exists,
(iii) $T_\alpha = \bigcap_{\beta<\alpha} T_\beta$, if $\alpha - 1$ does not exist,

where T_α is a hyperconvex hull of $T(A)$ or $T(A \cap T_{\alpha-1})$ so that the sequence (T_α) is nonincreasing.

Remark 4.11. It is easy to deduce from the properties of the hyperconvex hull studied in Sect. 4.4 that, given A, M and T as in the previous definition, there always exists a transfinite sequence associated with T on A.

Next we show some properties of these transfinite sequences.

Lemma 4.7. *Let (T_α) be a transfinite sequence associated to T on A. Then:*

(a) Each T_α is closed.
(b) $f(A \cap T_\alpha) \subseteq T_{\alpha+1}$ for all α.
(c) If $\beta < \alpha$, then $T_\alpha \subseteq T_\beta$.
(d) $f(A \cap T_\alpha) \subseteq T_\alpha$ for all α.
(e) There exists an ordinal number η such that $T_\alpha = T_\eta$ for all $\alpha \geq \eta$.

Proof. (a), (b), (c), and (d) are the direct consequence of Definition 4.26 and the properties of hyperconvex hulls from Sect. 4.4.

To prove (e), it suffices to note that if the cardinal of the ordinal δ is greater than the cardinal of the set of all subsets of M, then there must be repetitions in the transfinite sequence $\{T_\alpha\colon 0 \leq \alpha \leq \delta\}$. This implies that T_α remains constant from a certain index. □

Corollary 4.11. *Let (T_α) be a transfinite sequence associated with T on M, and suppose M is bounded. Then, the set T_η given by Lemma 4.7 is nonempty.*

Proof. It is sufficient to prove that for any ordinal number α, $T_\alpha \neq \emptyset$. We proceed by transfinite induction.

The case $\alpha = 0$ is trivial since $T(D) \subseteq T_0$, and therefore T_0 is nonempty. Now, we have to consider the following two cases: the ordinal α has a predecessor, in which case, from the inductive hypothesis, $T_{\alpha-1}$ is nonempty, and hence, $T_\alpha = h(f(M \cap T_{\alpha-1}))$ is nonempty, or the ordinal α has no predecessor, in which case $T_\alpha = \bigcap_{\beta<\alpha} T_\beta$. By using the inductive hypothesis and Theorem 4.8, we can deduce that T_α is nonempty. □

Next we give the definition of *ultimately compact* operator.

Definition 4.27. We say that $T^\infty(D)$ is an *ultimate range* of the operator T on the set M if it is the limit set of a transfinite sequence associated with T on D. The operator T is said to be *ultimately compact* (or *limit compact*) if it is continuous and there exists an ultimate range $T^\infty(D)$ of T such that $T(D \cap T^\infty(D))$ is relatively compact in M.

Remark 4.12. We will show later that γ-condensing mappings are ultimately compact. Moreover, from Corollary 4.11, if $D = M$ and M is bounded, any ultimate range of T is nonempty.

Lemma 4.8. *The following properties hold:*

(a) $T^\infty(D) = h\big(T(D \cap T^\infty(D))\big)$.
(b) *If* $D_1 \subseteq D$, *then* $T\colon D_1 \to M$ *has an ultimate range* $T^\infty(D_1)$ *contained in* $T^\infty(D)$.
(c) *If* T *is ultimately compact on* D *and* $D_1 \subseteq D$, T *is ultimately compact on* D_1 *too.*
(d) *The operator* $T\colon D \to M$ *is ultimately compact if and only if* $T^\infty(D)$ *is compact.*
(e) *The operator* $T\colon D \to M$ *is ultimately compact if and only if for any* $B \subset M$, $h(T(B \cap D)) = B$ *implies that* B *is compact.*

Proof. (a) It directly follows from the definition.

(b) It suffices to prove that for each transfinite sequence associated with T on D, (T_α), we may construct a transfinite sequence associated with T on D_1, (T'_α), such that $T'_\alpha \subseteq T_\alpha$ for all α. Suppose we fix (T_α). We now proceed by induction to construct (T'_α).

For $\alpha = 0$, we have $T(D_1) \subseteq T(D) \subseteq T_0$. Since T_0 is hyperconvex, by Corollary 4.7, there exists a hyperconvex hull of $T(D_1)$, $h(T(D_1))$, contained in T_0. Let $T'_0 = h(T(D_1))$.

If α has a predecessor, we proceed in the same way. In another case, the conclusion follows after applying the inductive hypothesis.

(c) Follows from (b).

(d) Let $T^\infty(D)$ be compact. By definition we have $T(D \cap T^\infty(D)) \subseteq T^\infty(D)$. Thus, $T(D \cap T^\infty(D))$ is relatively compact.

Conversely, suppose that $T(D \cap T^\infty(D))$ is relatively compact. Since

$$T^\infty(D) = h(T(D \cap T^\infty(D)))$$

we deduce that $T^\infty(D)$ is compact.

(e) Let T be ultimately compact on D then, by (c), T is also ultimately compact on $B \cap D$. If $h(T(B \cap D)) = B$ then, $T^\infty(B \cap D) = B$ and so B is compact. The converse implication is a direct consequence of (a) and (c). □

Theorem 4.18. *Let* M *be a bounded hyperconvex set, and suppose* $D \subseteq M$ *is closed. If* $T\colon D \to M$ *is* γ*-condensing, then* T *is ultimately compact on* D.

Proof. Let $T^\infty(D)$ be any ultimate range of T on D. By the previous lemma

$$h\big(T(D\cap T^\infty(D))\big) = T^\infty(D).$$

From this,

$$h\big(T(D\cap T^\infty(D))\big) \supseteq D\cap T^\infty(D).$$

Since γ is monotone, we obtain

$$\gamma\big(h(T(D\cap T^\infty(D))\big) \geq \gamma(D\cap T^\infty(D)).$$

But, from Corollary 4.10,

$$\gamma\big(h(T(D\cap T^\infty(D))\big) = \gamma(T(D\cap T^\infty(D))).$$

Hence,

$$\gamma\big(T(D\cap T^\infty(D))\big) \geq \gamma(D\cap T^\infty(D)).$$

Bearing in mind that T is condensing, we may conclude $D\cap T^\infty(D)$ is relatively compact. It follows hence that T is ultimately compact on D. □

Theorem 4.19. *Let M be a bounded hyperconvex metric space, and suppose $T: M\to M$ is ultimately compact on M. Then, T has a fixed point in M.*

Proof. The proof is straightforward if we note that $T^\infty(M)$ is nonempty, compact, and hyperconvex. Then, it is enough to apply Corollary 4.9 (Schauder theorem) to the map $T: T^\infty(M)\to T^\infty(M)$. □

Remark 4.13. A very fruitful line of arguments in Fixed Point Theory has been to obtain abstract versions of already existing results. One of the most successful examples of this is Penot abstract formulation on Kirk fixed point theorem [69]. Following a similar idea to Penot's one and the approach we find in [23] to Darbo–Sadovskii theorem, Khamsi presented in [46] an abstract approach to Darbo–Sadovskii theorem.

The following theorem, which we offer without proof, was given in [13] and is very closely related to Theorem 4.19 although they are different as it was shown in [14].

Theorem 4.20. *Let M be a hyperconvex metric space and $x_0\in M$, let $T: M\to M$ continuous. If every $D\subseteq M$ such that D is isometric to $h(T(D))$ or $D = T(D)\cup\{x_0\}$ is relatively compact, then T has a fixed point in M.*

Remark 4.14. It is not hard to see that any γ-condensing mapping satisfies the hypothesis of this theorem. Moreover, notice that M is not required to be bounded here.

4.6 Metric Fixed Point Property and Hyperconvexity

First results on existence of fixed points for nonexpansive mappings in hyperconvex spaces are due to Sine [78] and Soardi [82]. Properly speaking neither Sine nor Soardi speak about hyperconvexity in their works as they were rather working with some Banach spaces which had been shown to be reluctant to have good properties about existence of fixed point for nonexpansive mappings. However, these Banach spaces happened to be $\mathscr{P}_1$-spaces and so, ultimately, hyperconvex metric spaces. In fact, their proofs take over to the metric context without very much changes. Here we will give the proof based on Penot [69] formulation of Kirk fixed point theorem. This same result was also proved in [6].

Theorem 4.21. *Let M be a nonempty bounded hyperconvex metric space and $T\colon M \to M$ a nonexpansive mapping. Then, T has a fixed point. Moreover, the set of fixed points of T, that is,*

$$Fix(T) = \{x \in M\colon T(x) = x\}$$

is a hyperconvex subset of M.

Proof. Consider $\mathscr{A}(M)$ the family of admissible subsets of M and set

$$\mathscr{F} = \{A \in \mathscr{A}(M)\colon A \neq \emptyset \text{ and } T(A) \subseteq A\}.$$

This family is nonempty because $M \in \mathscr{F}$. From Theorem 4.7, it is immediate that we are under the assumptions to apply Zorn's lemma in the class $\mathscr{F}$. So, there is a minimal element in $\mathscr{F}$, let it be A_0. First thing we note is that $\text{cov}(T(A_0)) = A_0$. This follows since $T(A_0) \subseteq A_0$ and $\text{cov}(T(A_0)) \subseteq A_0$. Using this, we have

$$T\left(\text{cov}(T(A_0))\right) \subseteq T(A_0) \subseteq \text{cov}(T(A_0)),$$

which clearly implies that $\text{cov}(T(A_0)) \in \mathscr{F}$. The minimality of A_0 then implies our claim.

Now note that, from Corollary 4.2, to see that

$$C(A_0) = \{x \in M\colon r_x(A_0) = r(A_0)\}$$

belongs to $\mathscr{F}$ it is enough to check that it is T-invariant. Let $x \in C(A_0)$ which exists also from Corollary 4.2. Then, we have that $A_0 \subseteq B(x, r(A_0))$. Since T is nonexpansive, we get that $T(A_0) \subseteq B(T(x), r(A_0))$, which implies that

$$A_0 = \mathrm{cov}(T(A_0)) \subseteq B(T(x), r(A_0)).$$

Hence, $T(x) \in C(A_0)$, and so $C(A_0) \in \mathscr{F}$. The minimality of A_0 implies then that $A_0 = C(A_0)$ which is impossible from Corollary 4.2 unless A_0 is singleton. Therefore, A_0 is singleton and its element is a fixed point of T.

It rests to prove that $\mathrm{Fix}(T)$ is hyperconvex. Let $\{x_i\}_{i\in I}$ be a collection of points in $\mathrm{Fix}(T)$ and $\{r_i\}_{i\in I} \subseteq [0,\infty)$ such that

$$d(x_i, x_j) \le r_i + r_j, \qquad \text{for all } i, j \in I.$$

Set $H_0 = \bigcap_{i\in I} B(x_i, r_i)$, the hyperconvexity of M implies that H_0 is nonempty. Moreover, since the centers are points of $\mathrm{Fix}(T)$ and T is nonexpansive, we further have that H_0 is T-invariant. Moreover H_0 is bounded so T has a fixed point in H_0 and the proof is completed. □

Remark 4.15. The above proof follows the same pattern as the original proof of Kirk in [51] or Penot [69] for the abstract case. This pattern consists basically in two steps. In the first one we show that decreasing families of a certain collection of sets are compact, meaning that their intersections are nonempty and still a set in the same family. Then, we apply Zorn's lemma to obtain a minimal set and, as a second step, a normal structure argument is applied to show that minimal sets must be a singleton and so a fixed point for the mapping.

The next corollary says that any nonexpansive self-mapping defined on a bounded metric space has a fixed point, the problem is that it may not be where we would expect.

Corollary 4.12. *Let M be a bounded metric space and $T : M \to M$ a nonexpansive mapping, then T can be extended in a nonexpansive way to εM and its extension has a fixed point.*

Proof. Denote by M the isometric copy of M in its hyperconvex hull εM. Then, $T : M \to \varepsilon M$ can be extended to a nonexpansive mapping $\tilde{T} : \varepsilon M \to \varepsilon M$. Now, since εM is bounded and hyperconvex, $\tilde{T}$ has a fixed point. □

The fact that the set of fixed points of a given mapping has good properties has been very studied in Fixed Point Theory, especially in relation to the existence of fixed points of commuting mappings. We see next why.

Corollary 4.13. *Let M be a bounded hyperconvex metric space and $T_1, \dots, T_n$ a collection of nonexpansive commuting (that is, $T_i \circ T_j = T_j \circ T_i$ for $i, j \in \{1, \dots, n\}$) self-mappings on M. Then, the set of common fixed points is nonempty and hyperconvex.*

Proof. Since the mappings T_i are commuting then, by induction,

$$T_i : \mathrm{Fix}(T_{i-1}) \cap \cdots \cap \mathrm{Fix}(T_1) \to \mathrm{Fix}(T_{i-1}) \cap \cdots \cap \mathrm{Fix}(T_1)$$

is nonexpansive and $\mathrm{Fix}(T_{i-1}) \cap \cdots \cap \mathrm{Fix}(T_1)$ is hyperconvex. Therefore, T_i has a fixed points in $\mathrm{Fix}(T_{i-1}) \cap \cdots \cap \mathrm{Fix}(T_1)$ and its set of fixed points in $\mathrm{Fix}(T_{i-1}) \cap \cdots \cap \mathrm{Fix}(T_1)$ is hyperconvex and coincides with $\mathrm{Fix}(T_i) \cap \cdots \cap \mathrm{Fix}(T_1)$. □

This corollary was extended to any collection of nonexpansive commuting mappings defined on a bounded hyperconvex metric space by Baillon [6].

Theorem 4.22. *Let M be a bounded hyperconvex metric space. Any commuting family of nonexpansive self-mappings $\{T_i\}_{i\in I}$ on M has a common fixed point. Moreover, the common fixed points set $\bigcap_{i\in I} Fix(T_i)$ is hyperconvex.*

Proof. Let $\Gamma = 2^I = \{\beta\colon \beta \subseteq I\}$. Γ is downward directed by the set inclusion. Applying the same argument as in Corollary 4.13, we can deduce that for every $\beta \in \Gamma$, the set F_β of common fixed points of the mappings T_i, $i \in \beta$, is nonempty and hyperconvex. Clearly the family $\{F_\beta\}_{\beta\in\Gamma}$ is decreasing. Applying Baillon intersection theorem (Theorem 4.8) the conclusion follows.

Remark 4.16. As it has been recorded in different places, Baillon asked whether boundedness can be relaxed in Theorem 4.21. More precisely he asked whether the conclusion holds if the nonexpansive mapping has bounded orbits, that is, sets $\{T^n(x)\}_{n\in\mathbb{N}}$ are bounded. In the classical Kirk‘s fixed point theorem, having bounded orbit implies the existence of fixed points [31, p. 87]. This is not the case after an example given to Prus [25, p. 412] which the reader may find in Remark 1.4 in Chap. 1. The reason behind this difference may be that asymptotic center technique [85, p. 77] does not apply in hyperconvex spaces.

The interested reader can find further readings on hyperconvexity and metric fixed point theory in [25,50,59,79–81] and references therein. We finish this section with some results on boundary conditions and fixed points which were already announced in Sect. 4.3 with Proposition 4.5.

Sine applied the notion of proximinal nonexpansive retract to the study of Ky Fan best approximation results in hyperconvex spaces. We now exhibit some of the existing results in this direction. We begin with an extension of Darbo–Sadovskii theorem (Corollary 4.9).

Theorem 4.23. *Let D be an admissible subset of a hyperconvex metric space M, and let $T\colon D \to M$ be a γ-condensing mapping for which $T(\partial D) \subset D$. Then, T has a fixed point.*

Proof. Let R be the nonexpansive proximinal retraction given by Theorem 4.12. It is easy to see that the mapping $R \circ T\colon D \to D$ is condensing, and, since D is hyperconvex, $R \circ T$ has a fixed point, say $x_0 \in D$. If $T(x_0) \in D$, then

$$R \circ T(x_0) = T(x_0) = x_0.$$

If $T(x_0) \notin D$, then $x_0 \notin \partial D$, but since $R(x) \in \partial D$ for any $x \in M \setminus D$ and x_0 is a fixed point for $R \circ T$, we have that x_0 must be at the same time at ∂D and at the interior of D which is a contradiction. □

Remark 4.17. In this theorem γ stands for any measure of noncompactness which is invariant under passage to convex hull in linear spaces.

Next we show a Ky Fan's approximation principle for compact admissible subsets of a hyperconvex space.

Theorem 4.24. *Let D be a compact admissible subset of a hyperconvex metric space M and $T : D \to M$ a continuous mapping. Then there exists $x \in D$ such that*

$$d(x, T(x)) = \inf\{d(y, T(x)) : y \in D\}.$$

Proof. Consider R the proximinal nonexpansive retraction onto D, then $R \circ T : D \to D$ has a fixed point which is the solution to the Ky-Fan approximation problem. □

The same holds for nonexpansive mappings after adequately modifying the hypothesis.

Theorem 4.25. *Let D be an admissible subset of a hyperconvex metric space M and $T : D \to M$ a nonexpansive mapping. Then, there exists $x \in D$ such that*

$$d(x, T(x)) = \inf\{d(y, T(x)) : y \in D\}.$$

Proof. Same reasoning as in the previous theorem. □

Remark 4.18. After Remark 4.5 it follows, with no change in proofs, that these boundary fixed point theorems hold true if D is weakly externally hyperconvex instead of admissible.

4.7 Metric Fixed Point Property for Multivalued Mappings and Nonexpansive Selections

Among all properties proved for hyperconvex metric spaces one of the most surprising ones is about the behavior of nonexpansive multivalued mappings. In this chapter we are going to present this property along with pertinent results about selection of multivalued mappings and some applications of it.

The theory of finding *nice* selections of multivalued mappings is a very important one which finds multiple applications. The most well-known result in this subject is Michael's selection theorem (see Chap. 6 in this monograph). We are going to deal only with metric conditions, for a very fruitful theory under topological conditions and applications on hyperconvex spaces the reader may consult [85].

The next result was first discovered by Sine in [80] for mappings with admissible values and, some years later, re-discovered by Khamsi–Kirk–Martínez [49] for mappings with externally hyperconvex values. The interested reader should look in these two references for more information and applications. For simplicity in the exposition, we will present the version for admissible values.

Theorem 4.26. *Let M be a hyperconvex space and $T: M \to M$ a nonexpansive multivalued mapping with admissible values. Then, there exists a single-valued mapping $f: M \to M$ with $f(x) \in T(x)$ for $x \in M$, for which*

$$d(f(x), f(y)) \leq d_H(T(x), T(y)), \quad \textit{for each } x, y \in M,$$

where d_H stands for the Pompeiu–Hausdorff metric (Definition 4.20).

Proof. Let $\mathscr{F}$ denote the collection of all pairs (D, f) with $D \subseteq M$,

$$f: D \to M, \;\; f(d) \in T(d) \text{ for } d \in D,$$

and

$$d(f(x), f(y)) \leq d_H(T(x), T(y)) \text{ for each } x, y \in D.$$

$\mathscr{F} \neq \emptyset$ because for $x_0 \in D$ we can take $(\{x_0\}, f)$ with $f(x_0)$ any choice in $T(x_0)$. Define an order relation on $\mathscr{F}$ by setting

$$(D_1, f_1) \preceq (D_2, f_2) \text{ if and only if } D_1 \subseteq D_2 \text{ and } f_2|_{D_1} = f_1.$$

Let $\{(D_\alpha, f_\alpha)\}$ be a decreasing chain in $(\mathscr{F}, \preceq)$. Then, it follows that $(\cup_\alpha D_\alpha, f) \in \mathscr{F}$ where f is defined on each D_α as $f|_{D_\alpha} = f_\alpha$. By Zorn's lemma, $(\mathscr{F}, \preceq)$ has a maximal element, say (D, f). Assume $D \neq M$ and select $x_0 \in M \setminus D$. Set $\tilde{D} = D \cup \{x_0\}$ and consider the set

$$J = \bigcap_{x \in D} B(f(x), d_H(T(x), T(x_0))) \bigcap T(x_0).$$

By construction, J is an intersection of balls each two of them with nonempty intersection, therefore, J is nonempty. Choose $y_0 \in J$ and define

$$\tilde{f}(x) = \begin{cases} y_0, & \text{if } \;\; x = x_0, \\ f(x), & \text{if } \;\; x \in D. \end{cases}$$

Since $d(\tilde{f}(x_0), \tilde{f}(x)) = d(y_0, f(x)) \leq d_H(T(x), T(x_0))$, we conclude that $(D \cup \{x_0\}, \tilde{f}) \in \mathscr{F}$, contradicting the maximality of (D, f). Therefore, $D = M$ and the proof is completed. □

Remark 4.19. It may be one of the most challenging questions in metric hyperconvexity, already raised by Sine in [80], to show whether this result may be extended to mappings with hyperconvex values instead of admissible. As it was explained earlier, we can assume values are externally hyperconvex [49] but the answer is even unknown for mappings with weakly externally hyperconvex values.

Surely the attentive reader has noticed that hyperconvexity is needed in the proof of Theorem 4.26 only at the target space. This gives the following corollary.

Corollary 4.14. *Theorem 4.26 remains true if $T: M \to N$ where M is any metric space and N is hyperconvex.*

From Theorem 4.26, we also obtain a fixed point result for multivalued mappings.

Definition 4.28. Let $T: D \subseteq M \to M$ be a multivalued mapping, then x is a *fixed point* for T if $x \in T(x)$.

Corollary 4.15. *Let M be a bounded hyperconvex metric space and $T: M \to M$ a nonexpansive multivalued mapping with admissible values. Then, T has a fixed point.*

Proof. From Theorem 4.26, we can find $f: M \to M$ a single-valued nonexpansive selection of T. From Theorem 4.21, f has a fixed point x_0 in M. Then,

$$x_0 = f(x_0) \in T(x_0),$$

and so the corollary follows. □

One of the most immediate applications that Sine gave to Theorem 4.26 was to study the property of nonexpansive proximinal retract enjoyed by admissible subsets of a hyperconvex metric space (see Theorem 4.12 in this chapter).

A bit more involved is to study the structure of the fixed point set of nonexpansive multivalued mappings with admissible values in a hyperconvex space. Still the following strong property holds.

Theorem 4.27. *Let M be a hyperconvex metric space and $T: M \to M$ a nonexpansive multivalued mapping with admissible values and such that $Fix(T) \neq \emptyset$. Then there exists a nonexpansive selection f of T such that $Fix(f) = Fix(T)$. In particular, $Fix(T)$ is hyperconvex.*

Proof. The proof of this theorem follows the same lines as that of Theorem 4.26. We just need to be a little bit more careful about the selection we construct. The key now is to consider the family $\mathscr{F}$ of pairs (D, f) such that f is a nonexpansive selection of T on D, $\mathrm{Fix}(T) \subseteq D$ and $f(x) = x$ for every $x \in \mathrm{Fix}(T)$. Notice that this class is nonempty because $(\mathrm{Fix}(T), Id)$, with Id the identity map, is in it. Then, the proof follows the same steps and arguments as that of Theorem 4.26. □

We close this section with a last application of Theorem 4.26.

Theorem 4.28. *Let M be a hyperconvex metric space and $\lambda > 0$. Let L_λ be the family of all bounded λ-Lipschitzian mappings from M into M endowed with the supremum metric. Then, L_λ is a hyperconvex metric space.*

Proof. Let $\{f_\alpha\}_{\alpha \in \mathscr{A}} \subseteq L_\lambda$ and $\{r_\alpha\}_{\alpha \in \mathscr{A}} \subseteq [0, \infty)$ with $d(f_\alpha, f_\beta) \leq r_\alpha + r_\beta$ as usual. From the hyperconvexity of M, it directly follows that

$$J(x) = \bigcap_{\alpha \in \mathscr{A}} B(f_\alpha(x), r_\alpha) \neq \emptyset, \quad \text{for each } x \in M.$$

Then, $J\colon M \to M$ is a multivalued mapping with admissible nonempty values. Let us see next that J is λ-Lipschitz, that is,

$$d_H(J(x),J(y)) \le \lambda d(x,y), \quad \text{for each } x,y \in M.$$

To prove this it suffices to see that $J(x) \subseteq N_{\lambda d(x,y)}(J(y))$ (see Definition 4.20) for each $x,y \in M$. Take $z \in J(x)$, then, for each α,

$$\begin{aligned} d(z,f_\alpha(y)) &\le d(z,f_\alpha(x)) + d(f_\alpha(x),f_\alpha(y)) \\ &\le d(z,f_\alpha(x)) + \lambda d(x,y) \\ &\le r_\alpha + \lambda d(x,y). \end{aligned}$$

From Lemma 4.3, we have that

$$z \in \bigcap_{\alpha \in \mathscr{A}} B(f_\alpha(y), r_\alpha + \lambda d(x,y)) = N_{\lambda d(x,y)_{J(y)}}.$$

Now, from Theorem 4.26, it is possible to select $f(x) \in J(x)$ for each $x \in M$, so that $f \in L_\lambda$. This f guarantees that L_λ is hyperconvex. □

4.8 KKM Theory and Hyperconvex Spaces

One of the most deeply studied and earliest equivalent formulation to Brouwer fixed point theorem [31, p. 187] is the *theorem of Knaster, Kuratowski, and Mazurkiewicz* (the *KKM theorem* for short) firstly given in 1929. The KKM theory is the study of applications of various equivalent formulations of the KKM theorem and their generalizations.

In this section we are presenting some basic results on KKM theory in hyperconvex spaces. For a very exhaustive exposition on this topic the reader may consult [85]. This is a still developing subject, for updates on the topic the interested reader may also consult [16, 68].

KKM theory falls in the topological side of fixed point theory with a very strong dependence on the existence of a certain convexity structure. That explains in part why this theory needed so much time to develop in a metric context. It was Horvath, in 1991, who focused on the idea of working with abstract convexity structures [36], which were called *generalized convexity structures*, to give abstract versions of some results which had been linked to the notion of linear convexity. In fact, Horvath began a very fruitful line of research with multiple applications in problems related to economic mathematics [76]. Horvath stated KKM results for such structures to prove two years later in [37] that hyperconvex metric spaces were within those spaces which enjoyed the so-called generalized convex structures. Actually, first results on KKM theory for hyperconvex metric spaces were given in [37].

A few years later, in 1996, in an independent way Khamsi gave a new approach to KKM theory for hyperconvex metric spaces [44]. Again the idea goes through a certain convexity structure although this time it is the Kuratowski embedding (Theorem 4.13) the one that provides hyperconvex spaces with that convex structure. This approach got very popular among researchers working on hyperconvex spaces and fixed point in metric spaces, see, for instance, [60]. We will show in this section how this approach goes. Next we describe the convex structure inherited by hyperconvex spaces via Kuratowski embedding.

Definition 4.29. Let M be a metric space and $x,y \in M$. A *metric segment* from x to y is a mapping $c\colon [0,1] \to M$ such that $c(0) = x$, $c(1) = y$, $d(x,c(t)) = td(x,y)$ and $d(y,c(t)) = (1-t)d(x,y)$. In particular, $c([0,1])$ is isometric to the real interval $[0,d(x,y)]$.

Proposition 4.8. *Let M be a hyperconvex metric space. Then, for each x,y, there exists at least a metric segment joining both points.*

Proof. By Kuratowski embedding we can see M embedded into $(\ell_\infty M)$. Since M is hyperconvex there exists a nonexpansive retraction $R\colon \ell_\infty(M) \to M$. Take now $x,y \in M$ and its linear segment I in ℓ_∞. Then, $R(I)$ is isometric to the real interval $[0,d(x,y)]$ and it is a metric segment joining x and y. □

The reader may consult [44] or [25] for a deeper description of this convex structure. Now we present a set of definitions that will be needed in this section.

Definition 4.30. Let M be a metric space. A subset A of M is said to be *finitely closed* if for every finite collection $\{x_1,\ldots,x_n\}$ of points of M, the $\mathrm{cov}(\{x_i\}) \cap A$ is closed.

Definition 4.31. A collection of sets $\{A_\alpha\}_{\alpha\in\mathscr{A}}$ of a metric space M is said to have the *finite intersection property* if the intersection of each finite subfamily is nonempty.

The following technical result will be needed.

Lemma 4.9. *Let M be a metric space and $\{A_\alpha\}_{\alpha\in\mathscr{A}}$ a family of closed subsets of M with the finite intersection property. If there exists a compact set A_{α_0} in the collection, then $\bigcap_{\alpha\in\mathscr{A}} A_\alpha$ is nonempty.*

Proof. We write the proof for a numerable family of sets, the general case follows after applying transfinite induction. Index the numerable family of sets given as $\{A_n\}_{n\in\mathbb{N}}$ with A_1 compact. Then, since finite intersections are nonempty, we have that $B_n = \cap_{i=1}^n A_i$ is a decreasing family of nonempty compact sets. Then,

$$\bigcap_{n\in\mathbb{N}} A_n = \bigcap_{n\in\mathbb{N}} B_n \neq \emptyset,$$

and the proof is completed. □

The central concept in KKM theory is that of *KKM mapping* that we give next.

Definition 4.32. Let M be a metric space and $A \subseteq M$. A multivalued mapping $G\colon A \to M$ is said to be a *Knaster–Kuratowski–Mazurkiewicz mapping*, or a *KKM mapping* for short, if

$$\operatorname{cov}(\{x_1,\ldots,x_n\}) \subseteq \bigcup_{i=1}^{n} G(x_i), \quad \text{for all } x_1,\ldots,x_n \in M,$$

where $\operatorname{cov}(\{x_1,\ldots,x_n\})$ stands for the admissible hull of $\{x_1,\ldots,x_n\}$.

Then, we have the following result known as the *KKM principle*.

Theorem 4.29. *Let M be a hyperconvex metric space and A a nonempty subset of M. Let $G\colon A \to M$ be a KKM mapping such that each $G(x)$ is finitely closed. Then, the family $\{G(x)\colon x \in A\}$ has the finite intersection property.*

Proof. Suppose that what we want to prove is false, that is, that there exist $x_1,\ldots,x_n \in A$ such that $\bigcap_{i=1}^{n} G(x_i) = \emptyset$. Set $L = \operatorname{cov}(\{x\})$ the admissible hull of $\{x_1,\ldots,x_n\}$ in M. Let M_∞ be the admissible hull of M in $\ell_\infty(M)$ and C the convex hull of $\{x_1,\ldots,x_n\}$ in M_∞. Let R be a nonexpansive retraction from M_∞ onto M. Then, in particular, $R(C) \subseteq L$. Our assumptions imply that $L \cap G(x_i)$ is closed for every $i = 1,2,\ldots,n$. Since

$$\left(\bigcap_{i=1}^{n} G(x_i)\right) \bigcap L = \emptyset,$$

then, for every $c \in C$ there exists i_0 such that $R(c)$ does not belong to $L \cap G(x_{i_0})$. Hence, $\operatorname{dist}(R(c), L \cap G(x_{i_0})) > 0$ because $L \cap G(x_{i_0})$ is closed. Therefore,

$$\alpha(c) = \sum_{i=1}^{n} \operatorname{dist}(R(c), L \cap G(x_i))$$

is not zero for any $c \in C$. Define the mapping $F\colon C \to C$ as a linear combination in $\ell_\infty(M)$ by

$$F(c) = \frac{1}{\alpha(c)} \sum_{i=1}^{n} \operatorname{dist}(R(c), L \cap G(x_i)) x_i.$$

Clearly, F is continuous. Moreover, since C is compact, then it has a fixed point c_0, that is, there exists $c_0 \in C$ such that $F(c_0) = c_0$. Set $I = \{i\colon \operatorname{dist}(R(c_0), L \cap G(x_i)) > 0\}$, then we have

$$c_0 = \frac{1}{\alpha(c_0)} \sum_{i \in I} \operatorname{dist}(R(c_0), L \cap G(x_i))\, x_i.$$

Therefore, $R(c_0) \notin \bigcup_{i\in I} G(x_i)$ and $R(c_0) \in \mathrm{cov}(\{x_i : i \in I\})$, contradicting the assumption that $\mathrm{cov}(\{x_i : i \in I\}) \subseteq \bigcup_{i\in I} G(x_i)$. □

As an immediate consequence we obtain the following intersection result.

Theorem 4.30. *Let M be a hyperconvex metric space and $A \subseteq M$ nonempty. Let $G: A \to M$ be a KKM mapping such that $G(x)$ is closed for any $x \in A$ and $G(x_0)$ is compact for some $x_0 \in A$. Then,*

$$\bigcap_{x\in A} G(x) \neq \emptyset.$$

Proof. This is a direct consequence of Lemma 4.9 and Theorem 4.29. □

Consequences and applications of KKM theorems are usually connected with Min-Max inequalities, fixed point theorems for multivalued mappings, saddle points, and Nash equilibria (see, for instance, [85] or any of the references [37, 44, 60, 68, 76]). The following theorem is among these consequences. Note that this theorem is a multivalued version of Ky Fan approximation principle which has already been treated in this chapter in Theorems 4.24 and 4.25.

Theorem 4.31. *Let M be a compact hyperconvex metric space and A a nonempty admissible subset of M. Suppose that $T: A \to M$ is a multivalued continuous mapping with admissible values. Then, there exists $x_0 \in A$ such that*

$$dist(x_0, T(x_0)) = \inf_{x\in A} dist(x, T(x_0)).$$

Proof. Define the multivalued mapping $G: A \to M$ by

$$G(x) = \{y \in A : \mathrm{dist}(y, T(y)) \leq \mathrm{dist}(x, T(y))\}, \quad \text{for each } x \in A.$$

Notice that $G(x)$ is nonempty because $x \in G(x)$, and, since T is continuous, $G(x)$ is closed for each $x \in A$. We want to prove that G is a KKM mapping. Suppose it is not, then there exist a nonempty and finite subset $\{x_1, \dots, x_n\}$ of A and $y \in \mathrm{cov}(\{x_1, \dots, x_n\})$ such that

$$\mathrm{dist}(x_i, T(y)) < \mathrm{dist}(y, T(y))$$

for $i = 1, \dots, n$. Let $\varepsilon > 0$ such that

$$\mathrm{dist}(x_i, T(y)) < \mathrm{dist}(y, T(y)) - \varepsilon, \quad \text{for } i = 1, \dots, n.$$

Let $r = \mathrm{dist}(y, T(y)) - \varepsilon$. Then, $x_i \in T(y) + r$ for $i = 1, \dots, n$. From Lemma 4.3, $T(y) + r \in \mathscr{A}(M)$, thus $\mathrm{cov}(\{x_1, \dots, x_n\}) \subseteq T(y) + r$. This in turn implies that $y \in T(y) + r$, and hence

$$\mathrm{dist}(y, T(y)) \leq r = \mathrm{dist}(y, T(y)) - \varepsilon,$$

which is not possible by assumption. Therefore, G is a KKM mapping. Now, since M is compact, we can apply Theorem 4.30 and the point $x_0 \in \bigcap_{x\in A} G(x)$ is the solution to our problem. □

4.9 Fixed Point Theory and ℝ-Trees

Complete ℝ-trees can be regarded as a subclass of hyperconvex metric spaces, this fact was made clear by W.A. Kirk in [53]. However, ℝ-trees are important elements on their own. In fact they have been extensively applied to graph discrete collections of data and very specifically have been applied in phylogenetics where they are one of the most relevant tools for modeling. Fixed point results for ℝ-tree metric spaces have existed since long ago, see, for instance, [66], but the interest of theorists of metric fixed point theory on these spaces did not begin until 2006 with [25]. That was the initial point for a burst in a systematic study of both topological and metric fixed point properties of ℝ-trees. The very peculiar structure of these spaces has allowed researchers to find results much more powerful than their counterparts for other kinds of spaces as normed spaces. In this chapter we give a very up-to-date exposition on fixed points results for ℝ-trees.

We take as our starting point results given in [26] to continue with the amazing sequence of results obtained by different authors after it. Results inspiring this section can be found in some of these references [1,3,8,19,53,55–57,70]. However, the fact that compact ℝ-trees have the fixed point property for continuous maps goes back to Young [84]. Some familiarity with geodesic spaces is assumed in this section; for details, see [10].

Definition 4.33. Let M be a metric space. A *geodesic* in M is an isometry from $\mathbb{R}$ into M (we may also refer to the image of this isometry as a geodesic). For $x,y \in M$, a *geodesic path* from x to y is a mapping $c:[0,l]\to M$, where $[0,l]\subseteq\mathbb{R}$, such that $c(0)=x, c(l)=y$ and $d(c(t),c(t'))=|t-t'|$ for every $t,t'\in[0,l]$. The image $c([0,l])$ of c forms a *geodesic segment* which joins x and y and is not necessarily unique. If no confusion arises, we will use $[x,y]$ to denote a geodesic segment joining x and y.

Definition 4.34. M is a *geodesic space* if every two points $x,y\in M$ can be joined by a geodesic path. M is said to be *uniquely geodesic* if it is geodesic and for each $x,y\in M$ there exists a unique geodesic path joining them.

There are different ways to define a convex subset of a geodesic space. The next one works perfectly for ℝ-trees.

Definition 4.35. A subset A of a geodesic metric space M is said to be *convex* if any geodesic segment that joins every two points of A is contained in A.

The definition of ℝ-*trees*, also known as *real trees*, was given in Example 4.1. The following theorem can be found in [53].

Theorem 4.32. *A metric space is a complete $\mathbb{R}$-tree if and only if it is hyperconvex and has unique metric segments joining its points. Therefore, $\mathbb{R}$-trees are uniquely geodesic metric spaces.*

The main goal of [26] was to suggest a new metric approach to the classical fixed edge theorem of Nowakowski and Rival [66]. The first fact that it is noticed in [26] is the relation between $\mathbb{R}$-trees and *gated sets*.

Definition 4.36. Let M be a metric space and $A \subseteq M$. A is said to be *gated* if for any point $x \notin A$ there exists a unique point $x_A \in A$ (called the *gate of x in A*) such that for any $z \in A$,

$$d(x,z) = d(x,x_A) + d(x_A,z).$$

It is immediate to see that gated sets in a complete geodesic space are always closed and convex. (Remember that a *convex set* in a uniquely geodesic metric space is any set which contains any segment with endpoints in the same set.) Moreover, it was noticed in [21] that gated subsets of a complete geodesic space X are proximinal nonexpansive retracts of X. The next lemma is not hard to prove.

Lemma 4.10. *Gated subsets of an $\mathbb{R}$-tree are precisely its closed and convex subsets.*

Next we present a surprising fact from gated sets which is at the heart of many results obtained for $\mathbb{R}$-trees. Usually results asserting that a certain collection of descending closed sets has nonempty intersection require some compactness conditions and, of course, that the sets are bounded. That will not be the case for gated sets.

Proposition 4.9. *Let M be a complete geodesic space, and let $\{H_\alpha\}_{\alpha\in\mathscr{A}}$ be a collection of nonempty gated subsets of M which is directed downward by set inclusion. If M (or more generally, some H_α) does not contain a geodesic ray (that is, a subset isometric to a real half-line), then*

$$\bigcap_{\alpha\in\mathscr{A}} H_\alpha \neq \emptyset.$$

Proof. Let $H_0 \in \{H_\alpha\}_{\alpha\in\mathscr{A}}$, select $x_0 \in H_0$ and let

$$r_0 = \sup\{\operatorname{dist}(x_0, H_0 \cap H_\alpha) : \alpha \in \mathscr{A}\}.$$

If $x_0 \in \cap_{\alpha\in\mathscr{A}} H_\alpha$ we are finished. Otherwise choose $H_1 \in \{H_\alpha\}_{\alpha\in\mathscr{A}}$ so that, $H_1 \subset H_0$, $x_0 \notin H_1$, and

$$\operatorname{dist}(x_0, H_1) \geq \begin{cases} r_0 - 1, & \text{if } r_0 < \infty, \\ 1, & \text{if } r_0 = \infty \end{cases}.$$

Now take x_1 to be the gate of x_0 in H_1. Having defined x_n, let

$$r_n = \sup\{\operatorname{dist}(x_n, H_n \cap H_\alpha) : \alpha \in \mathscr{A}\}.$$

Now choose $H_{n+1} \in \{H_\alpha\}_{\alpha\in\mathscr{A}}$ so that $x_n \notin H_{n+1}$ (if possible), $H_{n+1} \subset H_n$, and

$$\operatorname{dist}(x_n, H_{n+1}) \geq \begin{cases} r_n - \dfrac{1}{n}, & \text{if } r_n < \infty, \\ 1, & \text{if } r_n = \infty \end{cases}.$$

Now take x_{n+1} to be the gate of x_n in H_{n+1}. Either this process terminates after a finite number of steps, yielding a point $x_n \in \bigcap_{\alpha\in\mathscr{A}} H_\alpha$, or we have sequences $\{x_n\}$, $\{H_n\}$ for which $i < j \Rightarrow x_j$ is the gate of x_i in H_j. Since X does not contain a geodesic ray, it must be the case that $r_n < \infty$ for some n (and hence for all n). By transitivity of gated sets the sequence $\{x_n\}$ is linear and thus lies on a geodesic in X. Since X does not contain a geodesic ray, the sequence $\{x_n\}$ must in fact be Cauchy. Let $x_\infty = \lim_n x_n$. Since each of the sets H_n is closed, clearly $x_\infty \in \bigcap_{n=1}^\infty H_n$. Also, $\sum_{n=1}^\infty r_n < \infty$, so $\lim_n r_n = 0$.

Now let P_α, $\alpha \in \mathscr{A}$, be the nearest point projection of X onto H_α, and for each $n \in \mathbb{N}$, let $y_n = P_\alpha(x_n)$. Then, $d(y_n, x_n) \leq r_n$, and since P_α is nonexpansive, for any $m, n \in \mathbb{N}$, $d(y_n, y_m) \leq d(x_n, x_m)$. It follows that $P_\alpha(x_\infty) = x_\infty$ for each $\alpha \in \mathscr{A}$. Therefore, $x_\infty \in \bigcap_{\alpha\in\mathscr{A}} H_\alpha$. □

Proposition 4.10. *Let M be a complete geodesic space, and let $\{H_n\}$ be a descending sequence of nonempty gated subsets of M. If $\{H_n\}$ has a bounded selection, then*

$$\bigcap_{n=1}^{\infty} H_n \neq \emptyset.$$

Proof. Here we simply describe the step-by-step procedure. Let $\{z_n\}$ be a bounded selection for $\{H_n\}$. Let $x_0 = z_0$. Then, let n_1 be the smallest integer such that $x_0 \notin H_{n_1}$. Let x_1 be the gate of x_0 in H_{n_1} and take $x_2 = z_{n_1}$. Now take n_2 to be the smallest integer such that $x_2 \notin H_{n_2}$ and take x_3 to be the gate of x_2 in H_{n_2}. Continuing this procedure inductively it is clear that one generates a sequence $\{x_n\}$ which is isometric to an increasing sequence of positive numbers on the real line. Since $\{x_{2n}\}$ is a subsequence of the bounded sequence $\{z_n\}$ it must be the case that $\{x_n\}$ is also bounded. Therefore, $\lim_n x_n$ exists and lies in $\bigcap_{n=1}^\infty H_n$. □

The next result also stands for gated sets and was noticed by Markin in [63].

Theorem 4.33. *Let M be a complete geodesic space with a convex metric (so, in particular, balls contain geodesic segments with endpoints in the given ball) and T a multivalued mapping with values that are bounded gated subsets of M. Then there is a mapping $f\colon M \to M$ such that $f(x) \in Tx$ for each $x \in M$ and $d(f(x), f(y)) \leq H(Tx, Ty)$ for each $x, y \in M$, where H stands for the Pompeiu–Hausdorff distance.*

Proof. For each $z \in M$ define the mapping $f\colon M \to M$ such that $f(x)$ is the unique closest point to z in Tx, that is, $f(x) = P_{Tx}z$. Take $\alpha = d(z, f(x))$ and $\beta = d(z, f(y))$ and assume that $\alpha \geq \beta$. Therefore $f(y) \in B(z, \alpha)$. Let p be the gate of $f(y)$ in Tx. Since $d(f(y), f(x)) = d(f(y), p) + d(p, f(x))$ the point p lies on the geodesic segment connecting $f(x)$ and $f(y)$. By convexity of the metric, this segment is contained in $B(z, \alpha)$. This implies that $p \in B(z, \alpha)$, and so it must be the case that $p = f(x)$. The conclusion trivially follows now. □

The main result in [26] requires the following lemma given in [52].

Lemma 4.11. *Suppose M is uniquely geodesic with a convex metric, suppose $T\colon M \to M$ is nonexpansive, and suppose $x_0 \in M$ satisfies*

$$d\left(x_0, T\left(x_0\right)\right) = \inf\left\{d\left(x, T\left(x\right)\right) : x \in M\right\} > 0.$$

Then, the sequence $\{T^n\left(x_0\right)\}$ is unbounded and lies on a geodesic ray.

Theorem 4.34. *Let M be a complete $\mathbb{R}$-tree, and suppose K is a closed convex subset of M which does not contain a geodesic ray. Then, every commuting family $\mathfrak{F}$ of nonexpansive self-mappings on K has a nonempty common fixed point set.*

Proof. Let $T \in \mathfrak{F}$. We first show that the fixed point set of T is nonempty. Let

$$d = \inf\left\{d\left(x, T\left(x\right)\right) : x \in K\right\}$$

and let

$$F_n := \left\{x \in K : d\left(x, T\left(x\right)\right) \leq d + \frac{1}{n}\right\}.$$

Since K is a closed convex subset of a complete $\mathbb{R}$-tree, K itself is hyperconvex and $\{F_n\}$ is a descending sequence of nonempty closed convex (hence gated) subsets of K. Since K does not contain a geodesic ray, Proposition 4.9 implies $F := \bigcap_{n=1}^{\infty} F_n \neq \emptyset$. Therefore there exists $z \in K$ such that

$$d\left(z, T\left(z\right)\right) = d.$$

Since K does not contain a geodesic ray, in view of Lemma 4.11, $d = 0$.

Because $\mathbb{R}$-trees are uniquely geodesic, the fixed point set F of T is closed and convex, and hence again an $\mathbb{R}$-tree. Now suppose $G \in \mathfrak{F}$. Since G and T commute it follows that $G : F \to F$, and by applying the preceding argument to G and F we conclude that G has a nonempty fixed point set in F. In particular the fixed point set of T and the fixed point set of G intersect. Since these are gated sets in X, by the Helly property of gated sets we conclude that every finite subcollection of $\mathfrak{F}$ has a nonempty common fixed point set (which is itself gated). Now let $\mathscr{A}$ be the collection of all finite subcollections of $\mathfrak{F}$, and for $\alpha \in \mathscr{A}$, let H_α be the common fixed point set of α. Then given $\alpha, \beta \in \mathscr{A}$, $H_{\alpha \cup \beta} \subseteq H_\alpha \cap H_\beta$, so clearly the family

$\{H_\alpha\}_{\alpha\in\mathscr{A}}$ is directed downward by set inclusion. Since these are all gated sets, we again apply Proposition 4.9 to conclude that $\bigcap_{\alpha\in\mathscr{A}} H_\alpha \neq \emptyset$, and thus that $\mathfrak{F}$ has a nonempty common fixed point set. □

Remark 4.20. The significance of this result is the fact that K itself is not assumed to be bounded.

Surprisingly enough W.A. Kirk noticed in [54] that when considered Theorem 4.34 on just one mapping the nonexpansiveness condition can be replaced by continuity.

Theorem 4.35. *Let M be a geodesically bounded complete $\mathbb{R}$-tree. Then, every continuous mapping $T: X \to X$ has a fixed point.*

Proof. For $u,v \in M$, we let $[u,v]$ denote the unique metric segment joining both points and let $[u,v) = [u,v] \setminus \{v\}$. To each $x \in M$ associate $\phi(x)$ as follows. For each $t \in [x,Tx]$, let $\xi(t)$ be the point in M for which

$$[x,Tx] \cap [x,Tt] = [x,\xi(t)].$$

Such a point always exists since M is an $\mathbb{R}$-tree. If $\xi(Tx) = Tx$, take $\phi(x) = Tx$. Otherwise it must be the case that $\xi(Tx) \in [x,Tx)$. Let

$$A = \{t \in [x,Tx] : \xi(t) \in [x,t]\};$$
$$B = \{t \in [x,Tx] : \xi(t) \in [t,Tx]\}.$$

Now a connectedness reasoning yields that there exists $\phi(x) \in A \cap B$. If $\phi(x) = x$, then $Tx = x$ and we are fare. Otherwise, $x \neq \phi(x)$ and

$$[x,Tx] \cap [x,T\phi(x)] = [x,\phi(x)].$$

Now let $x_0 \in M$ and let $x_n = \phi^n(x_0)$. Assuming that the process does not terminate upon reaching a fixed point of T, by construction, the points $\{x_0,x_1,x_2,\cdots\}$ are linear, and thus lie on a subset of M which is isometric with a subset of the real line. Since M does not contain a geodesic of infinite length, it must be the case that $\{x_n\}$ is a Cauchy sequence. Let z be the limit of $\{x_n\}$, then, by continuity, Tz is the limit of $\{Tx_n\}$. Now, by construction,

$$d(Tx_n, Tx_{n+1}) = d(Tx_n, x_{n+1}) + d(x_{n+1}, Tx_{n+1}),$$

from where it finally follows that $z = Tz$. □

The technique developed to prove Theorem 4.35 had a big impact on the theory and was applied by many other authors in different versions and extensions of this theorem. Among these new versions those ones dealing with multivalued mappings were to surprise us by the constantly and unexpected weakening of the required conditions.

Definition 4.37. Let $T\colon X \to 2^Y$ be a multivalued mapping with nonempty values, then

- T is said to be *upper semicontinuous* at $x_0 \in X$ if for each open $V \subseteq Y$ such that $T(x_0) \subseteq Y$ there exists an open set $U \subseteq X$ which contains x_0 such that
$$T(U) \subseteq V.$$
- T is said to be almost lower semicontinuous at $x_0 \in X$ if for each $\varepsilon > 0$ there is an open neighborhood $U \subseteq X$ of x_0 such that
$$\bigcap_{x\in U} N_\varepsilon(T(x)) \neq \emptyset.$$
- T is said to be ε-semicontinuous at $x_0 \in X$ if for each $\varepsilon > 0$ there is an open neighborhood $U \subseteq X$ of x_0 such that
$$T(x) \cap N_\varepsilon(T(x_0)) \neq \emptyset, \quad \text{for all } x \in U.$$

A best approximation result was given for upper semicontinuous mappings by W.A. Kirk and B. Panyanak in [56] while a similar one for almost lower semicontinuous mappings was obtained by J.T. Markin in [63]. However both were unified by B. Piątek in [70]. In fact the concept of ε-semicontinuity was introduced in [70] with the goal of unifying both results. The following proposition, not very hard to prove, was shown in [70].

Proposition 4.11. *If a multivalued mapping is either almost lower semicontinuous or upper semicontinuous, then it is ε-semicontinuous too.*

Then the following result, which contains those of [56, 63] as particular cases, was proved in [70].

Theorem 4.36. *Let M be a complete $\mathbb{R}$-tree and let K be a nonempty convex closed and geodesically bounded subset of M. If $F\colon K \to 2^K$ is an ε-semicontinuous mapping with nonempty convex closed values, then F has a fixed point.*

Proof. Let $x \in X$ and let $r(x) = P_{F(x)}(x)$. If x is not a fixed point, then $d(x, r(x)) > 0$. For each $t \in [x, r(x)]$ we define $\xi(t)$ as
$$[x, r(x)] \cap [x, r(t)] = [x, \xi(t)].$$

Let
$$A = \{t \in [x, r(x)] \mid \xi(t) \in [x, t]\},$$
$$B = \{t \in [x, r(x)] \mid \xi(t) \in [t, r(x)]\}.$$

Clearly, $r(x) \in A$ and $x \in B$ and A and B are closed. Indeed, let (t_n) be a sequence of elements of B such that $t_n \to t$. Assume that $t \in A \setminus B$. Then $d(t, \xi(t)) = \varepsilon > 0$. Let $n \in \mathbb{N}$ be such that $d(t, t_n) < \varepsilon/2$. For each $u \in F(t)$ and $v \in F(t_n)$ we obtain

$$r(t) \in [u, \xi(t)], \quad \xi(t) \in [r(t), \xi(t_n)], \quad \xi(t_n) \in [\xi(t), r(t_n)], \quad r(t_n) \in [\xi(t_n), v].$$

Then, we have

$$[\xi(t), \xi(t_n)] \subset [u, v]$$

and finally $\inf_{z \in F(t_n)} d(z, F(t)) \geq d(\xi(t), \xi(t_n)) > \varepsilon/2$ for each $n \in \mathbb{N}$ sufficiently large which contradicts the ε-semicontinuity of F.
Since A is compact there is $\varphi(x) \in [x, r(x)]$ such that $d(x, \varphi(x)) = \inf_{t \in A} d(x, t)$. Moreover, $\varphi(x) \in A \cap B$ which implies that

$$[x, r(x)] \cap [x, r(\varphi(x))] = [x, \varphi(x)]. \tag{4.2}$$

Now suppose that F has not a fixed point in X. Therefore, we have

$$d(x, \varphi(x)) > 0, \qquad x \in X. \tag{4.3}$$

Let us choose any $x_0 \in X$. We define a transfinite sequence $(x_\alpha)_{\alpha < \Omega}$ such that

$$d(x_0, x_\alpha) = \sum_{\beta < \alpha} d_\beta \tag{4.4}$$

and

$$d(x_0, \varphi(x_\alpha)) = d(x_0, x_\alpha) + d(x_\alpha, \varphi(x_\alpha)) \tag{4.5}$$

where Ω is the order type of the set $\{\alpha \mid \bar{\alpha} \leq \aleph_0\}$ and $d_\beta = d(x_\beta, \varphi(x_\beta))$.

Let α be a limit number. By the geodesically boundedness of X and (4.4) the countable sum $\sum_{\beta < \alpha} d_\beta$ is bounded. So, there is a sequence of points x_{α_n} such that $\lim_{n \to \infty} \sum_{\beta < \alpha_n} = \sum_{\beta < \alpha}$ and $x_{\alpha_n} \to \bar{x} \in X$. Let us define $x_\alpha := \bar{x}$. Clearly (4.4) and (4.5) are satisfied. The proof of (4.5) is not different from the proof of the closedness of B.

If $\alpha = \beta + 1$, we define $x_\alpha := \varphi(x_\beta)$. By (4.2) with $x = x_\beta$, (4.4) and (4.5) we obtain $d(x_0, x_\alpha) = d(x_0, x_\beta) + d(x_\beta, x_\alpha) = \sum_{\gamma < \alpha} d_\gamma$ and $d(x_0, \varphi(x_\alpha)) = d(x_0, x_\alpha) + d_\alpha$.
Now let us define

$$m := \sup_{\alpha < \Omega} \sum_{\beta < \alpha} d_\beta. \tag{4.6}$$

If m were equal to infinity, points x_α would lie on a geodesic ray. Hence $m < \infty$ and one can find a sequence α_n for which $d(x_0, x_{\alpha_n}) \to m$. Clearly, there is $\alpha < \Omega$ such that $\alpha_n < \alpha$ for each $n \in \mathbb{N}$. Moreover, $d(x_0, x_\alpha) = m$ which implies that $d(x_0, x_{\alpha+1}) = d(x_0, \varphi(x_\alpha)) > m$. This contradicts (4.6). □

4.10 New Trends in Hyperconvexity

Metric hyperconvexity was introduced in 1956 [4] in relation to the study of extension of uniformly continuous mappings. In 1979 [78, 82], we find the first results on existence of fixed points for spaces with a *hyperconvex geometry*, but it was not until the very end of the eighties [6, 79] that fixed point properties were not studied in a more systematic way for hyperconvex metric spaces. As we have seen in this chapter, it was from that moment that many works have appeared in the literature where metric fixed point theory, topological fixed point theory, and KKM theory got very extensively developed in hyperconvex metric spaces, see [25,28,85] and the references therein. Metric hyperconvexity and Fixed Point Theory, however, is a still developing topic. In this last section we will address some of these more recent trends in hyperconvexity and fixed points.

4.10.1 Two Long-Standing Open Problems

If we look at Metric Fixed Point Theory and hyperconvex spaces we find two very challenging problems which have not been solved yet.

One of the most influential works for hyperconvex metric spaces is Sine's [80]. We have addressed main results of [80] in Sect. 4.7. In particular, Theorem 4.26 asserts that any multivalued nonexpansive mapping with admissible values in a hyperconvex metric space admits a nonexpansive single-valued selection. Sine raised the following problem in [80].

Problem 4.1. Can the hypothesis that the multivalued mapping has admissible values be relaxed in Theorem 4.26? Is the result still true if the values are hyperconvex subsets of a hyperconvex metric space?

Very little advances have been made regarding this question. In [49], as it was explained in Sect. 4.7, the condition of admissible values was relaxed to the one of externally hyperconvex values. However, one of the main consequences of Theorem 4.26 studied by Sine in [81] was that admissible subsets were proximinal nonexpansive retracts. The problem of proximinal nonexpansive retracts has been described in Sect. 4.3 and, as it was explained in Remark 4.5, it was completely solved in [24] obtaining as a solution that proximinal nonexpansive retracts were, precisely, weakly externally hyperconvex subsets. Therefore, the best candidate for a positive answer to Problem 4.1 are these later sets.

The second problem is about *asymptotically nonexpansive mappings*.

Definition 4.38. Let M be a metric space, then a mapping $T\colon M \to M$ is said to be *asymptotically nonexpansive* if there exists $\{\lambda_n\}_{n\in\mathbb{N}} \subseteq \mathbb{R}$ with $\lim_{n\to\infty} \lambda_n = 1$ such that

$$\|T^n(x) - T^n(y)\| \leq \lambda_n d(x,y), \quad \text{for all } x,y \in M.$$

Proving that asymptotically nonexpansive mappings have fixed points when defined on spaces with nice asymptotic centers of sequence (see [31, Sect. 9]) is a very easy fact. So, since nonexpansive mappings behave so well in hyperconvex spaces, many authors have wondered about the next problem.

Problem 4.2. Let M be a bounded hyperconvex metric space and $T\colon M \to M$ an asymptotically nonexpansive mapping, is it true that T must have a fixed point?

The only positive approach so far to this problem was made by Khamsi in [47], where it was shown that for any such mapping as in Problem 4.2 it must be the case that

$$\inf_{x \in M} d(x, T(x)) = 0,$$

that is, such mappings have approximate fixed point sequences, i.e. sequences $\{x_n\}_{n \in \mathbb{N}}$ such that

$$\lim_{n \to \infty} d(x_n, T(x_n)) = 0.$$

4.10.2 Ultrametrics and Hyperconvex Metric Spaces

Ultrametric spaces [62] are interesting objects which find many applications in different fields (see, for instance, [73]).

Definition 4.39. A metric space M is said to be an *ultrametric space* if the metric satisfies the strong triangle inequality, that is, if

$$d(x,y) \le \max\{d(x,z), d(y,z)\}, \quad \text{for all } x,y,z \in M.$$

Existence of fixed points for ultrametric spaces has been studied by different authors, see, for instance, [72, 77]. Relations between ultrametrics and trees have also been explored [38] and hyperconvex ultrametric spaces were defined in [9] where also their fixed point properties were studied. This topic seems still to be unexplored and more connections between both notions in relation to fixed point properties have been recently shown in [58].

4.10.3 Diversities and Hyperconvexity

A general theory on *diversities* has recently been proposed by Bryant and Tupper in [12]. The authors introduce diversities in this work as a sort of multi-way metrics which inherits its name after some special appearances in works on phylogenetic and ecological diversity [30, 64, 67, 83]. In [12] the authors aim to develop a theory of

Tight Span Diversities (Hyperconvex Hulls, see Sect. 4.4) parallel to the theory of Tight Span for metric spaces independently given by Dress [20] and Isbell [41].

Definition 4.40. Let X be a set and denote $\langle X\rangle$ as the set of its finite subsets, then a diversity is a pair (X,δ) where $\delta\colon \langle X\rangle \to \mathbb{R}$ such that

(i) $\delta(A) \geq 0$, and $\delta(A) = 0$ if and only if $|A| \leq 1$, where $|A|$ stands for the cardinality of A.
(ii) If $B \neq \emptyset$, then $\delta(A\cup C) \leq \delta(A\cup B) + \delta(B\cup C)$.

A diversity will be said bounded if there exists $M \geq 0$ such that $\delta(A) \leq M$ for each $A \in \langle X\rangle$.

A very interesting thing from the definition of diversity is that any diversity (X,δ) induces a metric on X given by

$$d(x,y) = \delta(\{x,y\}).$$

Therefore, diversities carry with them a natural metric structure which has been very little studied so far. Most interesting examples so far of diversities are:

Example 4.4. (a) *Diameter diversity.* Let (X,d) be a metric space. For all $A \in \langle X\rangle$ let

$$\delta(A) = \max_{a,b\in A} d(a,b) = \operatorname{diam}(A).$$

Then, (X,δ) is a diversity which is called the diameter diversity generated by (X,d). Therefore, any metric space generates a diameter diversity.

(b) *Phylogenetic diversity.* Consider (T,d) a real tree, let μ be the one-dimensional Hausdorff measure on it. Notice that in this case $\mu([a,b]) = d(a,b)$ for any $a,b \in T$. If $A \subseteq T$, then the convex hull of A is defined as

$$\operatorname{conv}(A) = \bigcup_{a,b\in A} [a,b]$$

and we say that A is convex if $A = \operatorname{conv}(A)$ (see [25] for details). Then, it happens that

$$\delta_t(A) = \mu(\operatorname{conv}(A))$$

defines a diversity on T which is called the *real-tree* diversity (T,δ_t) for (T,d). Finally, always following [12], a diversity (X,δ) is a *phylogenetic diversity* if it can be embedded in a real-tree diversity for some complete real tree (T,d).

Hyperconvexity for diversities is defined as follows in [12].

Definition 4.41. A diversity (X,δ) is said to be *hyperconvex* if for all $r : \langle X\rangle \to \mathbb{R}$ such that

$$\delta\left(\bigcup_{A\in\mathscr{A}} A\right) \leq \sum_{A\in\mathscr{A}} r(A) \tag{4.7}$$

for all $\mathscr{A} \subseteq \langle X\rangle$ finite, with $r(\emptyset) = 0$, there is $z \in X$ such that $\delta(\{z\} \cup Y) \leq r(Y)$ for all finite $Y \subseteq X$.

These notions have been considered for the first time in relation to fixed point theory in [71] where some intriguing facts, among other results, have been found, as, for instance:

Lemma 4.12. *Let (X,δ) be a hyperconvex diversity then its induced metric space need not be hyperconvex.*

Theorem 4.37. *If (X,d) is a metric space induced by a bounded hyperconvex diversity, then it has* abstract *normal structure.*

Theorem 4.38. *If (X,d) is a bounded metric space induced by a hyperconvex diversity, then it need not have the fixed point property for nonexpansive mapping. However, if (X,d) has been induced by a bounded hyperconvex diversity, then it has the fixed point property for nonexpansive mappings, that is, any nonexpansive mapping $T: X \to X$ has a fixed point.*

4.10.4 Q-hyperconvexity

Another recent research line in hyperconvexity is the one for the so-called *q-hyperconvex quasi-metric spaces*. This notion was introduced by Salbany in [75].

Definition 4.42. A quasi-pseudometric space (M,d) is said to be *Isbell convex* if for each family $\{x_i\}_{i\in I}$ of points in M and families of nonnegative real numbers $\{r_i\}_{i\in I}$ and $\{s_i\}_{i\in I}$ the conditions $d(x_i,x_j) \leq r_i + s_j$ whenever $i,j \in I$ imply that

$$\bigcap_{i\in I}(B_d(x_i,r_i) \cap B_{d^t}(x_i,s_i)) \neq \emptyset,$$

where $d^t(x,y) = d(y,x)$.

In [42] the notion of Isbell hull for these structures is studied. In [61] the authors propose a successful study paralleling those of [49, 80] for nice selections of multivalued mappings with applications to fixed point theory.

Acknowledgments The first author wishes to acknowledge The University of Tabuk for its kind invitation to take part in *The International Mathematical Workshop (Fixed Point Theory and its Applications) May 26–27, 2012.* This work is motivated by that event.

Rafa Espínola and Aurora Fernández León were supported by DGES, MTM2012-34847C02-01 and Junta de Andalucía, Grant FQM-127.

References

1. Aksoy, A.G., Khamsi, M.A.: A selection theorem in metric trees. Proc. Am. Math. Soc. **134**, 2957–2966 (2006)
2. Ambrosio, A., Tilli, P.: Topics on Analysis in Metric Spaces. Oxford University Press, Oxford (2004)
3. Amini-Harandi, A., Farajzadeh, A.P.: Best approximation, coincidence and fixed point theorems for set-valued maps in $\mathbb{R}$-trees. Nonlinear Anal. **71**, 1649–1653 (2009)
4. Aronszajn, N., Panitchpakdi, P.: Extensions of uniformly continuous transformations and hyperconvex metric spaces. Pac. J. Math. **6**, 405–439 (1956)
5. Ayerbe, J.M., Domínguez. T., López, G.: Measures of Noncompactness on Metric Fixed Point Theory. Birkhäuser, Basel (1997)
6. Baillon, J.B.: Nonexpansive mappings and hyperconvex spaces. Contemp. Math. **72**, 11–19 (1988)
7. Bollobás, B.: Problem 29, Intersecting Convex Sets: Helly's Theorem. In: The Art of Mathematics: Coffee Time in Memphis, pp. 91–93. Cambridge University Press, Cambridge (2006)
8. Borkowski, M., Bugajewski, D., Phulara D.: On some properties of hyperconvex spaces. Fixed Point Theor. Appl. **2010**, Article ID 213812, 19 (2010)
9. Bouamama, S., Misane, D.: Hyperconvex ultrametric spaces and fixed point theory. New Zeal. J. Math. **34**, 25–29 (2005)
10. Bridson, M.R., Haefliger, A.: Metric Spaces of Non-positive Curvature. Springer, Berlin, Heidelberg (1999)
11. Brown, A. L., Deutsch, F., Indumathi, V., Kenderov, P.S.: Lower semicontinuity concepts, continuous selections, and set valued metric projections. J. Approx. Theor. **115**, 120–143 (2002)
12. Bryant, D., Tupper, P.F.: Hyperconvexity and tight-span theory for diversities. Adv. Math. **231**, 3172–3198 (2012)
13. Bugajewski, D., Espínola, R.: Remarks on some fixed point theorems for hyperconvex metric spaces and absolute retracts. Lecture Note in Pure and Applied Mathematics, vol. 213, pp. 85–92. Marcel-Dekker, New York (2000)
14. Bugajewski, D., Grzelaczyk, E.: A fixed point theorem in hyperconvex spaces. Arch. Math. **75**, 395–400 (2000)
15. Cerdà, J.: Linear Functional Analysis. Graduate Studies in Mathematics, vol. 116, American Mathematical Society, Providence, Thode Island (2010)
16. Chang, T.-H., Chen, C.-M., Chen, C.H.: Generalized 2-gKKM property in a hyperconvex metric space and its applications. Nonlinear Anal. **72**, 50–56 (2010)
17. Day, M.M.: Normed Linear Spaces, 3rd edn. Springer, New York (1973)
18. Dugundji, J., Granas, A.: Fixed Point Theory. Springer, New York (2003)
19. Dhompongsa, S., Kirk, W.A., Panyanak, B.: Nonexpansive set-valued mappings in metric and Banach spaces. J. Nonlinear Convex Anal. **8**, 35–45 (2007)
20. Dress, A.: Trees, tight extensions of metric spaces, and the cohomological dimension of certain groups: a note on combinatorial properties of metric spaces. Adv. Math. **53**, 321–402 (1984)
21. Dress, A., Scharlau, R.: Gated sets in metric spaces. Aequationes Math. **34**, 112–120 (1987)
22. Dress, A., Huber, K.T., Moulton V.: Metric Spaces in Pure and Applied Mathematics. (English summary) Proceedings of the Conference on Quadratic Forms and Related Topics (Baton Rouge, LA, 2001). Doc. Math. 2001, Extra Vol., pp. 121–139.
23. Espínola, R.: Darbo-Sadovski's theorem in hyperconvex metric spaces. Supplemento Rendiconti Circolo Mat. Palermo, Serie II **40**, 129–137 (1996)
24. Espínola, R.: On selections of the metric projection and best proximity pairs in hyperconvex spaces. Ann. Univ. Mariae Curie-Skłodowska Sect. A **59**, 9–17 (2005)

25. Espínola, R., Khamsi, M.A.: Introduction to hyperconvex spaces. In: Kirk, W.A., Sims, B. (eds.) Handbook of Metric Fixed Point Theory, pp. 391–435. Kluwer Academic, Dordrecht (2001)
26. Espínola, R., Kirk, W.A.: Fixed point theorems in $\mathbb{R}$-trees with applications to graph theory. Topology Appl. **153**, 1046–1055 (2006)
27. Espínola, R., López, G.: Ultimately compact operators in hyperconvex metric spaces. In: Takahashi, W., Tamaka, T.T. (eds.) Nonlinear Analysis and Convex Analysis, pp. 142–149. World Scientific Publishing, Singapore (1999)
28. Espínola, R., Lorenzo, P.: Metric fixed point theory on hyperconvex spaces: Recent progress. Arabian J. Math. **1**, 439–463 (2012)
29. Espínola, R., Kirk, W.A., López, G.: Nonexpansive retracts in hyperconvex spaces. J. Math. Anal. Appl. **251**, 557–570 (2000)
30. Faith, D.: Conservation evaluation and phylogenetic diversity. Biol. Conserv. **61**, 1–10 (1992)
31. Goebel, K., Kirk, W.A.: Topics in Metric Fixed Point Theory. Cambridge University Press, Cambridge (1990)
32. Goodner, D.B.: Projections in normed linear spaces. Trans. Am. Math. Soc. **69**, 89–108 (1950)
33. Grothendieck, A.: Une caractérisation vectorielle-métrique des espaces L^1. Canad. J. Math. **7**, 552–561 (1955)
34. Hasumi, M.: The extension property of complex Banach spaces. Tohoku Math. J. **10**, 135–142 (1958)
35. Heinonen, J.: Geometric Embeddings of Metric Spaces. University of Jyvaskyla, Finland (2003)
36. Horvath, C.D.: Contractibility and generalized convexity. J. Math. Anal. Appl. **156**, 341–357 (1991)
37. Horvath, C.D.: Extension and selection theorems in topological spaces with a generalized convexity structure. Ann. Fac. Sci. Toulouse Math. **2**, 253–269 (1993)
38. Hughes, B.: Trees and ultrametric spaces: a categorical equivalence. (English summary) Adv. Math. **189**, 148–191 (2004)
39. Hustad, O.: Intersection properties of balls in complex Banach spaces whose duals are L_1-spaces. Acta Math. **132**, 283–313 (1974)
40. Justad, O.: A noteo on complex $\mathscr{P}_1$-spaces. Israel J. Math. **16**, 117–119 (1973)
41. Isbell, J.R.: Six theorems about injective metric spaces. Comment. Math. Helvetici **39**, 439–447 (1964)
42. Kemajou, E., Künzi, H.-P. A., Olela Otafudu, O.: The Isbell-hull of a di-space. Topology Appl. **159**, 2463–2475 (2012)
43. Kelley, J.L.: Banach spaces with the extension property. Trans. Am. Math. Soc. **72**, 323–326 (1952)
44. Khamsi, M.A.: KKM and Ky Fan Theorems in hyperconvex metric spaces. J. Math. Anal. Appl. **204**, 298–306 (1996)
45. Khamsi, M.A.: One-local retract and common fixed point for commuting mappings in metric spaces. Nonlinear Anal. **27**, 1307–1313 (1996)
46. Khamsi, M.A.: Sadovskii's fixed point theorem without convexity. Nonlinear Anal. **53**, 829–837 (2003)
47. Khamsi, M. A.: On asymptotically nonexpansive mappings in hyperconvex metric spaces. Proc. Am. Math. Soc. **132**, 365–373 (2004)
48. Khamsi, M.A., Kirk, W.A.: An Introduction to Metric Spaces and Fixed Point Theory. Wiley Interscience, New York (2001)
49. Khamsi, M.A., Kirk, W.A., Martínez-Yáñez, C.: Fixed point and selection theorems in hyperconvex spaces. Proc. Am. Math. Soc. **128**, 3275–3283 (2000)
50. Khamsi, M.A., Lin, M., Sine, R., On the fixed points of commuting nonexpansive maps in hyperconvex spaces. J. Math. Anal. Appl. **168**, 372–380 (1992)
51. Kirk, W.A.: A fixed point theorem for mappings which do not increase distances. Am. Math. Mon. **72**, 1004–1006 (1965)

52. Kirk, W.A.: Krasnoselskii's iteration process in hyperbolic space. Numer. Funct. Anal. Optim. **4**, 371–381 (1981–1982)
53. Kirk, W.A.: Hyperconvexity of $\mathbb{R}$-trees. Fund. Math. **156**, 67–72 (1998)
54. Kirk, W.A.: Fixed point theorems in CAT(0) spaces and $\mathbb{R}$-trees. Fixed Point Theor. Appl. **2004**, 309–316 (2004)
55. Kirk, W.A.: A note on geodesically bounded $\mathbb{R}$-trees. Fixed Point Theor. Appl. **2010**, Article ID 393470, 4 (2010)
56. Kirk, W.A., Panyanak, B.: Best approximation in $\mathbb{R}$-trees. Numer. Funct. Anal. Optim. **28**, 681–690 (2007)
57. Kirk, W.A., Panyanak, B.: Remarks on best approximation in R-trees. Ann. Univ. Mariae Curie-Skłodowska Sect. A **63**, 133–138 (2009)
58. Kirk, W.A., Shahzad, N.: Some fixed point results in ultrametric spaces. (English summary) Topology Appl. **159**, 3327–3334 (2012)
59. Kirk, W.A., Shin, S.S.: Fixed point theorems in hyperconvex spaces. Houston J. Math. **23**, 175–187 (1997)
60. Kirk, W.A., Sims, B., Xian-Zhi Yuan, G.: The Khaster, Kuratowski and Mazurkiewicz theory in hyperconvex metric spaces and some of its applications. Nonlinear Anal. **39**, 611–627 (2000)
61. Künzi, H.-P.A., Olela Otafudu, O.: q-hyperconvexity in quasipseudometric spaces and fixed point theorems. J. Funct. Spaces Appl. **2012**, Article ID 765903, 18 (2012)
62. Lemin, A.: On ultrametrization of general metric spaces. Proc. Am. Math. Soc. **131**, 979–989 (2003)
63. Markin, J.T.: Fixed points, selections, and best approximation for multivalued mappings in $\mathbb{R}$-trees. Nonlinear Anal. **67**, 2712–2716 (2007)
64. Minh, B., Klaere, S., von Haeseler, A.: Taxon selection under split diversity. Syst. Biol. **58**, 586–594 (2009)
65. Nachbin, L.: A theorem of the Hahn-Banach type for linear transformations. Trans. Am. Math. Soc. **68**, 28–46 (1950)
66. Nowakowski, R., Rival, I.: Fixed-edge theorem for graphs with loops. J. Graph Theor. **3**, 339–350 (1979)
67. Pachter, L., Speyer, D.: Reconstructing trees from subtree weights. Appl. Math. Lett. **17**, 615–621 (2004)
68. Park, S.: Evolution of the 1984 KKM theorem of Ky Fan. Fixed Point Theor. Appl. **2012**(146), 14 (2012)
69. Penot, J.P.: Fixed point theorems without convexity. Bull. Soc. Math. France Mem. **60**, 129–152 (1979)
70. Piątek, B.: Best approximation of coincidence points in metric trees. Ann. Univ. Mariae Curie-Skłodowska Sect. A **62**, 113–121 (2008)
71. Piątek, B., Espínola, R.: Diversities, hyperconvexity and fixed points. arXiv:1303.7146 [math.MG]
72. Priess-Crampe, S., Ribenboim, P.: The common point theorem for ultrametric spaces. Geometriae Dedicata **71**, 105–110 (1998)
73. Priess-Crampe, S., Ribenboim, P.: Ultrametric spaces and logic programming. J. Logic Program. **42**, 59–70 (2000)
74. Sakai, S.: C^*-Algebras and W^*-Algebras. Springer, Berlin, New York (1971)
75. Salbany, S.: Injective objects and morphisms. In: Categorical Topology and Its Relation to Analysis, Algebra and Combinatorics, pp. 394–409, Prague, 1988. World Scientific Publication, Teaneck, NJ (1989)
76. Sánchez, M.C., Llinares, J.V., Subiza, B.: A KKM-result and an application for binary and non-binary choice functions. Econom. Theor. **21**, 185–193 (2003)
77. Schörner, E.: Ultrametric fixed point theorems and applications. Fields Institute Communication, vol. 33. American Mathematical Society, Providence, RI (2003)
78. Sine, R.: On nonlinear contraction semigroups in sup norm spaces. Nonlinear Anal. **3**, 885–890 (1979)
79. Sine, R.: Hyperconvexity and approximate fixed points. Nonlinear Anal. **13**, 863–869 (1989)

80. Sine, R.: Hyperconvexity and nonexpansive multifunctions. Trans. Am. Math. Soc. **315**, 755–767 (1989)
81. Sine, R.: Hyperconvexity and approximate fixed points. Nonlinear Anal. **13**, 863–869 (1989)
82. Soardi, P.: Existence of fixed points of nonexpansive mappings in certain Banach lattices. Proc. Am. Math. Soc. **73**, 25–29 (1979)
83. Steel, M.A.: Phylogenetic diversity and the Greedy algorithm. Syst. Biol. **54**, 527–529 (2005)
84. Young, G.S., The introduction of local connectivity by change of topology. Am. J. Math. **68**, 479–494 (1946)
85. Xian-Zhi Yuan, G.: Fixed point and related theorems for set-valued mappings. In: Kirk, W.A., Sims, B. (eds.) Handbook of Metric Fixed Point Theory, pp. 643–690. Kluwer Academic, Dordrecht (2001)

Chapter 5
An Introduction to Fixed Point Theory in Modular Function Spaces

W.M. Kozlowski

5.1 Introduction

Before we introduce the definition and the fundamental properties of modular function spaces, let us discuss a historical background which will provide also a motivation for the introduction of the methods of modular function spaces into the fixed point theory.

The results and methods of fixed point theory, applied to spaces of measurable functions, have been used extensively in the field of integral equations and integral inequalities. Since the 1930s many prominent mathematicians like Orlicz and Birnbaum recognized that using the methods of L^p-spaces alone created many complications and in some cases did not allow to solve some non-power type integral equations, see [8]. They considered spaces of functions with some growth properties different from the power type growth control provided by the L^p-norms. Orlicz and Birnbaum considered, for instance, function spaces defined as follows:

$$L^{\varphi} = \left\{ f : \mathbb{R} \to \mathbb{R};\ \exists \lambda > 0 : \int_{\mathbb{R}} \varphi(\lambda |f(x)|)\, dm(x) < \infty \right\},$$

where $\varphi : [0,\infty) \to [0,\infty)$ was assumed to be a convex function increasing to infinity, that is, the function which to some extent behaves similarly to power functions $\varphi(t) = t^p$. Let us mention two typical examples of such functions: $\varphi_1(t) = e^t - t - 1$ or $\varphi_2(t) = e^{t^2} - 1$. The possibility of introducing the structure of a linear metric in L^{φ} as well as the interesting properties of these spaces, later named Orlicz spaces, and many applications to differential and integral equations with kernels of nonpower types were among the reasons for the development of the theory of

W.M. Kozlowski (✉)
School of Mathematics and Statistics, University of New South Wales, Sydney, NSW 2052, Australia
e-mail: w.m.kozlowski@unsw.edu.au

S. Almezel et al. (eds.), *Topics in Fixed Point Theory*, DOI 10.1007/978-3-319-01586-6_5,

Orlicz spaces, their applications and generalizations. Consider, for example, the following Hammerstein nonlinear integral equation which plays an important role in the elasticity theory:

$$f(x) = \int_0^1 k(x,y)\varphi(f(y))dy,$$

where $\varphi(u)$ is a function which increases more rapidly than an arbitrary power function. Krasnosel'skii and Rutickii [55] showed that the Hammerstein operator defined by the right member of this integral equation does not operate in any of the L^p spaces. And yet, they showed how to find an Orlicz space where the Hammerstein operator is well defined and possesses properties allowing to use some fixed point theorems for solving the corresponding integral equation.

Using the apparatus of the modular function spaces we can go much further: the operator itself is used for the construction of a function modular and hence of a space in which this operator has required properties. This technique together with a relevant modular function space fixed point theorem can be applied to solving integral equations, like Urysohn integral equation, see Example 5.9.

An additional importance for applications of modular function spaces consists in the richness of structure of modular function spaces, that—besides being Banach spaces (or F-spaces in a more general settings)—are equipped with modular equivalents of norm or metric notions, and also are equipped with almost everywhere convergence and convergence in submeasure. As the above-mentioned example of the Urysohn operator vividly demonstrated, in many situations in integral equations, approximation and fixed point theory, modular type conditions are much more natural and modular type assumptions can be more easily verified than their metric or norm counterparts. There are also important results that can be proved only using the apparatus of modular function spaces. From this perspective, the fixed point theory in modular function spaces should be considered as complementary to the fixed point theory in normed and metric spaces.

The theory of contractions and nonexpansive mappings defined on convex subsets of Banach spaces has been well developed since the 1960s (see, for example, [9, 15, 20, 21, 23, 41]), and generalized to metric spaces (see, for example, [4, 22, 33]), and modular function spaces (see, for example, [30, 31, 36, 37]). The corresponding fixed point results were then extended to larger classes of mappings like asymptotic mappings [32, 42], pointwise contractions [40] and asymptotic pointwise contractions and nonexpansive mappings [25, 34, 43, 44].

While Banach Contraction Principle and its generalizations usually provide a fixed point construction method as limits of orbits, the proofs of the fixed point existence results for nonexpansive mappings and their generalizations are of the existential nature and do not describe any algorithms for constructing fixed points of an asymptotic pointwise ρ-nonexpansive mapping. It is well known that the fixed point construction iteration processes for generalized nonexpansive mappings have been successfully used to develop efficient and powerful numerical methods for solving various nonlinear equations and variational problems, often of great importance for applications in various areas of pure and applied science. Kozlowski proved convergence to fixed points of some iterative algorithms applied to asymptotic pointwise nonexpansive mappings in Banach spaces [48, 53] and the existence of common fixed points of semigroups of pointwise Lipschitzian mappings in Banach spaces [49]. Recently the weak and strong convergence of such processes to common fixed points of semigroups of mappings in Banach spaces was demonstrated in [52, 54]. The pioneering work on the convergence of Mann and Ishikawa algorithms in modular function spaces can be found in [13]. Our approach is based on this paper. We would like to emphasize that all convergence theorems presented in this chapter define constructive algorithms that can be actually implemented. When dealing with specific applications of these theorems, one should take into consideration how additional properties of the mappings, sets and modulars involved, can influence the actual implementation of the algorithms defined in this paper.

The existence of common fixed points for families of contractions and nonexpansive mappings in Banach spaces has been investigated since the early 1960s, see, for example, DeMarr [14], Browder [9], Belluce and Kirk [6, 7], Lim [56], and Bruck [10]. The asymptotic approach for finding common fixed points of semigroups of Lipschitzian (but not pointwise Lipschitzian) mappings has also been investigated for some time, see, for example, Tan and Xu [72]. It is worthwhile mentioning the recent studies on the special case, when the parameter set for the semigroup is equal to $\{0,1,2,3,\ldots\}$ and $T_n = T^n$, the nth iterate of an asymptotic pointwise nonexpansive mapping, that is, such a $T : C \to C$ that there exists a sequence of functions $\alpha_n : C \to [0,\infty)$ with $\|T^n(x) - T^n(y)\| \leq \alpha_n(x)\|x - y\|$. Kirk and Xu [44] proved the existence of fixed points for asymptotic pointwise contractions and asymptotic pointwise nonexpansive mappings in Banach spaces, while Hussain and Khamsi extended this result to metric spaces [25], and Khamsi and Kozlowski to modular function spaces [34, 35]. In the context of modular function spaces with Δ_2-property, Khamsi discussed in [29] the existence of nonlinear semigroups in Musielak–Orlicz spaces and considered some applications to differential equations. The existence of common fixed points for general semigroups of mappings acting in modular function spaces was proved by Kozlowski in [50].

5.2 Modular Function Spaces and Modular Geometry

Let Ω be a nonempty set and Σ be a nontrivial σ-algebra of subsets of Ω. Let $\mathscr{P}$ be a nontrivial δ-ring of subsets of Ω, recall this means that $\mathscr{P}$ is closed with respect to forming of countable intersections, and finite unions and differences. Assume further that $E \cap A \in \mathscr{P}$ for any $E \in \mathscr{P}$ and $A \in \Sigma$. Let us assume that there exists an increasing sequence of sets $K_n \in \mathscr{P}$ such that $\Omega = \bigcup K_n$. By $\mathscr{E}$ we denote the linear space of all simple functions with supports from $\mathscr{P}$. By $\mathscr{M}_\infty$ we will denote the space of all extended measurable functions, that is, all functions $f : \Omega \to [-\infty,\infty]$ such that there exists a sequence $\{g_n\} \subset \mathscr{E}$, $|g_n| \leq |f|$ and $g_n(\omega) \to f(\omega)$ for all $\omega \in \Omega$. By 1_A we denote the characteristic function of the set A.

5.2.1 Introduction to Modular Function Spaces

Definition 5.1. Let $\rho : \mathcal{M}_\infty \to [0,\infty]$ be a nontrivial, convex, and even function. We say that ρ is a *regular convex function pseudomodular* if:

(a) $\rho(0) = 0$;
(b) ρ is monotone, that is, $|f(\omega)| \leq |g(\omega)|$ for all $\omega \in \Omega$ implies $\rho(f) \leq \rho(g)$, where $f,g \in \mathcal{M}_\infty$;
(c) ρ is orthogonally subadditive, that is, $\rho(f1_{A\cup B}) \leq \rho(f1_A) + \rho(f1_B)$ for any $A,B \in \Sigma$ such that $A \cap B \neq \emptyset$, where $f \in \mathcal{M}_\infty$;
(d) ρ has the Fatou property: $|f_n(\omega)| \uparrow |f(\omega)|$ for all $\omega \in \Omega$ implies $\rho(f_n) \uparrow \rho(f)$, where $f \in \mathcal{M}_\infty$;
(e) ρ is order continuous in $\mathcal{E}$, that is, $g_n \in \mathcal{E}$ and $|g_n(\omega)| \downarrow 0$ implies $\rho(g_n) \downarrow 0$.

Similarly as in the case of measure spaces, we say that a set $A \in \Sigma$ is ρ-*null* if $\rho(g1_A) = 0$ for every $g \in \mathcal{E}$. We say that a property holds ρ-*almost everywhere* if the exceptional set is ρ-null. As usual we identify any pair of measurable sets whose symmetric difference is ρ-null as well as any pair of measurable functions differing only on a ρ-null set. With this in mind we define

$$\mathcal{M} = \{f \in \mathcal{M}_\infty : |f(\omega)| < \infty \ \rho - a.e\},$$

where each $f \in \mathcal{M}$ is actually an equivalence class of functions equal ρ-a.e. rather than an individual function.

Definition 5.2. Let ρ be a regular convex function pseudomodular.

(a) We say that ρ is a *regular convex function semimodular* if $\rho(\alpha f) = 0$ for every $\alpha > 0$ implies that $f = 0 \ \rho - a.e.$;
(b) We say that ρ is a *regular convex function modular* if $\rho(f) = 0$ implies that $f = 0 \ \rho - a.e.$

The class of all nonzero regular convex function modulars defined on Ω will be denoted by $\mathfrak{R}$.

Remark 5.1. Let us denote $\rho(f,E) = \rho(f1_E)$ for $f \in \mathcal{M}$, $E \in \Sigma$. Also, by convention for $\alpha > 0$ we will write $\rho(\alpha,E)$ instead of $\rho(\alpha 1_E)$. We will use these notations when convenient. It is easy to prove that $\rho(f,E)$ is a convex function modular in the sense of Definition 2.1.1 in [47] (more precisely, it is a convex function modular with the Fatou property). Therefore, we can use all results of the standard theory of modular function spaces as per the framework defined by Kozlowski in [45–47, 51], see also [30] for an early exposition of the fixed point theory in the modular function spaces, and [61–64] for the basics of the general modular theory.

Remark 5.2. Note that if ρ is a regular convex function modular then to verify that a set E is ρ-null it suffices to prove that there exists $\alpha > 0$ such that $\rho(\alpha,E) = 0$.

Definition 5.3 ([45–47]). Let ρ be a convex function modular.

(a) A *modular function space* is the vector space $L_\rho(\Omega,\Sigma)$, or briefly L_ρ, defined by

$$L_\rho = \{f \in \mathcal{M} : \rho(\lambda f) \to 0 \ as\ \lambda \to 0\}.$$

(b) The following formula defines a norm in L_ρ (frequently called the *Luxemburg norm*):

$$\|f\|_\rho = \inf\{\alpha > 0 : \rho(f/\alpha) \le 1\}.$$

Remark 5.3. It is not difficult to prove that $\|\cdot\|_\rho$ defines actually a norm such that $\|f\|_\rho \le \|g\|_\rho$ whenever $|f)| \le |g|$ ρ-a.e. It is also straightforward to demonstrate that $\|f_n\|_\rho \to 0$ if and only if $\rho(\alpha f_n) \to 0$ for every $\alpha > 0$. While it is interesting to investigate the Luxemburg norm related properties of modular function spaces, we focus our attention on those fixed point results that can be formulated purely in terms of function modulars. The relationship between modulars and respective Luxemburg norms is discussed in Propositions 5.5, 5.6, and 5.7 below, see also Example 5.12 in the section "Nonexpansive Mappings".

Let us start with some examples of modular function spaces. See [47] for a more exhaustive list.

Example 5.1. L^p-space, $p \ge 1$, is a modular function space generated by

$$\rho(f) = \int_{\mathbb{R}} |f(t)|^p dm(t).$$

Example 5.2. l^p-space, $p \ge 1$, is a modular function space generated by

$$\rho(f) = \sum_{i=1}^{\infty} |f_i|^p.$$

Example 5.3. Orlicz space is a modular function space generated by

$$\rho(f) = \int_{\mathbb{R}} \varphi(|f(t)|)dm(t),$$

provided φ is an Orlicz function, that is, it is a nonnegative, convex function such that $\varphi(0) = 0$.

Example 5.4. Musielak–Orlicz space is a modular function space generated by

$$\rho(f) = \int_{\mathbb{R}} \phi(t, |f(t)|)dm(t),$$

provided φ is nonnegative, convex function in the second variable such that $\varphi(x,0)=0$, and φ is a measurable, locally integrable function in the first variable such that $\varphi(.,u)>0$ $m-a.e.$ for every $u>0$.

Example 5.5. Lorentz p-space is a modular function space generated by

$$\rho(f)=\sup_{\tau\in\mathscr{T}}\int_{\mathbb{R}}|f(t)|^p d\mu_\tau(t)$$

where μ is a σ-finite measure, $\mathscr{T}$ is a group of all measure preserving transformations $\tau:\mathbb{R}\to\mathbb{R}$ and $\mu_\tau(E)=\mu(\tau^{-1}(E))$.

Example 5.6. Orlicz–Lorentz space is a modular function space generated by

$$\rho(f)=\sup_{\tau\in\mathscr{T}}\int_{\mathbb{R}}\varphi(|f(t)|)d\mu_\tau(t)$$

where φ is an Orlicz function, μ is a σ-finite measure, $\mathscr{T}$ is a group of all measure preserving transformations $\tau:\mathbb{R}\to\mathbb{R}$ and $\mu_\tau(E)=\mu(\tau^{-1}(E))$.

Example 5.7. Let $\Omega=[0,1]$, $\{\Omega_n\}$ be a countable disjoint partition of Ω such that $m(\Omega_p)=2^{-p}$ where m is the Lebesgue measure on $[0,1]$. Let $\mathscr{P}$ be a δ-ring generated by the sets of the form $A\cap\Omega_p$ for all $p\in\mathbb{N}$ and all measurable sets $A\subset[0,1]$. For a measurable function $f:\Omega\to\mathbb{R}$ we put

$$\rho(f)=\sum_{p=1}^{\infty}\left(\int_{\Omega_p}|f(t)|^p dm(t)\right)^{\frac{1}{p}}+\sup_{p\in\mathbb{N}}\int_{\Omega_p}|f(t)|^p dm(t).$$

In the fixed point theory in modular function spaces we use often a notion of ρ-convergence. It is defined, among other important terms, in the following definition.

Definition 5.4. Let $\rho\in\mathfrak{R}$.

(a) We say that $\{f_n\}$ is ρ-*convergent* to f and write $f_n\to f$ (ρ) if $\rho(f_n-f)\to 0$.
(b) A sequence $\{f_n\}$ where $f_n\in L_\rho$ is called a ρ-*Cauchy* sequence if $\rho(f_n-f_m)\to 0$ as $n,m\to\infty$.
(c) A set $B\subset L_\rho$ is called ρ-*closed* if for any sequence of $f_n\in B$, the convergence $f_n\to f$ (ρ) implies that f belongs to B.
(d) A set $B\subset L_\rho$ is called ρ-*bounded* if its ρ-diameter is finite; the ρ-diameter of B is defined as $\delta_\rho(B)=\sup\{\rho(f-g):f\in B,g\in B\}$.
(e) A set $B\subset L_\rho$ is called *strongly* ρ-*bounded* if there exists $\beta>1$ such that $M_\beta(B)=\sup\{\rho(\beta(f-g)):f\in B,g\in B\}<\infty$.
(f) A set $B\subset L_\rho$ is called ρ-*compact* if for any $\{f_n\}$ in B, there exists a subsequence $\{f_{n_k}\}$ and an $f\in B$ such that $\rho(f_{n_k}-f)\to 0$.
(g) A set $B\subset L_\rho$ is called ρ-*a.e. closed* if for any $\{f_n\}$ in B which ρ-a.e. converges to some f, then we must have $f\in B$.

(h) A set $B \subset L_\rho$ is called *ρ-a.e. compact* if for any $\{f_n\}$ in B, there exists a subsequence $\{f_{n_k}\}$ which ρ-a.e. converges to some $f \in B$.

(i) Let $f \in L_\rho$ and $B \subset L_\rho$. The *ρ-distance* between f and B is defined as

$$d_\rho(f,B) = \inf\{\rho(f-g) : g \in B\}.$$

Let us note that ρ-convergence does not necessarily imply ρ-Cauchy condition. Also, $f_n \to f\ (\rho)$ does not imply in general $\lambda f_n \to \lambda f\ (\rho)$, $\lambda > 1$.

The following useful results is a direct consequence of the order continuity property of function modulars, see part (e) of Definition 5.1.

Proposition 5.1. *If $\{f_n\}$ converges uniformly to f on a set $E \in \mathscr{P}$, then for every $\alpha > 0$*

$$\lim_{n\to\infty} \rho\left(\alpha(f_n - f), E\right) = 0.$$

The next result follows from the Fatou property of ρ.

Proposition 5.2. *Let $\rho \in \Re$ and $f,\ f_n \in \mathscr{M}_\infty$. If $f_n \to f\ \rho-a.e.$ then*

$$\rho(f) \leq \liminf_{n\to\infty} \rho(f_n).$$

Proof. Set $g_n = \inf_{k\geq n} |f_k|$. Hence

$$\liminf_{n\to\infty} \rho(g_n) \leq \liminf_{n\to\infty} \rho(f_n).$$

Note that, in virtue of the Fatou property, the left-hand side of the above inequality is equal to $\rho(f)$ since $|g_n| \uparrow |f|$ almost everywhere with respect to ρ. □

The next theorem describes and an important relationship between the ρ-convergence and the convergence ρ-a.e.

Theorem 5.1. *Let $\rho \in \Re$. Assume that $\{f_n\}$ satisfies ρ-Cauchy condition. Then, there exists a subsequence $\{f_{n_k}\}$ of $\{f_n\}$ and a function $f \in \mathscr{M}$ such that $f_{n_k} \to f$ ρ-a.e.*

Proof. Denoting $E_{n,m}(\varepsilon) = \{\omega \in \Omega : |f_n(\omega) - f_m(\omega)| \geq \varepsilon\}$, we get

$$\rho(\varepsilon, E_{n,m}(\varepsilon)) \leq \rho(f_n - f_m, E_{n,m}(\varepsilon)) \leq \rho(f_n - f_m) \to 0.$$

Hence, for any $k \in \mathbb{N}$ there exists $n_0(k) \in \mathbb{N}$ such that if $n \geq n_0(k)$ and $m \geq n_0(k)$, then

$$\rho\left(2^{-k}, E_{n,m}(2^{-k})\right) < 2^{-k}.$$

Taking $n_1 = n_0(1)$, $n_k = \max\{n_{k-1}, n_0(k)\}$, we get the sequence $g_k = f_{n_k}$. Define

$$E_k = \{\omega \in \Omega : |g_k(\omega) - g_{k+1}(\omega)| \geq 2^{-k}\},$$

and observe that $\rho(2^{-k}, E_k) < 2^{-k}$. Denote $G_j = \bigcup_{m=j}^{\infty} E_m$ and observe that for every $j \geq i \geq k$ and every $\omega \in \Omega \setminus G_k$ there holds

$$|g_i(\omega) - g_j(\omega)| \leq \sum_{m=i}^{\infty} |g_m(\omega) - g_{m+1}(\omega)| \leq 2^{1-i}.$$

Define a measurable function $h = \sum_{j=1}^{\infty} 2^{-j} 1_{G_j \setminus G_{j+1}}$ and calculate

$$\rho(h, G_k) = \rho\left(h, \bigcup_{n=k}^{\infty} (G_n \setminus G_{n+1})\right) \leq$$

$$\sum_{n=k}^{\infty} \rho(h_n, G_n \setminus G_{n+1}) \leq \sum_{n=k}^{\infty} \rho(2^{-n}, E_n) \leq 2^{1-k}.$$

Hence, $\rho(h, G) = 0$, where $G = \bigcap_{j=1}^{\infty} G_j$. Observe that if ω does not belong to G, then there exists k such that for any $j \geq i \geq k$ there holds $|g_i(\omega) - g_j(\omega)| < 2^{1-i}$. Therefore, there exists a real number $f(\omega)$ such that $f_{n_i}(\omega) = g_i(\omega) \to f(\omega)$. Clearly, the function f is measurable. It remains to be shown that G is ρ-null. Note that $G = \bigcup_{n=1}^{\infty} D_n$ where $D_n = G \cap \{\omega \in \Omega : h(\omega) \geq n^{-1}\}$. Therefore,

$$0 = \rho(h, G) \geq \rho(h, D_n) \geq \rho(n^{-1}, D_n).$$

Hence, D_n is ρ-null for every $n \in \mathbb{N}$ which implies that G is ρ-null. □

We will use the above result to prove the fundamental completeness theorem.

Theorem 5.2. *Let $\rho \in \mathfrak{R}$. L_ρ is complete with respect to ρ-convergence.*

Proof. Let a sequence $\{f_n\} \subset L_\rho$ be ρ-Cauchy. From Theorem 5.1, there exists a subsequence $\{f_{n_k}\}$ and a function $f \in \mathcal{M}$ such that $f_{n_k} \to 0$ ρ-a.e. Fix $\varepsilon > 0$ and note that there exists $n_0 \in \mathbb{N}$ such that $\rho(f_m - f_n) < \varepsilon$ provided $m, n \geq n_0$. Using Proposition 5.2, we get

$$\rho(f - f_n) \leq \liminf_{k \to \infty} \rho(f_{n_k} - f_n) \leq \varepsilon$$

for $n \geq n_0$. It remains to be proved that $f \in L_\rho$. Let $\lambda_n \to 0$. Fix $\varepsilon > 0$ and take $k \in \mathbb{N}$ such that $\rho(f - f_k) < \varepsilon$. For n sufficiently large we obtain

$$\rho(\lambda_n(f - f_k)) \leq \rho(f - f_k) < \varepsilon,$$

which implies that $\rho(\lambda_n(f - f_k)) \to 0$ as $n \to \infty$. Thus, $f_k - f$ belongs to L_ρ. Since $f_k \in L_\rho$ and L_ρ is a linear space, then f belongs to L_ρ as claimed. □

Let us follow [47] and prove a version of the Egoroff Theorem which plays an important role in the theory of modular function spaces.

Theorem 5.3. *Let $\rho \in \mathfrak{R}$ and let $f_n \to f$ $\rho-a.e.$ where $f,\ f_n \in \mathcal{M}$. There exists a nondecreasing sequence of sets $H_k \in \mathcal{P}$ such that $H_k \uparrow \Omega$ and $\{f_n\}$ converges uniformly to f on every H_k.*

Proof. Let us fix $\alpha > 0$ and recall that there exists an nondecreasing sequence of sets $K_n \in \mathcal{P}$ such that $\Omega = \bigcup K_n$. Let us fix temporarily $n \in \mathbb{N}$. We shall prove that there exist sets $F_n \in \mathcal{P}$ such that $F_n \subset K_n$ and

$$\rho(\alpha, K_n \setminus F_n) < \frac{1}{2^n} \tag{5.1}$$

and $\{f_m\}$ converges uniformly to f on each F_n. By omitting, if necessary, a ρ-null set from Ω, we may assume that the sequence $\{f_n\}$ pointwise converges to f everywhere. Set

$$E_k^m = \bigcap_{i=k}^{\infty} \left\{ \omega \in K_n : \ |f_i(\omega) - f(\omega)| < \frac{1}{m} \right\}$$

and observe that $E_1^m \subset E_2^m \subset \ldots$ and, since $\{f_n\}$ pointwise converges to f everywhere, that

$$K_n \subset \bigcup_{k=1}^{\infty} E_k^m$$

for every $m \in \mathbb{N}$. Then there exists a positive integer $n_0 = n_0(m)$ such that

$$\rho(\alpha, K_n \setminus E_{n_0(m)}^m) < \frac{1}{2^{m+n}}.$$

Denoting

$$F_n = \bigcap_{m=1}^{\infty} E_{n_0(m)}^m$$

we have

$$\rho(\alpha, K_n \setminus F_n) = \rho\left(\alpha, K_n \setminus \bigcap_{m=1}^{\infty} E^m_{n_0(m)}\right)$$
$$= \rho\left(\alpha, \bigcup_{m=1}^{\infty}\left(K_n \setminus E^m_{n_0(m)}\right)\right) \le \sum_{m=1}^{\infty} \rho\left(\alpha, K_n \setminus E^m_{n_0(m)}\right) < \frac{1}{n},$$

proving (5.1). It follows from the definition of F_n that to every $m \in \mathbb{N}$ there corresponds $k_m \in \mathbb{N}$ such that for $i \ge k_m$

$$|f_i(\omega) - f(\omega)| < \frac{1}{m},$$

for all $\omega \in F_n$, which means that $\{f_i\}$ converges uniformly to f on F_n. Let us define now

$$H_k = \bigcup_{n=1}^{k} F_n,\ H = \bigcup_{k=1}^{\infty} H_k.$$

Clearly, $\{H_k\}$ is a nondecreasing sequence of sets from $\mathscr{P}$ such that

$$\rho(\alpha, K_k \setminus H) \le \rho(\alpha, K_k \setminus H_k) \le \rho(\alpha, K_k \setminus F_k) < \frac{1}{2^k}$$

and $\{f_m\}$ converges uniformly to f on every H_k. It remains to prove that $\Omega \setminus H$ is ρ-null. Indeed,

$$\rho(\alpha, \Omega \setminus H) \le \rho(\alpha, K_k \setminus H) + \sum_{n=k+1}^{\infty} \rho(\alpha, K_n \setminus H) < \frac{1}{2^k} + \sum_{n=k+1}^{\infty} \frac{1}{2^n}$$

for every $k \in \mathbb{N}$. Thus $\rho(\alpha, \Omega \setminus H) = 0$. The proof is complete. □

The following useful result is an easy corollary from Theorems 5.1 and 5.3.

Proposition 5.3. *Let $\rho \in \Re$. Assume that $\rho(f_n - f) \to 0$, then there exists a subsequence $\{f_{n_k}\}$ of $\{f_n\}$ such that $f_{n_k} \to f$ ρ-a.e.*

Proof. By Theorem 5.1, there exists a subsequence $\{f_{n_k}\}$ of $\{f_n\}$ and a measurable function g such that $f_{n_k} \to g$ ρ-a.e. We need to show that $f = g$ ρ-a.e. This is not obvious as the modular ρ does not need to satisfy the triangle property. By the Egoroff Theorem 5.3 there exists an increasing sequence of sets $\{H_m\}$ from $\mathscr{P}$ such that $\bigcup_{m=1}^{\infty} H_m = \Omega$ and f_n converges to g uniformly on each H_m. Hence,

$$\rho\left(\frac{f-g}{2}, H_m\right) \le \rho(f - f_{n_k}, H_m) + \rho(f_{n_k} - g, H_m) \to 0, \quad \text{as } k \to \infty.$$

Hence, $\rho(f-g,H_m)=0$ for every m which implies that $f=g$ ρ-a.e. □

As an immediate consequence from Propositions 5.2 and 5.3 we get the following important result.

Proposition 5.4. *Let $\rho \in \Re$. Then ρ-balls $B_\rho(f,r)=\{g \in L_\rho : \rho(f-g) \le r\}$ are ρ-closed and ρ-a.e. closed.*

Let us come back to the question of the relationship between the modular and norm convergence in modular function spaces.

Proposition 5.5. *Let $\rho \in \Re$. The following assertions are true:*

(a) If $\|f\|_\rho < 1$ then $\rho(f) \le \|f\|_\rho$
(b) $\|f\|_\rho \le 1$ if and only if $\rho(f) \le 1$

Proof. Part (a): Let α be any number such that $\|f\|_\rho < \alpha < 1$. Then it follows from the definition of the Luxemburg norm that

$$\rho\left(\frac{f}{\alpha}\right) \le 1,$$

hence

$$\rho(f) = \rho\left(\alpha\frac{f}{\alpha}\right) \le \alpha\rho\left(\frac{f}{\alpha}\right) \le \alpha.$$

Since α was chose arbitrarily it follows that $\rho(f) \le \|f\|_\rho$, as claimed.
Part (b): Assume first that $\|f\|_\rho \le 1$ and fix arbitrarily $0 < \lambda < 1$. Hence $\|\lambda f\|_\rho < 1$. By Part (a) we get

$$\rho(\lambda f) \le \|\lambda f\|_\rho = \lambda\|f\|_\rho.$$

Taking $\lambda \to 1$ and using the Fatou property, we obtain the desired inequality. Let us prove now the other implication. Assume now that $\rho(f) \le 1$. If $\rho(f)=1$, then by definition $\|f\|_\rho \le 1$, so we may assume that $\rho(f) < 1$. Assume to the contrary that $\|f\|_\rho > 1$. There exists then a sequence of positive numbers $\{\lambda_n\}$ such that $\lambda_n \uparrow 1$ and $\|\lambda_n f\|_\rho > 1$ which, by the definition of the Luxemburg norm, implies that $\rho(\lambda_n f) > 1$. By passing with $n \to \infty$ and using the Fatou property we get $\rho(f) \ge 1$ which contradicts our assumption. □

The following definition plays an important role in the theory of modular function spaces.

Definition 5.5. Let $\rho \in \Re$. We say that ρ has the *Δ_2-property* if $\rho(2f_n) \to 0$ whenever $\rho(f_n) \to 0$.

The next result is an immediate consequence from the above definition and from Proposition 5.5.

Proposition 5.6. *Let $\rho \in \mathfrak{R}$. The convergence with respect to the function modular ρ is equivalent to the convergence with respect to the Luxemburg norm $\|\cdot\|_\rho$ if and only if ρ has the Δ_2-property.*

The following comparison of various notion of compactness is an immediate consequence of the above results.

Proposition 5.7. *Let $\rho \in \mathfrak{R}$. The following relationships hold for sets $C \subset L_\rho$:*

(a) If C is ρ-compact, then C is ρ-a.e. compact.
(b) If C is $\|.\|_\rho$-compact, then C is ρ-compact.
(c) If ρ satisfies Δ_2, then $\|.\|_\rho$-compactness and ρ-compactness are equivalent in L_ρ.

5.2.2 Geometrical Properties of Modular Function Spaces

In this section, we discuss the general geometrical methods for the consideration of fixed point properties, similarly as is the case in the Banach space setting. We believe that recent results, see, for example, [2, 13, 34, 35], provide further evidence for the existence of such a general theory. Indeed, the most common approach in the Banach space fixed point theory for generalized nonexpansive mappings is to assume the uniform convexity of the norm which implies the reflexivity, and—via the Milman Theorem—guarantees the weak compactness of the closed bounded sets. As we will see, the notion of a uniform convexity of function modulars in conjunction with the property (R) being the modular equivalence of the Banach space reflexivity, [34, 35, 37], equips us with the powerful tools for proving the fixed point property in modular function spaces. We shall see that the property (R) represents the most important, from the fixed point theory viewpoint, geometric characterization of reflexive spaces: every nonincreasing sequence of nonempty, convex, bounded sets has a nonempty intersection. The property (R) also aligns well to the metric equivalents of reflexivity defined by the notions of compact convexity structures [25]. This idea has been further developed in [2] to introduce notions of admissible sets and related modular versions of normal and compact convexity structures. All of this provides a set of powerful techniques for proving existence of common fixed points for commutative families of mappings acting in modular function spaces and for investigating the topological properties of the set of common fixed points.

Let us start by reminding ourselves that the modulus of convexity of a Banach space X is defined as

$$\delta(\varepsilon) = \inf\left\{1 - \left\|\frac{x+y}{2}\right\| : x, y \in X,\ \|x\| \le 1,\ \|y\| \le 1,\ \|x-y\| \ge \varepsilon\right\}.$$

A Banach space X is called *uniformly convex* if $\delta(\varepsilon) > 0$ for every $\varepsilon > 0$. We will investigate notions of the modular equivalents of uniform convexity of ρ. As demonstrated below, one concept of uniform convexity in normed spaces generates several different types of uniform convexity in modular function spaces. This is due primarily to the fact that in general modulars are not homogeneous.

Definition 5.6. Let $\rho \in \Re$. We define the following uniform convexity type properties of the function modular ρ:

(a) Let $r > 0, \varepsilon > 0$. Define

$$D_1(r,\varepsilon) = \{(f,g) : f,g \in L_\rho, \rho(f) \le r, \rho(g) \le r, \rho(f-g) \ge \varepsilon r\}.$$

Let

$$\delta_1(r,\varepsilon) = \inf\left\{1 - \frac{1}{r}\rho\left(\frac{f+g}{2}\right) : (f,g) \in D_1(r,\varepsilon)\right\}, \text{ if } D_1(r,\varepsilon) \neq \emptyset,$$

and $\delta_1(r,\varepsilon) = 1$ *if* $D_1(r,\varepsilon) = \emptyset$. We say that ρ satisfies $(UC1)$ if for every $r > 0, \varepsilon > 0$, $\delta_1(r,\varepsilon) > 0$. Note, that for every $r > 0$, $D_1(r,\varepsilon) \neq \emptyset$, for $\varepsilon > 0$ small enough.

(b) We say that ρ satisfies $(UUC1)$ if for every $s \ge 0, \varepsilon > 0$, there exists

$$\eta_1(s,\varepsilon) > 0$$

depending on s and ε such that

$$\delta_1(r,\varepsilon) > \eta_1(s,\varepsilon) > 0 \text{ for } r > s.$$

(c) Let $r > 0, \varepsilon > 0$. Define

$$D_2(r,\varepsilon) = \left\{(f,g) : f,g \in L_\rho, \rho(f) \le r, \rho(g) \le r, \rho\left(\frac{f-g}{2}\right) \ge \varepsilon r\right\}.$$

Let

$$\delta_2(r,\varepsilon) = \inf\left\{1 - \frac{1}{r}\rho\left(\frac{f+g}{2}\right) : (f,g) \in D_2(r,\varepsilon)\right\}, \text{ if } D_2(r,\varepsilon) \neq \emptyset,$$

and $\delta_2(r,\varepsilon) = 1$ *if* $D_2(r,\varepsilon) = \emptyset$. We say that ρ satisfies $(UC2)$ if for every $r > 0, \varepsilon > 0$, $\delta_2(r,\varepsilon) > 0$. Note, that for every $r > 0$, $D_2(r,\varepsilon) \neq \emptyset$, for $\varepsilon > 0$ small enough.

(d) We say that ρ satisfies $(UUC2)$ if for every $s \ge 0, \varepsilon > 0$ there exists

$$\eta_2(s,\varepsilon) > 0$$

depending on s and ε such that

$$\delta_2(r,\varepsilon) > \eta_2(s,\varepsilon) > 0 \; for \; r > s.$$

(e) We say that ρ is Strictly Convex, (SC), if for every $f,g \in L_\rho$ such that $\rho(f) = \rho(g)$ and

$$\rho\left(\frac{f+g}{2}\right) = \frac{\rho(f)+\rho(g)}{2}$$

we have $f = g$.

Remark 5.4. Let us observe that for $i = 1,2$, $\delta_i(r,0) = 0$, and $\delta_i(r,\varepsilon)$ is an increasing function of ε for every fixed r. Note also that

$$\delta_1(r,\varepsilon) = \inf\left\{\delta'(r,h) : \; h \in L_\rho, \; \rho(h) \geq r\varepsilon\right\},$$

$$\delta_2(r,\varepsilon) = \inf\left\{\delta'(r,h) : \; h \in L_\rho, \; \rho\left(\frac{h}{2}\right) \geq r\varepsilon\right\},$$

where

$$\delta'(r,h) = \inf\left\{1 - \frac{1}{r}\rho\left(f + \frac{h}{2}\right) : \; f \in L_\rho, \rho(f) \leq r, \rho(f+h) \leq r\right\}.$$

Proposition 5.8. *The following conditions characterize the relationship between the above defined notions:*

(a) $(UUCi)$ *implies* (UCi) *for* $i = 1,2.$;
(b) $\delta_1(r,\varepsilon) \leq \delta_2(r,\varepsilon)$;
(c) $(UC1)$ *implies* $(UC2)$;
(d) $(UC2)$ *implies* (SC);
(e) $(UUC1)$ *implies* $(UUC2)$;

Proof. The proofs of these implications follow immediately from Definition 5.6. □

Remark 5.5. Observe that, denoting $\rho_\alpha(u) = \alpha\rho(u)$, and the corresponding moduli of convexity by $\delta_{\rho_\alpha,i}$, where $i = 1,2$, we have

$$\delta_{\rho_\alpha,i}(r,\varepsilon) = \delta_{\rho,i}\left(\frac{r}{\alpha},\varepsilon\right),$$

or

$$\delta_{\rho,i}(r,\varepsilon) = \delta_{\rho_\alpha,i}(r\alpha,\varepsilon).$$

Hence, ρ is (UCx), where (UCx) is any of the conditions from Definition 5.6, if and only if there exists $\alpha > 0$ such that ρ_α is (UCx). In particular, taking $\alpha = \frac{1}{r}$, it is enough to prove any of the conditions defining (UCx) with $r = 1$.

Example 5.8. It is known that in Orlicz spaces, the Luxemburg norm is uniformly convex if and only φ is uniformly convex and Δ_2 property holds; this result can be traced to early papers by Luxemburg [57], Milnes [59], Akimovic [1], and Kaminska [27]. It is also known that, under suitable assumptions, $(UC2)$ in Orlicz spaces is equivalent to the very convexity of the Orlicz function [12, 37]. Remember that the function φ is called very convex if or every $\varepsilon > 0$ and any $x_0 > 0$, there exists $\delta > 0$ such that

$$\varphi\left(\frac{1}{2}(x-y)\right) \geq \frac{\varepsilon}{2}\left(\varphi(x)+\varphi(y)\right) \geq \varepsilon\varphi(x_0),$$

implies

$$\varphi\left(\frac{1}{2}(x+y)\right) \leq \frac{1}{2}(1-\delta)\left(\varphi(x)+\varphi(y)\right).$$

Typical examples of Orlicz functions that do not satisfy the Δ_2 condition but are very convex are: $\varphi_1(t) = e^{|t|} - |t| - 1$ and $\varphi_2(t) = e^{t^2} - 1$, [55, 59]. Therefore, these are the examples of Orlicz spaces that are not uniformly convex in the norm sense and hence the classical Kirk theorem cannot be applied. However, these spaces are uniformly convex in the modular sense, and respective modular fixed point results can be applied.

Let us recall a notion of bounded away sequences of real numbers.

Definition 5.7. A sequence $\{t_n\} \subset (0,1)$ is called *bounded away from* 0 if there exists $0 < a < 1$ such that $t_n \geq a$ for every $n \in \mathbb{N}$. Similarly, $\{t_n\} \subset (0,1)$ is called *bounded away from* 1 if there exists $0 < b < 1$ such that $t_n \leq b$ for every $n \in \mathbb{N}$.

We need to prove few technical results that will play an important role in proving our main theorems. First let us mention the following result whose proof is elementary. Note that for $t = \frac{1}{2}$, this result follows directly from Definition 5.6.

Lemma 5.1. *Let $\rho \in \Re$ be $(UUC1)$ and let $t \in (0,1)$. Then for every $s > 0, \varepsilon > 0$ there exists $\eta_1^t(s,\varepsilon) > 0$ depending only on s and ε such that*

$$\delta_1^t(r,\varepsilon) > \eta_1^t(s,\varepsilon) > 0 \; for\ any\ r > s.$$

The next lemma introduces a useful technique which is used extensively for investigating convergence to fixed points in the $(UUC1)$ modular function spaces. It was introduced in [35] for the case $t_n = \frac{1}{2}$ and extended to more general case in [13], see, for example, [70] for the Banach space equivalent.

Lemma 5.2. *Let $\rho \in \Re$ be $(UUC1)$ and let $\{t_n\} \subset (0,1)$ be bounded away from* 0 *and* 1. *If there exists $R > 0$ such that*

$$\limsup_{n\to\infty} \rho(f_n) \le R, \ \limsup_{n\to\infty} \rho(g_n) \le R,$$

$$\lim_{n\to\infty} \rho(t_n f_n + (1-t_n)g_n) = R,$$

then,

$$\lim_{n\to\infty} \rho(f_n - g_n) = 0.$$

Proof. Assume to the contrary that this is not the case and fix an arbitrary $\gamma > 0$. Passing to a subsequence if necessary we may assume that there exists an $\varepsilon > 0$ such that

$$\rho(f_n) \le R+\gamma, \ \rho(g_n) \le R+\gamma \tag{5.2}$$

while

$$\rho(f_n - g_n) \ge (R+\gamma)\varepsilon. \tag{5.3}$$

Since $\{t_n\}$ is bounded away from 0 and 1 there exist $0 < a < b < 1$ such that $a \le t_n \le b$ for all natural n. Passing to a subsequence if necessary we can assume that $t_n \to t_0 \in [a,b]$. For every $t \in [01]$ and $f,g \in D_1(R+\gamma,\varepsilon)$ let us define $\lambda_{f,g}(t) = \rho(tf + (1-t)g)$. Observe that the function $\lambda_{f,g} : [0,1] \to [0,R+\gamma]$ is a convex function. Hence that the function

$$\lambda(t) = \sup\{\lambda_{f,g}(t) : f,g \in D_1(R+\gamma,\varepsilon)\}$$

is also convex on $[0,1]$, and consequently it is a continuous function on $[a,b]$. Noting that

$$\delta_1^t(R+\gamma,\varepsilon) = 1 - \frac{1}{r}\lambda_{f,g}(t),$$

we conclude that $\delta_1^t(R+\gamma,\varepsilon)$ is a continuous function of $t \in [a,b]$. Hence

$$\lim_{n\to\infty} \delta_1^{t_n}(R+\gamma,\varepsilon) = \delta_1^{t_0}(R+\gamma,\varepsilon). \tag{5.4}$$

By (5.2) and (5.3)

$$\delta_1^{t_n}(R+\gamma,\varepsilon) \le 1 - \frac{1}{R+\gamma}\rho(t_n f_n + (1-t_n)g_n). \tag{5.5}$$

From (5.4) we deduce that the left-hand side of (5.5) tends to $\delta_1^{t_0}(R+\gamma,\varepsilon)$ as $n\to\infty$ while the right-hand side tends to $\frac{\gamma}{R+\gamma}$ in view of (5.2). Hence

$$\delta_1^{t_0}(R+\gamma,\varepsilon)\leq\frac{\gamma}{R+\gamma}. \tag{5.6}$$

By $(UUC1)$ and by Lemma 5.1 there exists $\eta_1^{t_0}(R,\varepsilon)>0$ satisfying

$$0<\eta_1^{t_0}(R,\varepsilon)\leq\delta_1^{t_0}(R+\gamma,\varepsilon). \tag{5.7}$$

Combining (5.6) with (5.7) we get

$$0<\eta_1^{t_0}(R,\varepsilon)\leq\frac{\gamma}{R+\gamma}.$$

Letting $\gamma\to 0$ we get a contradiction which completes the proof. □

Lemma 5.3. *Let ρ satisfies $(UC2)$. Let $f,g\in L_\rho$ and $r>0$ be such that $f\neq g$, $\rho(f)\leq r$ and $\rho(g)\leq r$. Then,*

$$\rho\left(\frac{f+g}{2}\right)<r.$$

Proof. Set $h=f-g$. Since $f\neq g$ then $\rho(\frac{h}{2})>0$. By the definition of δ_2 it follows from

$$\rho\left(\frac{f-g}{2}\right)=\rho\left(\frac{h}{2}\right)=\left(\frac{1}{r}\rho\left(\frac{h}{2}\right)\right)r$$

that

$$\rho\left(\frac{f+g}{2}\right)\leq r\left[1-\delta_2\left(r,\frac{1}{r}\rho\left(\frac{h}{2}\right)\right)\right].$$

Notice that

$$\delta_2\left(r,\frac{1}{r}\rho\left(\frac{h}{2}\right)\right)>0$$

because ρ satisfies $(UC2)$ and $\frac{1}{r}\rho\left(\frac{h}{2}\right)>0$. Hence

$$\rho\left(\frac{f+g}{2}\right)<r,$$

as claimed. □

Lemma 5.4. *Let ρ satisfies $(UC2)$. Then, for every $r > 0$, $\delta_2(r,1) = 1$.*

Proof. Fix $r > 0$. Let $(x,y) \in D_2(r,1)$, i.e. $\rho(x) \le r$, $\rho(y) \le r$, and

$$\rho\left(\frac{x-y}{2}\right) \ge r.$$

Observe that

$$r \le \rho\left(\frac{x-y}{2}\right) \le \frac{\rho(x)+\rho(y)}{2} \le r.$$

Hence

$$\rho\left(\frac{x-y}{2}\right) = r.$$

Setting $z = -y$ we get

$$r = \rho\left(\frac{x-z}{2}\right) = \rho\left(\frac{x+y}{2}\right) \le r.$$

By Lemma 5.3 applied to $f = x$ and $g = z$ we obtain that $x = z$, that is, $x = -y$, which implies that

$$\rho\left(\frac{x+y}{2}\right) = 0.$$

Hence, by the definition of δ_2 we conclude that $\delta_2(r,1) = 1$ as claimed. □

In the next theorem, we investigate the relationship between the uniform convexity of function modulars and the unique best approximant property (for other results on best approximation in modular function spaces, see, for example, [38]). This result, Theorem 5.4 below, is used in the proof of Theorem 5.5 to establish relationship between the modular uniform convexity and the property (R) which is a modular equivalent of the Milman–Pettis theorem stating that uniform convexity of a Banach space implies its reflexivity.

Theorem 5.4. *Assume $\rho \in \Re$ is $(UUC2)$. Let $C \subset L_\rho$ be nonempty, convex, and ρ-closed. Let $f \in L_\rho$ be such that $d = d_\rho(f,C) < \infty$. There exists then a unique best ρ-approximant of f in C, that is, a unique $g_0 \in C$ such that*

$$\rho(f - g_0) = d_\rho(f,C).$$

Proof. Uniqueness follows from the Strict Convexity (SC) of ρ; remember that $(UUC2)$ implies (SC). Let us prove the existence of the ρ-approximant. Since C

is ρ-closed, we may assume without loss of any generality that $d = d_\rho(f,C) > 0$. Clearly there exists a sequence $\{f_n\} \in C$ such that

$$\rho(f - f_n) \leq d\left(1 + \frac{1}{n}\right).$$

We claim that $\left\{\frac{1}{2}f_n\right\}$ is ρ-Cauchy. Assume to the contrary that this is not the case. There exists then an $\varepsilon_0 > 0$ and a subsequence $\{f_{n_k}\}$ of $\{f_n\}$ such that

$$\rho\left(\frac{f_{n_k} - f_{n_p}}{2}\right) \geq \varepsilon_0,$$

for any $p,k \geq 1$. Since ρ is (UUC2), then ρ is (UC2). Hence

$$\rho\left(f - \frac{f_{n_k} + f_{n_p}}{2}\right) \leq \left(1 - \delta_2\left(d(k,p), \frac{\varepsilon_0}{d(k,p)}\right)\right) d(k,p),$$

where $d(k,p) = \left(1 + \frac{1}{\min(n_p,n_k)}\right) d$. For $p,k \geq 1$, we have $d(k,p) \leq 2d$. Hence

$$\delta_2\left(d(k,p), \frac{\varepsilon_0}{d(k,p)}\right) \geq \delta_2\left(d(k,p), \frac{\varepsilon_0}{2d}\right)$$

Since ρ is (UUC2) then there exists $\eta > 0$ such that

$$\delta_2\left(r, \frac{\varepsilon_0}{2d}\right) \geq \eta$$

for any $r > d/3$. Since $d(k,p) \geq d > d/3$, we get

$$\rho\left(f - \frac{f_{n_k} + f_{n_p}}{2}\right) \leq (1-\eta)d(k,p),$$

for any $k,p \geq 1$. By the convexity of C, $\frac{f_{n_k} + f_{n_p}}{2} \in C$. Using the definition of d, we get

$$d \leq \rho\left(f - \frac{f_{n_k} + f_{n_p}}{2}\right) \leq (1-\eta)d(k,p),$$

for any $k,p \geq 1$. If we let k,p go to infinity, we get $d \leq (1-\eta)d$, which is impossible. Hence $\left\{\frac{1}{2}f_n\right\}$ is ρ-Cauchy. By Proposition 5.4, $\left\{\frac{1}{2}f_n\right\}$ ρ-converges to a $g \in L_\rho$. Fix $m \geq 1$. Since $\left\{\frac{f_m + f_n}{2}\right\} \in C$ and ρ-converges to $\frac{f_m}{2} + g$ and C is ρ-closed, then we have $\frac{f_m}{2} + g \in C$. Letting $m \to \infty$, we get $2g \in C$. By Propositions 5.2 and 5.3, passing to a subsequence if necessary, we get

$$\rho(f - 2g) \leq \liminf_{n\to\infty} \rho\left(f - g - \frac{f_n}{2}\right) \leq \liminf_{n\to\infty} \liminf_{m\to\infty} \rho\left(f - \frac{f_n + f_m}{2}\right).$$

Since ρ is convex, we get

$$\liminf_{n\to\infty} \liminf_{m\to\infty} \rho\left(f - \frac{f_n + f_m}{2}\right) \leq \liminf_{n\to\infty} \liminf_{m\to\infty} \frac{\rho(f - f_n) + \rho(f - f_m)}{2} \leq d.$$

Hence, $\rho(f-2g) \leq d$. Since $2g \in C$, we get $d \leq \rho(f-2g)$. Therefore, $\rho(f-2g)=d$. In other words, $g_0 = 2g$ is the ρ-approximant of f in C. □

The unique best approximant property of the $(UUC2)$ modular function spaces is used for establishing the property (R). As elaborated previously, this is parallel to the well-known fact that uniformly convex Banach spaces are reflexive. The property (R) will be essential for the proof of several fixed point theorems in modular function spaces. Following [37] let us formally define the property (R).

Definition 5.8. We say that L_ρ has property (R) if and only if every nonincreasing sequence $\{C_n\}$ of nonempty, ρ-bounded, ρ-closed, convex subsets of L_ρ has nonempty intersection.

Theorem 5.5. *Let $\rho \in \Re$ be $(UUC2)$, then L_ρ has property (R).*

Proof. Let $\{C_n\}$ be a nonincreasing sequence of nonempty, ρ-bounded, ρ-closed, convex subsets of L_ρ. According to Definition 5.8 we need to demonstrate that $\{C_n\}$ has nonempty intersection. Fix any $f \in C_1$. By the ρ-boundedness of C_1, there exists a finite constant $M > 0$ such that for any $n \geq 1$, $\rho(f - g) < M$ for any $g \in C_n \subset C_1$. Hence,

$$\sup_{n\geq 1} d_\rho(f, C_n) < \infty.$$

Using the proximinality of ρ-closed convex subsets of L_ρ (Theorem 5.4), for every $n \geq 1$ there exists $f_n \in C_n$ such that $\rho(f - f_n) = d_\rho(f, C_n)$. It is easy to show that $\{d_\rho(f, C_n)\}$ is nondecreasing and bounded. Hence $\lim_{n\to\infty} d_\rho(f, C_n) = d$ exists. Assume that $d = 0$, then $d_\rho(f, C_n) = 0$, for any $n \geq 1$. Since $\{C_n\}$ are ρ-closed, we get $f \in C_n$ for any $n \geq 1$, which implies $\bigcap_{n\geq 1} C_n \neq \emptyset$. Therefore, we can assume $d > 0$.

In this case we claim that $\left\{\frac{1}{2} f_n\right\}$ is ρ-Cauchy. Indeed if we assume the contrary, then there exists an $\varepsilon_0 > 0$ and a subsequence $\{f_{n_k}\}$ of $\{f_n\}$ such that

$$\rho\left(\frac{f_{n_k}-f_{n_p}}{2}\right) \geq \varepsilon_0,$$

for any $p,k \geq 1$. Since ρ is (UUC2), then ρ is (UC2). Hence

$$\rho\left(f-\frac{f_{n_k}+f_{n_p}}{2}\right) \leq \left(1-\delta_2\left(d,\frac{\varepsilon_0}{d}\right)\right)d,$$

for any $p,k \geq 1$. So

$$d_\rho(f,C_{\max(n_p,n_k)}) \leq \rho\left(f-\frac{f_{n_k}+f_{n_p}}{2}\right) \leq \left(1-\delta_2\left(d,\frac{\varepsilon_0}{d}\right)\right)d,$$

for any $p,k \geq 1$. If we let $p,k \to \infty$, we will get

$$d \leq \left(1-\delta_2\left(d,\frac{\varepsilon_0}{d}\right)\right)d,$$

which is a contradiction because $\delta_2\left(d,\frac{\varepsilon_0}{d}\right) > 0$ by $(UC2)$. Hence the sequence $\left\{\frac{1}{2}f_n\right\}$ is ρ-Cauchy and therefore it ρ-converges to some $g \in L_\rho$. Let us prove that $2g \in C_n$, for any $n \geq 1$. Indeed, we have $\frac{f_k+f_p}{2} \in C_n$, for any $p,k \geq n$. Fix any $k \geq n$. Since $\left\{\frac{f_k+f_p}{2}\right\}$ ρ-converges to $\frac{f_k}{2}+g$ as $p \to \infty$, and C_n is ρ-closed, then $\frac{f_k}{2}+g \in C_n$, for any $k \geq n$. If we let $k \to \infty$, we get $2g \in C_n$, for any $n \geq 1$. Hence $\bigcap\limits_{n\geq 1} C_n \neq \emptyset$, which completes the proof. □

We will establish now a modular version of the parallelogram inequality for uniformly convex modular function spaces, see [73] and Beg [5] for the norm version of this inequality.

Lemma 5.5. *For each $0 < s < r$ and $\varepsilon > 0$, set*

$$\Psi(r,s,\varepsilon) = \inf\left\{\frac{1}{2}\rho^2(f)+\frac{1}{2}\rho^2(g)-\rho^2\left(\frac{f+g}{2}\right)\right\},$$

where the infimum is taken over all $f,g \in L_\rho$ such that $\rho(f) \leq r$, $\rho(g) \leq r$, $\max(\rho(f),\rho(g)) \geq s$, and $\rho(f-g) \geq r\varepsilon$.

If $\rho \in \Re$ is $(UUC1)$, then $\Psi(r,s,\varepsilon) > 0$ for any $0 < s < r$ and $\varepsilon > 0$. Moreover, for a fixed $r,s > 0$, we have

(a) $\Psi(r,s,0) = 0$;
(b) $\Psi(r,s,\varepsilon)$ is a nondecreasing function of ε;
(c) if $\lim\limits_{n\to\infty} \Psi(r,s,t_n) = 0$, then $\lim\limits_{n\to\infty} t_n = 0$.

Proof. It is easy to see that $\Psi(r,s,\varepsilon) \geq 0$ for any $0 < s < r$ and $\varepsilon > 0$. We want to show that $\Psi(r,s,\varepsilon) > 0$. Assume to the contrary that there exist $0 < s < r$ and $\varepsilon > 0$ such that $\Psi(r,s,\varepsilon) = 0$. Then there exist $\{f_n\}$ and $\{g_n\}$ such that

$$\lim_{n\to\infty} \frac{1}{2}\rho^2(f_n) + \frac{1}{2}\rho^2(g_n) - \rho^2\left(\frac{f_n+g_n}{2}\right) = 0 \tag{5.8}$$

and $\rho(f_n) \leq r$, $\rho(g_n) \leq r$, $max\Big(\rho(f_n),\rho(g_n)\Big) \geq s$, and $\rho(f_n - g_n) \geq r\varepsilon$. Since ρ is convex, we get

$$\rho^2\left(\frac{f_n+g_n}{2}\right) \leq \left(\frac{\rho(f_n)+\rho(g_n)}{2}\right)^2 \leq \frac{\rho^2(f_n)+\rho^2(g_n)}{2},$$

where we used the inequality $2ab \leq a^2 + b^2$, for any $a,b \in \mathbb{R}$. Hence

$$\left(\frac{\rho(f_n)-\rho(g_n)}{2}\right)^2 \leq \frac{1}{2}\rho^2(f_n) + \frac{1}{2}\rho^2(g_n) - \rho^2\left(\frac{f_n+g_n}{2}\right),$$

which implies $\lim\limits_{n\to\infty}\Big(\rho(f_n) - \rho(g_n)\Big) = 0$. Without loss of any generality, we may assume $\lim\limits_{n\to\infty}\rho(f_n) = R$ exists. This also implies that $\lim\limits_{n\to\infty}\rho(g_n) = R$. By (5.8) we get then

$$\lim_{n\to\infty}\rho(g_n) = \lim_{n\to\infty}\rho^2\left(\frac{f_n+g_n}{2}\right) = R.$$

Observe that

$$R = \lim_{n\to\infty}\max\Big(\rho(f_n),\rho(g_n)\Big) \geq s > 0.$$

By Lemma 5.2 then $\rho(f_n - g_n) \to 0$ contradicting the fact that $\rho(f_n - g_n) \geq r\varepsilon > 0$. The proofs of (a), (b), and (c) are straightforward. □

Let us introduce a notion of a ρ-type, a powerful technical tool which will be used in the proofs of our fixed point results.

Definition 5.9. Let $C \subset L_\rho$ be convex and ρ-bounded. A function $\tau : C \to [0,\infty]$ is called a ρ-*type* (or shortly a type) if there exists a sequence $\{g_k\}$ of elements of C such that for any $f \in C$ there holds

$$\tau(f) = \limsup_{k\to\infty}\rho(g_k - f).$$

The following lemma establishes a crucial minimizing sequence property of uniformly convex modular function spaces. It will be used in conjunction with the parallelogram property in the proof of the main fixed point result for nonexpansive mappings acting in modular function spaces.

Lemma 5.6. *Assume that $\rho \in \Re$ is $(UUC1)$. Let C be a ρ-closed ρ-bounded convex nonempty subset. Let τ be a ρ-type defined on C. Then, any minimizing sequence of τ is ρ-convergent. Its limit is independent of the minimizing sequence.*

Proof. Let $\{f_n\} \subset C$ be such that $\tau(f) = \limsup_{n\to\infty} \rho(f_n - f)$, $\tau_0 = \inf\{\tau(h) : h \in C\}$. Let $\{g_k\}$ be a minimizing sequence of τ. Since C is ρ-bounded, there exists $R > 0$ such that $\rho(f-g) \le R$ for any $f, g \in C$. The rest of the proof is split into two cases.

Case 1: Assume that $\tau_0 > 0$. Let us choose a $\sigma > 0$ such that $\tau_0 - \sigma > 0$. Let us fix g_m *and* g_k and select a subsequence $\{h_n\}$ of $\{f_n\}$ such that

$$0 < \tau_0 \le \tau\left(\frac{g_m + g_k}{2}\right) = \lim_{n\to\infty} \rho\left(\frac{g_m + g_k}{2} - h_n\right).$$

Then, for n sufficiently large we have

$$0 < \tau_0 - \sigma \le \rho\left(\frac{g_m + g_k}{2} - h_n\right) \le \max\left(\rho(g_m - h_n), \rho(g_k - h_n)\right).$$

Using (5.8) from Lemma 5.5 then

$$\rho^2\left(\frac{g_m+g_k}{2} - h_n\right) \le \frac{1}{2}\rho^2(g_m - h_n) + \frac{1}{2}\rho^2(g_k - h_n) - \Psi\left(R, \tau_0 - \sigma, \frac{1}{R}\rho(g_m - g_k)\right),$$

and passing with n to infinity we get

$$\tau^2\left(\frac{g_k + g_m}{2}\right) \le \frac{1}{2}\tau^2(g_k) + \frac{1}{2}\tau^2(g_m) - \Psi\left(R, \tau_0 - \sigma, \frac{1}{R}\rho(g_k - g_m)\right).$$

Hence,

$$\tau_0^2 \le \frac{1}{2}\tau^2(g_k) + \frac{1}{2}\tau^2(g_m) - \Psi\left(R, \tau_0 - \sigma, \frac{1}{R}\rho(g_k - g_m)\right),$$

for any $k, m \ge 1$. So,

$$\Psi\left(R, \tau_0 - \sigma, \frac{1}{R}\rho(g_k - g_m)\right) \le \frac{1}{2}\tau^2(g_k) + \frac{1}{2}\tau^2(g_m) - \tau_0^2.$$

Hence, $\lim_{k,m\to\infty} \Psi\left(R, \tau_0 - \sigma, \frac{1}{R}\rho(g_k - g_m)\right) = 0$. The properties satisfied by Ψ imply that $\{g_k\}$ is ρ-Cauchy. Since L_ρ is ρ-complete and C is ρ-closed, then $\{g_k\}$ is ρ-convergent to some point $g \in C$. Let us prove that any other minimizing sequence also ρ-converges to g. Indeed, let $\{u_n\} \in C$ be any minimizing sequence of τ. Using the same argument as previously, we have

$$\tau_0^2 \le \tau^2\left(\frac{g_n+u_n}{2}\right) \le \frac{1}{2}\tau^2(g_n)+\frac{1}{2}\tau^2(u_n) - \Psi\left(R, \tau_0-\sigma, \frac{1}{R}\rho(g_n-u_n)\right),$$

for some $\sigma > 0$ such that $\tau_0 - \sigma > 0$ and

$$\Psi\left(R, \tau_0-\sigma, \frac{1}{R}\rho(g_n-u_n)\right) \le \frac{1}{2}\tau^2(g_n)+\frac{1}{2}\tau^2(u_n)-\tau_0^2,$$

for any $n \ge 1$. As before, we get $\lim_{n\to\infty} \rho(g_n - u_n) = 0$. Since ρ is convex, we have

$$\rho\left(\frac{u-g}{3}\right) \le \frac{1}{3}\rho(u-u_n)+\frac{1}{3}\rho(u_n-g_n)+\frac{1}{3}\rho(g_n-g),$$

where u is the ρ-limit of $\{u_n\}$. Clearly, our assumptions imply that $\rho\left(\frac{u-g}{3}\right) = 0$ or $u = g$. This completes the proof of the Lemma for Case 1.

Case 2: Assume that $\tau_0 = 0$. Let

$$K = \bigcap_{n\ge 1} \overline{\mathrm{conv}}_\rho\left(\{f_k; k \ge n\}\right),$$

which is nonempty in view of property (R); recall $(UUC1)$ implies (R) by Theorem 5.5. Let $f_\infty \in K$. Let $h \in C$, $\varepsilon > 0$. By definition of τ, there exists $n_0 > 0$ such that for every $n > n_0$

$$\rho(f_n - h) \le \tau(h) + \varepsilon.$$

Therefore, $f_n \in B_\rho(h, \tau(h)+\varepsilon)$ for $n > n_0$. This fact implies

$$K \subset \overline{\mathrm{conv}}_\rho(\{f_n; n \ge n_0\}) \subset B_\rho(h, \tau(h)+\varepsilon).$$

Hence, $f_\infty \in B_\rho(h, \tau(h)+\varepsilon)$. Since this is true for every $\varepsilon > 0$, there holds $f_\infty \in B_\rho(h, \tau(h))$, that is,

$$\rho(f_\infty - h) \le \tau(h). \tag{5.9}$$

Let $\{g_k\}$ be a minimizing sequence of τ. Using (5.9) with $h = g_k$, we get

$$\rho(f_\infty - g_k) \le \tau(g_k) \to \tau_0 = 0 \ as\ k \to \infty,$$

which means that $\{g_k\}$ is ρ-convergent to f_∞. Since this limit is independent of the sequence $\{g_k\}$, the proof of Case 2 and of the Lemma is complete. □

5.2.3 Nonlinear Mappings in Modular Function Spaces

Let us introduce definitions of modular contractions, pointwise contractions, asymptotic pointwise mappings, and associated notions. Frequently, they are collectively referenced in the literature as *generalized contractions* and *generalized nonexpansive mappings*, respectively.

Definition 5.10. Let $\rho \in \mathfrak{R}$ and let $C \subset L_\rho$ be nonempty and ρ-closed. A mapping $T : C \to C$ is called a *ρ-contraction* or shortly *contraction* if there exists a number $c \in [0,1)$ such that

$$\rho(T(f) - T(g)) \leq c\rho(f - g), \quad \text{for all } f, g \in C.$$

Definition 5.11. Let $\rho \in \mathfrak{R}$ and let $C \subset L_\rho$ be nonempty and ρ-closed. A mapping $T : C \to C$ is called a *pointwise ρ-contraction* or shortly *pointwise contraction* if there exists $\alpha : C \to [0,1)$ such that

$$\rho(T(f) - T(g)) \leq \alpha(f)\rho(f - g), \quad \text{for all } f, g \in C.$$

Definition 5.12. Let $\rho \in \mathfrak{R}$ and let $C \subset L_\rho$ be nonempty and ρ-closed. A mapping $T : C \to C$ is called an *asymptotic pointwise mapping* if there exists a sequence of mappings $\alpha_n : C \to [0,\infty)$ such that

$$\rho(T^n(f) - T^n(g)) \leq \alpha_n(f)\rho(f - g), \quad \text{for all } f, g \in L_\rho.$$

(a) If $\alpha_n(f) = 1$ for every $f \in L_\rho$ and every $n \in \mathbb{N}$, then T is called *ρ-nonexpansive* or shortly nonexpansive.
(b) If $\{\alpha_n\}$ converges pointwise to $\alpha : C \to [0,1)$, then T is called *asymptotic pointwise ρ-contraction* or shortly *asymptotic pointwise contraction.*
(c) If $\limsup_{n\to\infty} \alpha_n(f) \leq 1$ for any $f \in L_\rho$, then T is called *asymptotic pointwise ρ-nonexpansive* or shortly *asymptotic pointwise nonexpansive.*
(d) If α_n is a constant function for every n, and $\limsup_{n\to\infty} \alpha_n \leq 1$ for any $f \in L_\rho$, then T is called *asymptotically ρ-nonexpansive* or shortly *asymptotically nonexpansive.*

Remark 5.6. It follows immediately from the definition that every ρ-contraction is a pointwise ρ-contraction, and every pointwise ρ-contraction is an asymptotic pointwise ρ-contraction.

Remark 5.7. It follows immediately from the definition that every ρ-nonexpansive mapping is asymptotically ρ-nonexpansive, and every asymptotically ρ-nonexpansive mapping is an asymptotic pointwise ρ-nonexpansive mapping.

Let us give some examples which will illustrate the role of the above defined notions. We start with a typical application of the methods of modular function spaces to the theory of nonlinear integral equations.

Example 5.9. As suggested in the Introduction, an operator itself may be used for the construction of a function modular and hence a space in which this operator has required properties like ρ-nonexpansiveness or ρ-contraction. Let us consider, for instance, the following, important for the theory of integral equations, Urysohn operator:

$$T(f)(x) = \int_0^1 k(x,y,|f(y)|)dy + f_0(x),$$

where f_0 is a fixed function and $f : [0,1] \to \mathbb{R}$ is Lebesgue measurable. For the kernel k we assume that

(a) $k : [0,1] \times [0,1] \times \mathbb{R}_+ \to \mathbb{R}_+$ is Lebesgue measurable,
(b) $k(x,y,0) = 0$,
(c) $k(x,y,.)$ is continuous, convex and increasing to $+\infty$,
(d) $\int_0^1 k(x,y,t)dx > 0$ for $t > 0$ and $y \in (0,1)$,

Assume, in addition, that for almost all $t \in [0,1]$ and all measurable functions f, g there exists a constant $K > 0$ such that

$$\int_0^1 \left\{ \int_0^1 k(t,u,|k(u,v,|f(v)|) - k(u,v,|g(v)|)|)dv \right\} du \le K \int_0^1 k(t,u,|f(u) - g(u)|)du.$$

Setting $\rho(f) = \int_0^1 \left\{ \int_0^1 k(x,y,|f(y)|)dy \right\} dx$ and using Jensen's inequality it is easy to show that ρ is a function modular and that $\rho(T(f) - T(g)) \le K\rho(f-g)$ on L_ρ, that is, T is K-Lipschitzian with respect to ρ. Let us summarize what we have done: given an integral operator we have constructed a modular function space in which this operator is Lipschitzian with the constant K. Obviously, if $K < 1$, then T becomes a ρ-contraction, or if $K \le 1$, ρ-nonexpansive mapping. Hence by applying relevant fixed point theorems from section "Existence of Fixed Points" one can solve the corresponding Urysohn integral equation.

The next example, taken from [20], illustrates the role of the asymptotic assumptions. Please remember that the Hilbert space l^2 is a modular function space too.

Example 5.10. Let C denote the unit ball in l^2 and let T be defined as follows:

$$T : (x_1, x_2, x_3, \dots) \mapsto (0, x_1^2, A_2 x_2, A_3 x_3, \dots)$$

where $A_i \in (0,1)$ are numbers such that $\prod_{i=2}^{\infty} = \frac{1}{2}$. Then T is Lipschitzian with constant 2 but it is not nonexpansive. Moreover,

$$\|T^i(x) - T^i(y)\| \leq 2\prod_{j=2}^{i} A_j \|x-y\|$$

for $i = 2, 3, \ldots$, which implies that

$$\lim_{i\to\infty} k_i = 2\lim_{i\to\infty}\prod_{j=2}^{i} A_j = 1,$$

that is, T is asymptotically nonexpansive.

The following simple finite-dimensional example gives another view on asymptotic pointwise contractions and asymptotic pointwise nonexpansive mappings.

Example 5.11. Let a be any real number from $(0,1)$ and $C = [a,1]$. Let $T : [a,1] \to \mathbb{R}$ be defined simply as $T(x) = \sqrt{x}$. Note that if $x, y \in C$ with $x \leq y$ then

$$|\sqrt{x} - \sqrt{y}| \leq \frac{1}{2\sqrt{x}}|x-y|,$$

which makes T pointwise nonexpansive (and actually nonexpansive) but only if $a \geq \frac{1}{4}$. We know that T has an obvious fixed point at $x = 1$. However if we started with $a < \frac{1}{4}$ we would not be able to use fixed point results like Banach Contraction Principle or Browder/Goehde/Kirk theorem to prove its existence. On the other hand, since in the same situation

$$|T^n(x) - T^n(y)| \leq \frac{1}{2^n} x^{\frac{1}{2^n}-1}|x-y|,$$

it is easy to see that T is an asymptotic pointwise contraction in $C = [a,1]$ for any $a \in (0,1)$ and hence, by Theorem 5.9 below, it has a unique fixed point in C which can be found by taking the limit of orbits. An easy calculation will show that this limit is indeed equal to 1.

In the sequel we will be using the following notion of the ρ-lower semicontinuity.

Definition 5.13. A function $\lambda : C \to [0,\infty]$, where $C \subset L_\rho$ is nonempty and ρ-closed, is called *ρ-lower semicontinuous* if for any $\alpha > 0$, the set $C_\alpha = \{f \in C : \lambda(f) \leq \alpha\}$ is ρ-closed.

Remark 5.8. It can be proved that ρ-lower semicontinuity is equivalent to the condition

$$\lambda(f) \leq \liminf_{n\to\infty} \lambda(f_n) \text{ provided } f, f_n \in C, \text{ and } \rho(f - f_n) \to 0.$$

The ρ-lower semicontinuous real-valued convex functions play an important role in modular function spaces because they attain their infimum provided ρ has property (R), as shown in our next result.

Lemma 5.7. *Let us assume that $\rho \in \Re$ has property (R). Let $K \subset L_\rho$ be nonempty, convex, ρ-closed, and ρ-bounded. If $\varphi : K \to [0,\infty)$ is a ρ-lower semicontinuous convex function, then there exists $x_0 \in K$ such that*

$$\varphi(x_0) = \inf\{\varphi(x) : x \in K\}.$$

Proof. Let $m = \inf\{\varphi(x) : x \in K\}$. The assumptions on φ imply $m < \infty$. For any $n \geq 1$, set

$$K_n = \left\{x \in K : \varphi(x) \leq m + \frac{1}{n}\right\}.$$

Clearly K_n is not empty and is a convex set because φ is a convex function. Also, K_n is ρ-closed since φ is ρ-lower semicontinuous. Since ρ satisfies property (R), then

$$K_\infty = \bigcap_{n\geq 1} K_n \neq \emptyset.$$

For $x_0 \in K_\infty$ there holds

$$\varphi(x_0) = \inf\{\varphi(x) : x \in K\},$$

as claimed. □

5.3 Existence of Fixed Points

The following notation will be used throughout this chapter.

Definition 5.14. For any mapping $T : C \to C$, by $F(T)$ we will denote the set of all fixed points, that is $F(T) = \{f \in C : \ T(f) = f\}$.

5.3.1 Generalized Contractions in Modular Function Spaces

For ρ-contractions the most natural way seems to be to obtain a modular equivalent of the celebrated Banach contraction principle:

Theorem 5.6 (Banach Contraction Theorem). *Let $C \neq \emptyset$ be a closed subset of a Banach space X, and $T : C \to C$ be a contraction, that is, there exists a constant $\alpha < 1$ such that*

$$\|T(x) - T(y)\| \leq \alpha \|x - y\|, \quad \textit{for all } x, y \in C.$$

Then, there exists a unique $z \in C$ *such that* $T(z) = z$. *Moreover, for any* $x \in C$, *there holds* $\|T^n(x) - z\| \to 0$ *as* $n \to \infty$, *where* T^n *is the n-th iterate of* T.

Indeed, the following result holds for the modular function spaces.

Theorem 5.7. *Let* $\rho \in \mathfrak{R}$. *Let* $C \subset L_\rho$ *be nonempty,* ρ*-closed, and* ρ*-bounded. Let* $T : C \to C$ *be a* ρ*-contraction. Then,* T *has a unique fixed point* $f_0 \in C$. *Moreover, for any* $f \in C$, *there holds* $\rho\left(T^n(f) - f_0\right) \to 0$ *as* $n \to \infty$, *where* T^n *is the n-th iterate of* T.

Proof. Let us fix $f_0 \in C$. Let $\delta_\rho(C) < \infty$ be a ρ-diameter of C which is finite due to the ρ-boundedness of C. Observe that

$$\begin{aligned}
\rho\left(T^{n+k}(f_0) - T^n(f_0)\right) &\leq \alpha\rho\left(T^{n+k-1}(f_0) - T^{n-1}(f_0)\right) \\
&\leq \alpha^n\rho\left(T^k(f_0) - f_0\right) \\
&\leq \alpha^n\delta_\rho(C) \to 0,
\end{aligned}$$

as $n \to \infty$ because $\alpha < 1$ and $\delta_\rho(C) < \infty$. Hence, $\{T^n(f_0)\}$ is ρ-Cauchy and, because of the ρ-completeness of L^ρ (Theorem 5.2), there exists a function $\bar{f} \in L^\rho$ such that $\rho\left(T^n(f_0) - \bar{f}\right)$. Because C is ρ-closed, we get $\bar{f} \in C$. Since

$$\begin{aligned}
\rho\left(\frac{\bar{f} - T(\bar{f})}{2}\right) &\leq \rho\left(\bar{f} - T^n(f_0)\right) + \rho\left(T^n(f_0) - T(\bar{f})\right) \\
&\leq \rho\left(\bar{f} - T^n(f_0)\right) + \alpha\rho\left(T^{n-1}(f_0) - \bar{f}\right) \to 0, \text{ as } n \to \infty.
\end{aligned}$$

Therefore, $T(\bar{f}) = \bar{f}$, which means that $\bar{f}$ is a fixed point. To prove the uniqueness part observe that if $T(f_1) = f_1$ and $T(f_2) = f_2$, then

$$\rho(f_1 - f_2) = \rho\left(T(f_1) - T(f_2)\right) \leq \alpha\rho(f_1 - f_2). \tag{5.10}$$

Since $\alpha < 1$ and the right-hand side is finite, equality (5.10) can hold only if $f_1 = f_2$ ρ-a.e. □

The remaining of this section is devoted to extending Theorem 5.7 to the case of pointwise ρ-contractions or asymptotic pointwise ρ-contractions. Our first result demonstrates that we can focus on proving only existence of fixed points as the uniqueness and the ρ-convergence of orbits are automatic in this case.

Theorem 5.8. *Let us assume that* $\rho \in \mathfrak{R}$. *Let* $K \subset L_\rho$ *be nonempty,* ρ*-closed, and* ρ*-bounded. Let* $T : K \to K$ *be a pointwise* ρ*-contraction or asymptotic pointwise* ρ*-contraction. Then* T *has at most one fixed point* $x_0 \in K$. *Moreover if* x_0 *is a fixed point of* T, *then the orbit* $\{T^n(x)\}$ ρ*-converges to* x_0 *for any* $x \in K$.

Proof. Since every pointwise ρ-contraction is an asymptotic pointwise ρ-contraction we can assume that T is an asymptotic pointwise ρ-contraction. Let $u, v \in C$ be two fixed points of T. Then we have

$$\rho(u-v) = \rho\left(T^n(u) - T^n(v)\right) \leq \alpha_n(u)\rho(u-v),$$

for any $n \geq 1$. If we let $n \to \infty$, we will get

$$\rho(u-v) \leq \alpha(u)\rho(u-v).$$

Since $\alpha(u) < 1$, we conclude that $\rho(u-v) = 0$ and hence $u = v$. This proves the uniqueness part. To prove the convergence, assume that x_0 is a fixed point of T. Fix an arbitrary $x \in C$. Let us prove that $\{T^n(x)\}$ converges to x_0. Indeed we have

$$\rho\left(T^{n+m}(x) - T^n(x_0)\right) = \rho\left(T^{n+m}(x) - x_0\right) \leq \alpha_n(x_0)\rho\left(T^m(x) - x_0\right),$$

for any $n, m \geq 1$. Hence

$$\limsup_{m\to\infty} \rho\left(T^{n+m}(x) - x_0\right) \leq \limsup_{m\to\infty} \alpha_n(x_0)\, \rho\left(T^m(x) - x_0\right).$$

Since $\limsup_{m\to\infty} \rho\left(T^{n+m}(x) - x_0\right) = \limsup_{m\to\infty} \rho\left(T^m(x) - x_0\right)$, we get

$$\limsup_{m\to\infty} \rho\left(T^m(x) - x_0\right) \leq \alpha_n(x_0) \limsup_{m\to\infty} \rho\left(T^m(x) - x_0\right),$$

for any $n \geq 1$. If we let $n \to \infty$, we obtain

$$\limsup_{m\to\infty} \rho\left(T^m(x) - x_0\right) \leq \alpha(x_0) \limsup_{m\to\infty} \rho\left(T^m(x) - x_0\right).$$

Since $\alpha(x_0) < 1$, we get $\limsup_{m\to\infty} \rho\left(T^m(x) - x_0\right) = 0$. Clearly we can derive the same equality where lim sup is replaced by lim inf which implies the desired conclusion: $\lim_{m\to\infty} \rho\left(T^m(x) - x_0\right) = 0$. □

We will start our discussion with the fixed point results in the case of uniformly continuous function modulars.

Definition 5.15. We will say that the function modular ρ *is uniformly continuous* if for every $\varepsilon > 0$ and $L > 0$ there exists $\delta > 0$ such that

$$|\rho(g) - \rho(h+g)| \leq \varepsilon, \text{ if } \rho(h) \leq \delta \text{ and } \rho(g) \leq L.$$

Remark 5.9. Let us mention that uniform continuity holds for a large class of function modulars. For instance, it can be proved that in Orlicz spaces over a finite atomless measure [12] or in sequence Orlicz spaces [27] the uniform continuity of the Orlicz modular is equivalent to the Δ_2-type condition.

Lemma 5.8. *Let $\rho \in \Re$ be uniformly continuous. Let $K \subset L_\rho$ be nonempty, convex, ρ-closed, and ρ-bounded. Then, any ρ-type $\tau : K \to [0,\infty]$ is ρ-lower semicontinuous in K.*

Proof. Let τ be a ρ-type. Let $\alpha > 0$, denote $C_\alpha = \{ x \in K : \tau(x) \le \alpha \}$. We need to prove that C_α is ρ-closed. Without loss of generality we can assume that C_α is nonempty. Let a sequence $\{x_k\} \subset C_\alpha$ be such that $\rho(x_0 - x_k) \to 0$ with $x_0 \in K$. We need to prove that $x_0 \in C_\alpha$, i.e. that $\tau(x_0) \le \alpha$. Let $\{y_m\}$ be a sequence that defines τ. By ρ-boundedness of K, there exists $L = \sup_m \rho(x_0 - y_m) < \infty$. Let us fix an arbitrary $\varepsilon > 0$. By uniform continuity of ρ there exists $\delta > 0$ such that

$$|\rho(g) - \rho(h+g)| \le \varepsilon \ \ if \ \ \rho(h) \le \delta \ \ and \ \ \rho(g) \le L. \tag{5.11}$$

Since $\rho(x_0 - x_k) \to 0$, there exists $p \ge 1$ such that $\rho(x_0 - x_k) \le \delta$ for $k \ge p$. Using (5.11) with $h = x_0 - x_p$ and $g = y_n - x_0$ we have

$$|\rho(y_n - x_0) - \rho(y_n - x_p)| \le \varepsilon \ \ for\ every \ \ n \ge 1.$$

By the definition of τ, we have then that $\tau(x_0) \le \alpha + 2\varepsilon$ because $x_p \in C_\alpha$. Since ε was chosen arbitrarily we have finally $\tau(x_0) \le \alpha$ as claimed. □

We will utilize the above lemma to prove the following fixed point theorem.

Theorem 5.9. *Assume that $\rho \in \Re$ is uniformly continuous and has property (R). Let $K \subset L_\rho$ be nonempty, convex, ρ-closed, and ρ-bounded. Let $T : K \to K$ be an asymptotic pointwise ρ-contraction or asymptotic pointwise ρ-contraction. Then, T has a unique fixed point $x_0 \in K$. Moreover, the orbit $\{T^n(x)\}$ converges to x_0 for any $x \in K$.*

Proof. Since every pointwise ρ-contraction is an asymptotic pointwise ρ-contraction we can assume that T is an asymptotic pointwise ρ-contraction. In view of Theorem 5.8 it is enough to show that T has a fixed point. Let us fix $x \in K$ and define the ρ-type by

$$\tau(u) = \limsup_{n\to\infty} \rho(T^n(x) - u),$$

for $u \in K$. By Lemma 5.8 the ρ-type τ is ρ-lower semicontinuous in K. Then, by Lemma 5.7, there exists $x_0 \in K$ such that

$$\tau(x_0) = \inf\{\tau(x) : x \in K\}.$$

Let us prove that $\tau(x_0) = 0$. Indeed, for any $n,m \ge 1$ we have

$$\rho(T^{n+m}(x) - T^m(x_0) \le \alpha_m(x_0)\rho(T^n(x) - x_0).$$

If let n go to infinity, we get

$$\tau(T^m(x_0)) \leq \alpha_m(x_0)\tau(x_0),$$

which implies

$$\tau(x_0) = \inf\{\tau(x) : x \in K\} \leq \tau(T^m(x_0)) \leq \alpha_m(x_0)\tau(x_0).$$

Passing with m to infinity, we get $\tau(x_0) \leq \alpha(x_0)\tau(x_0)$ which forces $\tau(x_0) = 0$ as $\alpha(x_0) < 1$. Hence, $\rho(T^n(x) - x_0)) \to 0$ as $n \to \infty$. By the ρ-continuity of T, this forces x_0 to be a fixed point of T. □

Remark 5.10. The conclusions of Theorem 5.9 were originally proved in [34] using slightly different techniques.

In the remaining of this section, we develop techniques which will allow us to replace the assumption of uniform continuity of ρ by somewhat weaker assumptions of Strong Opial property. This will be possible at a price of the assumption of the ρ-a.e. compactness of C and of restricting our considerations to a subspace E_ρ of L_ρ defined as follows.

Definition 5.16. Denote

$$L^0_\rho = \{f \in L_\rho : \ \rho(f,\cdot) \ is \ order \ continuous\}$$

and define

$$E_\rho = \{f \in L_\rho : \ \lambda f \in L^0_\rho \ for \ every \ \lambda > 0\}.$$

Let us define the Opial property and the Strong Opial property in modular function spaces.

Definition 5.17. We say that L_ρ satisfies the *ρ-a.e. Opial property* if for every $\{f_n\} \in L_\rho$ which is ρ-a.e. convergent to 0 such that there exists a $\beta > 1$ for which

$$\sup_n \{\rho(\beta f_n)\} < \infty,$$

the following inequality holds for any $g \in E_\rho$ not equal to 0

$$\liminf_{n\to\infty} \rho(f_n) \leq \liminf_{n\to\infty} \rho(f_n + g).$$

Definition 5.18. We say that L_ρ satisfies the *ρ-a.e. Strong Opial property* if for every $\{f_n\} \in L_\rho$ which is ρ-a.e. convergent to 0 such that there exists a $\beta > 1$ for which

$$\sup_n \{\rho(\beta f_n)\} < \infty,$$

the following equality holds for any $g \in E_\rho$

$$\liminf_{n\to\infty} \rho(f_n + g) = \liminf_{n\to\infty} \rho(f_n) + \rho(g).$$

Remark 5.11. It is not difficult to prove that the ρ-a.e. Strong Opial property implies ρ-a.e. Opial property, see, for example, [31].

Remark 5.12. Also, note that, in virtue of Theorem 2.1 in [31], every convex, orthogonally additive function modular ρ has the ρ-a.e. Strong Opial property. Let us recall that ρ is called orthogonally additive if $\rho(f, A \cup B) = \rho(f, A) + \rho(f, B)$ whenever $A \cap B = \emptyset$. Therefore, all Orlicz and Musielak–Orlicz spaces must have the Strong Opial property.

Remark 5.13. Observe that the Opial property in the norm sense, [66], does not necessarily hold for several classical Banach function spaces. For instance, the norm Opial property does not hold for L^p spaces for $1 \leq p \neq 2$ while the modular Strong Opial property holds in L^p for all $p \geq 1$.

A typical method of proof for the fixed point theorems is to construct a fixed point by finding an element on which a specific type function attains its minimum. As a matter of fact, that is exactly how we proved our Theorem 5.9. To be able to proceed with this method, one has to know that such an element indeed exists. In case of Theorem 5.9 the assumption of uniform continuity of ρ and property (R) ensured such an existence because in this case ρ-types are ρ-lower semicontinuous which was proved in Lemma 5.8. In the case of ρ with the Strong Opial property, we cannot guarantee anymore that the ρ-type is ρ-lower semicontinuous and therefore one needs additional assumptions to ensure that ρ-types attain their minimum. Not surprisingly a weak form of compactness will do the job in this case, as demonstrated in our next result.

Lemma 5.9. *Let $\rho \in \Re$. Assume that L_ρ has the ρ-a.e. Strong Opial property. Let $C \subset E_\rho$ be a nonempty, ρ-a.e. compact, strongly ρ-bounded, convex set. Then any ρ-type defined in C attains its minimum in C.*

Proof. Let us fix a ρ-type τ defined by

$$\tau(u) = \limsup_{t\to\infty} \rho(u - u_t)$$

where $u_t \in C$ for every $t \geq 0$, and denote $\tau_0 = \inf\{\tau(u) : u \in C\}$. Note that by the strong ρ-boundedness of C, there exists $\beta > 1$ such that $M_\beta(C) < \infty$. Observing that

$$\tau_0 \leq \sup\{\rho(\beta(f - g)) : \; f, g \in C\} = M_\beta(C) < \infty.$$

we conclude that there exists a sequence $\{x_n\} \subset C$ such that

$$\tau_0 = \lim_{n\to\infty} \tau(x_n).$$

Since C is ρ-a.e. compact, by passing to a subsequence if necessary, we can assume that there exists an $x_0 \in C$ such that $x_n \to x_0$ ρ-a.e. Let us select a sequence $t_n \to \infty$ so that

$$\tau(x_0) = \limsup_{t\to\infty} \rho(x_0 - u_t) = \lim_{n\to\infty} \rho(x_0 - u_{t_n}), \tag{5.12}$$

and denote $y_n = u_{t_n}$. By the ρ-a.e. compactness of C again, there exists a subsequence $\{y_{\varphi(n)}\}$ of $\{y_n\}$ which ρ-a.e. converges to some $y_0 \in C$. By the ρ-a.e. Strong Opial property we get

$$\liminf_{n\to\infty} \rho(y_{\varphi(n)} - x_m) = \liminf_{n\to\infty} \rho(y_{\varphi(n)} - y_0) + \rho(y_0 - x_m), \tag{5.13}$$

for any $m \geq 0$. Since

$$\begin{aligned}\liminf_{n\to\infty} \rho(y_{\varphi(n)} - x_m) &\leq \limsup_{n\to\infty} \rho(y_n - x_m) = \limsup_{n\to\infty} \rho(x_m - u_{t_n})\\ &\leq \limsup_{t\to\infty} \rho(x_m - u_t) = \tau(x_m),\end{aligned}$$

we conclude via (5.13) that

$$\liminf_{m\to\infty} \tau(x_m) \geq \liminf_{n\to\infty} \rho(y_{\varphi(n)} - y_0) + \liminf_{m\to\infty} \rho(y_0 - x_m).$$

Using the ρ-a.e. Strong Opial property again, to $\{x_m - x_0\}$ this time, we get

$$\liminf_{m\to\infty} \rho(y_0 - x_m) = \liminf_{m\to\infty} \rho(x_m - x_0) + \rho(x_0 - y_0)$$

which implies

$$\liminf_{m\to\infty} \tau(x_m) \geq \liminf_{n\to\infty} \rho(y_{\varphi(n)} - y_0) + \liminf_{m\to\infty} \rho(x_m - x_0) + \rho(x_0 - y_0).$$

Hence

$$\liminf_{m\to\infty} \tau(x_m) \geq \liminf_{n\to\infty} \rho(y_{\varphi(n)} - x_0) + \liminf_{m\to\infty} \rho(x_m - x_0),$$

which implies

$$\liminf_{m\to\infty} \tau(x_m) \geq \liminf_{n\to\infty} \rho(y_{\varphi(n)} - x_0). \tag{5.14}$$

Using (5.12) we get

$$\liminf_{n\to\infty} \rho(y_{\varphi(n)} - x_0) = \lim_{n\to\infty} \rho(y_n - x_0) = \tau(x_0). \tag{5.15}$$

Combining (5.15) with (5.14), we get

$$\tau_0 = \lim_{m\to\infty} \tau(x_m) \geq \tau(x_0).$$

On the other hand,

$$\tau_0 = \inf\{\tau(u) : u \in C\} \leq \tau(x_0).$$

Finally $\tau(x_0) = \tau_0 = \inf\{\tau(u) : u \in C\}$, which completes the proof of the lemma. □

We are now ready to prove our next fixed point theorem.

Theorem 5.10. *Let $\rho \in \Re$. Assume that L_ρ has the ρ-a.e. Strong Opial property. Let $K \subset E_\rho$ be a nonempty, ρ-a.e. compact, strongly ρ-bounded, convex set. Then, any $T : K \to K$ asymptotic pointwise ρ-contraction or pointwise ρ-contraction has a unique fixed point $x_0 \in K$. Moreover, the orbit $\{T^n(x)\}$ converges to x_0, for any $x \in K$.*

Proof. Since every pointwise ρ-contraction is an asymptotic pointwise ρ-contraction we can assume that T is an asymptotic pointwise ρ-contraction. In view of Theorem 5.8 it is enough to show that T has a fixed point. Let us fix then $x \in K$ and define the ρ-type by

$$\tau(u) = \limsup_{n\to\infty} \rho(T^n(x) - u),$$

for $u \in K$. By Lemma 5.9 there exists $x_0 \in K$ such that

$$\tau(x_0) = \inf\{\tau(x) : x \in K\}.$$

Using the same argument as in the proof of Theorem 5.9, we will get $\tau(x_0) = 0$. Hence, $\rho(T^n(x) - x_0)) \to 0$ as $n \to \infty$. By the ρ-continuity of T, this forces x_0 to be a fixed point of T. □

Remark 5.14. The conclusions of Theorem 5.10 were proven by Khamsi and Kozlowski in [34] using different methods.

5.3.2 Generalized Nonexpansive Mappings

Similarly as in the Banach space setting, the fixed point existence theorems for the ρ-nonexpansive mappings (and for their pointwise and pointwise asymptotic generalizations) acting in modular function spaces are much harder to obtain. Let us first consider the following question: Are the ρ-nonexpansive mappings really different from the mappings nonexpansive with respect to the Luxemburg norm associated with the modular ρ? First we will show the following simple result.

Proposition 5.9. *Let $\rho \in \Re$. If for every $\lambda > 0$*

$$\rho(\lambda(T(f) - T(g))) \leq \rho(\lambda(f - g)), \tag{5.16}$$

then $\|T(f) - T(g)\|_\rho \leq \|f - g\|_\rho$.

Proof. Assume to the contrary that there exist $f, g \in L_\rho$ and $\alpha > 0$ such that

$$\|f - g\|_\rho < \alpha < \|T(f) - T(g)\|_\rho.$$

Then, $\left\|\frac{f-g}{\alpha}\right\|_\rho < 1$, which by Proposition 5.5 part (a) implies that $\rho\left(\frac{f-g}{\alpha}\right) < 1$. It also implies that

$$1 < \left\|\frac{T(f) - T(g)}{\alpha}\right\|_\rho,$$

which, by Proposition 5.5 part (b), yields $1 < \rho\left(\frac{T(f)-T(g)}{\alpha}\right)$. Finally, setting $\lambda = \alpha^{-1}$, we obtain

$$\rho(\lambda(f - g)) < 1 < \rho(\lambda(T(f) - T(g))).$$

Contradiction completes the proof. □

In view of Proposition 5.9, we need to ask whether the inequality (5.16) needs to hold for every $\lambda > 0$ in order to ensure the norm nonexpansiveness. If we knew that it sufficed to assume it merely for $\lambda = 1$, then there would be no real reason to consider ρ-nonexpansiveness. The answer to this question can be found in the following example of a mapping which is ρ-nonexpansive but it is not $\|.\|_\rho$-nonexpansive.

Example 5.12. Let $X = (0, \infty)$ and Σ be the σ-algebra of all Lebesgue measurable subsets of X. Let $\mathscr{P}$ denote the δ-ring of subsets of finite measure. Define a function modular by

$$\rho(f) = \frac{1}{e^2} \int_0^\infty |f(x)|^{x+1} dm(x).$$

Let B be the set of all measurable functions $f : (0, \infty) \to \mathbb{R}$ such that $0 \leq f(x) \leq 1/2$. Consider the map

$$T(f)(x) = \begin{cases} f(x-1), & \text{for } x \geq 1, \\ 0, & \text{for } x \in [0,1]. \end{cases}$$

Clearly, we have $T(B) \subset B$. For every $f, g \in B$ and $\lambda \leq 1$, we have

$$\rho\left(\lambda\left(T(f) - T(g)\right)\right) \leq \lambda \rho\left(\lambda(f-g)\right),$$

which implies that T is ρ-nonexpansive. On the other hand, if we take $f = 1_{[0,1]}$, then

$$\|T(f)\|_\rho > e \geq \|f\|_\rho,$$

which clearly implies that T is not $\|.\|_\rho$-nonexpansive. Note that T is linear.

For historical reasons, let us quote without proof an early example of a fixed point theorem for ρ-nonexpansive mappings acting in modular function spaces, see Theorem 2.13 [36].

Theorem 5.11. *Let $\rho \in \Re$ satisfy the regular growth condition, that is, $w_\rho(t) < 1$ for all $t \in [0,1)$, where*

$$w_\rho(t) = \sup\left\{\frac{\rho(tf)}{\rho(f)} : f \in L_\rho,\ 0 < \rho(f) < \infty\right\}.$$

Let $C \subset L_\rho$ be a convex, ρ-bounded and ρ-a.e. compact such that $C - C \subset L^0_\rho$. Assume in addition that for every sequence $f_n \in C$ such that $f_n \to f$ ρ-a.e. with $f \in C$, and for every sequence of sets $G_k \downarrow \emptyset$,

$$\lim_{k\to\infty}\left(\sup_{n\in\mathbb{N}} \rho(f_n - f, G_k)\right) = 0.$$

If $T : C \to C$ is ρ-nonexpansive, then T has a fixed point.

The above theorem sheds an interesting light on the following example by Alspach [3].

Example 5.13. Define the operator

$$T(f)(x) = \begin{cases} \min\{2, 2f(x)\}, & \text{for } x \in [0,1/2], \\ \max\{0, 2f(2x-1) - 2\}, & \text{for } x \in (1/2,1] \end{cases}$$

on C, a convex subset of $L^1[0,1]$, defined by

$$C = \left\{f \in L^1[0,1] : 0 \leq f(x) \leq 2 \text{ a.e. and } \int_0^1 f(x)dx = 1\right\}.$$

Let $L_\rho = L^1[0,1]$. The operator T is an L^1-isometry on C hence is ρ-nonexpansive, but it does not have any fixed points. It is easy to see that all assumptions of Theorem 5.11 are satisfied except for the ρ-a.e. compactness of C. To see that C is not ρ-a.e. compact take a sequence $f_n = 2g_n$ where $g_n = 1_{A_n}$ and $A_n = [2^{-1} - 2^{-n}, 1 - 2^{-n}]$. Obviously $f_n \in C$ but $f_n \to 0$ ρ-a.e. while the zero function does not belong to C. Note that Theorem 5.11 gives an intrinsic reason why T, defined on $B = \{f \in L^1[0,1]; 0 \leq f(x) \leq 2 \text{ a.e.}\}$, must have a fixed point while $T: C \to C$ does not have to. Moreover, we do not refer to any geometrical properties of L^1 or its subspaces.

A more advanced fixed point results for ρ-nonexpansive mappings started showing up in the early 2000s, see, for example, [16–18]. It was only after a proper modular function space geometry was established that it became possible to obtain an elegant modular fixed point theorem for ρ-nonexpansive mappings, presented below in a generalized version which covers also the asymptotic pointwise case, Theorem 5.12. On a high level, the working of this theory can be summarized as follows:

(1) The uniform convexity property implies the unique best approximant property (Theorem 5.4).
(2) The uniform convexity property via the unique best approximant property implies the property (R) (Theorem 5.5).
(3) The uniform convexity property implies the parallelogram property (Lemma5.5).
(4) The parallelogram property implies the minimizing sequence property for type functions when the minimum is strictly positive (Lemma 5.6, Case 1).
(5) The property (R) implies the minimizing sequence property for type functions when the minimum is equal to zero (Lemma 5.6, Case 2).
(6) The minimizing sequence property for type functions implies the fixed point property for asymptotic pointwise nonexpansive mappings; the modular limit of a minimizing sequence for a type function defined by an orbit is an obvious candidate for a fixed point. As Theorem 5.12 below shows, this is indeed the case.

Theorem 5.12. *Assume $\rho \in \Re$ is $(UUC1)$. Let C be a ρ-closed ρ-bounded convex nonempty subset of L_ρ. Then, any $T: C \to C$ pointwise asymptotically nonexpansive mapping has a fixed point. Moreover, the set of all fixed points $F(T)$ is ρ-closed and convex.*

Proof. Let $f \in C$. Define the ρ-type

$$\tau(h) = \limsup_{n\to\infty} \rho(T^n(f) - h), \text{ for any } h \in C.$$

Let $\tau_0 = \inf\{\tau(h) : h \in C\}$ and note that $\tau_0 < \infty$ due to the ρ-boundedness of C. Let $\{g_n\} \subset C$ be a minimizing sequence of τ and $g \in C$ its ρ-limit which exists in view of Lemma 5.6. We will prove that g is a fixed point of T. First notice that

$$\tau(T^m(h)) \leq \alpha_m(h)\tau(h),$$

for any $h \in C$ and $m \geq 1$. In particular, we have

$$\tau(T^m(g_n)) \leq \alpha_m(g_n)\tau(g_n),$$

for any $n,m \geq 1$. By induction, we will build an increasing sequence $\{m_k\}$ such that

$$\alpha_{m_k+m}(g_k) \leq 1+\frac{1}{k}, \tag{5.17}$$

for $k,m \geq 1$. Since T is pointwise asymptotically nonexpansive, we have

$$\limsup_{m\to\infty} \alpha_m(g_1) \leq 1.$$

Therefore, there exists $m_1 \geq 1$ such that for any $m \geq 1$ we have

$$\alpha_{m_1+m}(g_1) \leq 1+\frac{1}{1}.$$

Since $\limsup_{m\to\infty} \alpha_m(g_2) \leq 1$, there exists $m_2 > m_1$ such that for any $m \geq 1$, we have

$$\alpha_{m_2+m}(g_2) \leq 1+\frac{1}{2}.$$

Assume now that m_k has been built. Since

$$\limsup_{m\to\infty} \alpha_m(g_{k+1}) \leq 1,$$

then there exists $m_{k+1} > m_k$ such that for any $m \geq m_{k+1}$, we have

$$\alpha_m(g_{k+1}) \leq 1+\frac{1}{k+1},$$

which completes our induction claim and proves (5.17). Let $n \geq 1$, $k \geq 1$ and $p \geq 0$ be integers. From (5.17) it follows that

$$\rho\left(T^{n+m_k+p}(f) - T^{m_k+p}(g_k)\right) \leq \left(1+\frac{1}{k}\right)\rho\left(T^n(f) - g_k\right). \tag{5.18}$$

By taking $\limsup_{n\to\infty}$ of both sides of (5.18), we obtain

$$\tau_0 \leq \tau\left(T^{m_k+p}(g_k)\right) \leq \left(1+\frac{1}{k}\right)\tau(g_k). \tag{5.19}$$

Passing with $k \to \infty$ in (5.19) gives us

$$\lim_{k\to\infty} \tau\left(T^{m_k+p}(g_k\right) = \tau_0,$$

which forces $\{T^{m_k+p}(g_k)\}$ to be a minimizing sequence of τ, for any $p \geq 0$. Lemma 5.6 implies $\{T^{m_k+p}(g_k)\}$ is ρ-convergent to g, for any $p \geq 0$. In particular, taking $p = 1$, we conclude that

$$\rho\left(T^{m_k+1}(g_k) - g\right) \to 0. \tag{5.20}$$

From

$$\rho\left(T^{m_k+1}(g_k) - T(g)\right) \leq \alpha_1(g)\rho\left(T^{m_k}(g_k) - g\right),$$

we get the following

$$\rho\left(T^{m_k+1}(g_k) - T(g)\right) \to 0. \tag{5.21}$$

Since, by Lemma 5.6, the ρ-limit of any ρ-convergent sequence is unique then (5.20) together with (5.21) imply that $T(g) = g$.

To prove that $F(T)$ is ρ-closed, let $f_n \in F(T)$ and $\rho(f_n - f) \to 0$. Observe that

$$\begin{aligned}\rho\left(\frac{1}{3}(T(f) - f)\right) &\leq \rho(T(f) - T(f_n)) + \rho(T(f_n) - f_n) + \rho(f_n - f) \\ &\leq \alpha_1(f)\rho(f_n - f) + \rho(f_n - f) \to 0.\end{aligned}$$

Hence $f \in F(T)$ proving $F(T)$ is ρ-closed. It remains to prove that $F(T)$ is convex. Note that to this end it suffices to show that

$$h = \frac{f+g}{2} \in F(T)$$

for any $f, g \in F(T)$. Without loss of generality, we can assume that $f \neq g$. Let $k > 0$ be an integer. We have

$$\rho(f - T^k(h)) = \rho(T^k(f) - T^k(h)) \leq \alpha_k(f)\rho\left(f - h\right),$$

and

$$\rho(g - T^k(h)) = \rho(T^k(g) - T^k(h)) \leq \alpha_k(g)\rho\left(g - h\right).$$

Since $\rho(f-h) = \rho(g-h) = \rho\left(\frac{f-g}{2}\right)$, and

$$\rho\left(\frac{f-g}{2}\right) \leq \frac{1}{2}\rho(f - T^k(h)) + \frac{1}{2}\rho(g - T^k(h)),$$

we conclude that

$$\lim_{k\to\infty}\rho(f-T^k(h)) = \lim_{k\to\infty}\rho(g-T^k(h)) = \rho\left(\frac{f-g}{2}\right).$$

Similarly we have

$$\rho\left(f-\frac{h+T^k(h)}{2}\right) \le \frac{1}{2}\rho\left(f-h\right)+\frac{1}{2}\rho(f-T^k(h)),$$

and

$$\rho\left(g-\frac{h+T^k(h)}{2}\right) \le \frac{1}{2}\rho\left(g-h\right)+\frac{1}{2}\rho(g-T^k(h)).$$

Since

$$\rho\left(\frac{f-g}{2}\right) \le \frac{1}{2}\rho\left(f-\frac{h+T^k(h)}{2}\right)+\frac{1}{2}\rho\left(g-\frac{h+T^k(h)}{2}\right),$$

we conclude that

$$\lim_{k\to\infty}\rho\left(f-\frac{h+T^k(h)}{2}\right) = \lim_{k\to\infty}\rho\left(g-\frac{h+T^k(h)}{2}\right) = \rho\left(\frac{f-g}{2}\right).$$

Therefore we have

$$\lim_{k\to\infty}\rho(f-T^k(h)) = \lim_{k\to\infty}\rho\left(f-\frac{h+T^k(h)}{2}\right) = \rho(f-h).$$

Lemma 5.2 applied to $A_k = f - T^k(h)$ and $B_k = T^k(h) - g$ implies that $\rho(A_k - B_k) \to 0$. Hence

$$\lim_{k\to\infty}\rho(h-T^k(h)) = \lim_{k\to\infty}\rho\left(\frac{A_k-B_k}{2}\right) \le \lim_{k\to\infty}\rho(A_k-B_k) = 0.$$

Clearly we will get $\lim_{k\to\infty}\rho(h-T^{k+m}(h)) = 0$, for any integer $m \ge 0$. Since

$$\rho(T^m(h)-T^{k+m}(h)) \le \alpha_m(h)\rho(h-T^k(h))$$

we get $\lim_{k\to\infty}\rho(T^m(h)-T^{k+m}(h)) = 0$. Finally using the inequality

$$\rho\left(\frac{h-T^m(h)}{2}\right) \le \frac{1}{2}\rho(h-T^{k+m}(h))+\frac{1}{2}\rho(T^m(h)-T^{k+m}(h)),$$

and by letting $k \to \infty$, we get $T^m(h) = h$, for any integer $m \geq 0$. Taking $m = 1$ we get $h \in F(T)$, which completes the proof. □

Remark 5.15. Observe that the statement of the above theorem is parallel to the celebrated Browder, Gohde, Kirk fixed point theorem (see [9,23,39]) but formulated purely in terms of function modulars without any reference to norms. Also, note that Theorem 5.12 extends outside nonexpansiveness and assumes merely asymptotic pointwise ρ-nonexpansiveness of the mapping T. Therefore, Theorem 5.12 can be actually understood as parallel not only to the Browder, Gohde, Kirk fixed point theorem for nonexpansive mappings in Banach spaces but also to the result of Kirk and Xu [44] for asymptotic pointwise nonexpansive mappings in Banach spaces.

5.4 Convergence of Fixed Point Iterative Algorithms

Assume $\rho \in \Re$ is $(UUC1)$. Let C be a ρ-closed ρ-bounded convex nonempty subset of L_ρ. Let $T : C \to C$ be a pointwise asymptotically nonexpansive mapping. According to Theorem 5.12 the mapping T has a fixed point. The proof of this important theorem is of the existential nature and does not describe any algorithm for constructing a fixed point of an asymptotic pointwise ρ-nonexpansive mapping. This chapter aims at filling this gap. Therefore, we will define iterative processes for the fixed point construction in modular function spaces and we will prove their convergence. These algorithms will be based on classical iterative methods introduced originally by Mann in [58] and Ishikawa in [26].

Denoting $a_n(x) = \max(\alpha_n(x), 1)$, we note that without loss of generality we can assume that T is asymptotically pointwise nonexpansive if

$$\rho(T^n(f) - T^n(g)) \leq a_n(f)\rho(f-g), \quad \text{for all } f,g \in C,\ n \in \mathbb{N},$$
$$\lim_{n\to\infty} a_n(f) = 1,\ a_n(f) \geq 1, \quad \text{for all } f \in C, \textit{ and } n \in \mathbb{N}.$$

Define $b_n(f) = a_n(f) - 1$. In view of the above, we have

$$\lim_{n\to\infty} b_n(f) = 0.$$

The above notation will be consistently used throughout this section.

By $\mathscr{T}(C)$ we will denote the class of all asymptotic pointwise nonexpansive mappings $T : C \to C$.

In this section, we will impose some restrictions on the behavior of a_n and b_n.

Definition 5.19. Define $\mathscr{T}_r(C)$ as a class of all $T \in \mathscr{T}(C)$ such that a_n is a bounded function for every $n \geq 1$ and $\sum_{n=1}^{\infty} b_n(x) < \infty$ for every $x \in C$.

Remark 5.16. The restrictions described in Definition 5.19 are typical for controlling the convergence of iterative processes for asymptotically nonexpansive mappings, see, for example, [48].

Let us start with the following technical result.

Lemma 5.10. *Let $\rho \in \Re$. Let $C \subset L_\rho$ be a convex set, and let $T \in \mathscr{T}_r(C)$. If $\{x_k\}$ is a ρ-approximate fixed point sequence for T, that is, $\rho(T(x_k) - x_k) \to 0$ as $k \to \infty$, then for every fixed $m \in \mathbb{N}$ there holds*

$$\rho\left(\frac{T^m(x_k) - x_k}{m}\right) \to 0, \quad \text{as } k \to \infty.$$

Proof. It follows from the fact that a_n is a bounded function for every $n \geq 1$ that there exists a finite constant $M > 0$ such that

$$\sum_{j=1}^{m-1} sup\{a_j(x) : x \in C\} \leq M.$$

Using the convexity of ρ and the ρ-nonexpansiveness of T we get

$$\rho\left(\frac{T^m(x_k) - x_k}{m}\right) = \rho\left(\frac{1}{m}\sum_{j=0}^{m-1}(T^{j+1}(x_k) - T^j(x_k))\right)$$

$$\leq \frac{1}{m}\sum_{j=0}^{m-1}\rho(T^{j+1}(x_k) - T^j(x_k)) \leq \rho(T(x_k) - x_k)\left(\sum_{j=1}^{m-1}a_j(x_n) + 1\right)$$

$$\leq \frac{1}{m}(M+1)\rho(T(x_k) - x_k) \to 0,$$

as $k \to \infty$. □

Corollary 5.1. *If, under the hypothesis of Lemma 5.10, ρ satisfies additionally the Δ_2 condition, then $\rho\,(T^m(x_k) - x_k) \to 0$ as $k \to \infty$.*

The following modular version of the Demiclosedness Principle will be used in the proof of our convergence Theorem 5.14. Our proof of the Demiclosedness Principle uses the parallelogram inequality valid in the modular spaces with the $(UUC1)$ property (Lemma 5.5).

Theorem 5.13 (Demiclosedness Principle [13]). *Let $\rho \in \Re$. Assume that*

(a) ρ is $(UCC1)$,
(b) ρ has Strong Opial Property,
(c) ρ has Δ_2 property and is uniformly continuous.

Let $C \subset L_\rho$ be a nonempty, convex, strongly ρ-bounded and ρ-closed, and let $T \in \mathscr{T}_r(C)$. Let $\{x_n\} \subset C$, and $x \in C$. If $x_n \to x$ $\rho-a.e.$ and $\rho(T(x_n)-x_n) \to 0$ then $x \in F(T)$.

Proof. Let us recall that by definition of uniform continuity of ρ to every $\varepsilon > 0$ and $L > 0$, there exists $\delta > 0$ such that

$$|\rho(g) - \rho(g+h)| \leq \varepsilon, \tag{5.22}$$

provided $\rho(h) < \delta$ and $\rho(g) \leq L$. Fix any $m \in \mathbb{N}$. Noting that $\rho(x_n - x) \leq M < \infty$ due to the strong ρ-boundedness of C and that $\rho\left(T^m(x_n) - x_n\right) \to 0$ by Corollary 5.1, it follows then from (5.22) with $g = x_n - x$ and $h = T^m(x_n) - x_n$ that

$$\left|\rho(x_n - x) - \rho\left(x_n - x + T^m(x_n) - x_n\right)\right| \to 0, \quad \text{as } n \to \infty.$$

Hence,

$$\limsup_{n\to\infty} \rho(x_n - x) = \limsup_{n\to\infty} \rho\left(T^m(x_n) - x\right). \tag{5.23}$$

Define the ρ-type φ by

$$\varphi(x) = \limsup_{n\to\infty} \rho(x_n - x).$$

By (5.23) we get

$$\varphi(x) = \limsup_{n\to\infty} \rho\left(T^m(x_n) - x\right).$$

Hence, for every $y \in C$ there holds

$$\begin{aligned} \varphi(T^m(y)) &= \limsup_{n\to\infty} \rho\left(T^m(x_n) - T^m(y)\right) \\ &\leq a_m(y) \limsup_{n\to\infty} \rho(x_n - y) = a_m(y)\varphi(y) \end{aligned} \tag{5.24}$$

Using (5.24) with $y = x$ and by passing with m to infinity we conclude that

$$\limsup_{m\to\infty} \varphi(T^m(x)) \leq \varphi(x). \tag{5.25}$$

Since ρ satisfies the Strong Opial property, it also satisfies the Opial property. Since $x_n \to x$ $\rho-a.e.$, it follows via the Opial property that for any $y \neq x$

$$\varphi(x) = \limsup_{n\to\infty} \rho(x_n - x) < \limsup_{n\to\infty} \rho(x_n - y) = \varphi(y),$$

which implies that

$$\varphi(x) = \inf\{\varphi(y) : y \in C\}. \tag{5.26}$$

Combining (5.25) with (5.26), we have

$$\varphi(x) \leq \limsup_{m\to\infty} \varphi(T^m(x)) \leq \varphi(x),$$

that is,

$$\limsup_{m\to\infty} \varphi(T^m(x)) = \varphi(x).$$

We claim that

$$\lim_{m\to\infty} \rho\left(T^m(x) - x\right) = 0. \tag{5.27}$$

Assume to the contrary that (5.27) does not hold, that is

$$\rho\left(T^m(x) - x\right) \text{ does not tend to zero.} \tag{5.28}$$

By Δ_2, then it follows from (5.28) that $\rho\left(\frac{T^m(x)-x}{2}\right)$ does not tend to zero. By passing to a subsequence if necessary we can assume that there exists $0 < t < M$ such that

$$\rho\left(\frac{T^m(x) - x}{2}\right) > t > 0, \quad \text{for } m \in \mathbb{N},$$

which implies that

$$\rho(x_n - x) + \rho\left(x_n - T^m(x)\right) > \frac{t}{2}, \quad \text{for every } m, n \in \mathbb{N}.$$

Hence,

$$\max\left\{\rho(x_n - x), \rho\left(x_n - T^m(x)\right)\right\} \geq \frac{t}{4}, \quad \text{for every } m, n \in \mathbb{N}.$$

Applying the parallelogram inequality from Lemma 5.5,

$$\rho^2\left(\frac{z+y}{2}\right) \leq \frac{1}{2}\rho^2(z) + \frac{1}{2}\rho^2(y) - \Psi\left(r, s, \frac{1}{r}\rho(z-y)\right),$$

where $\rho(z) \leq r, \rho(y) \leq r$ and $\max\{\rho(z), \rho(y)\} \geq s$ for $0 < s < r$, with $r = M$, $s = \frac{t}{4}$, $z = x_n - x$, $y = T^m(x)$, we get

$$\rho^2\left(x_n - \frac{x+T^m(x)}{2}\right) \leq \frac{1}{2}\rho^2(x_n - x) + \frac{1}{2}\rho^2(x_n - T^m(x))$$
$$-\Psi\left(M, \frac{t}{4}, \frac{1}{M}\rho(x - T^m(x))\right).$$

Note that by (5.26)

$$\varphi^2(x) \leq \varphi^2\left(\frac{x+T^m(x)}{2}\right) = \limsup_{n\to\infty}\rho^2\left(x_n - \frac{x+T^m(x)}{2}\right).$$

Combining the last two formulas we obtain

$$\varphi^2(x) \leq \tfrac{1}{2}\limsup_{n\to\infty}\rho^2(x_n - x) + \tfrac{1}{2}\limsup_{n\to\infty}\rho^2(x_n - T^m(x))$$
$$-\Psi\left(M, \tfrac{t}{4}, \tfrac{1}{M}\rho(x - T^m(x))\right),$$

which implies

$$0 \leq \Psi\left(M, \frac{t}{4}, \frac{1}{M}\rho(x - T^m(x))\right) \leq \frac{1}{2}\varphi^2(T^m(x)) - \frac{1}{2}\varphi^2(x).$$

$$0 \leq \limsup_{m\to\infty}\Psi\left(M, \frac{t}{4}, \frac{1}{M}\rho(x - T^m(x))\right)$$
$$\leq \frac{1}{2}\limsup_{m\to\infty}\varphi^2(T^m(x)) - \frac{1}{2}\varphi^2(x) \leq 0.$$

Using the properties of Ψ, we conclude that $\rho(x - T^m(x))$ tends to zero itself, which contradicts our assumption (5.28). Hence, $\rho(x - T^m(x)) \to 0$ as $m \to \infty$. Clearly, then $\rho(x - T^{m+1}(x)) \to 0$ as $m \to \infty$, that is, $T^{m+1}(x) \to x(\rho)$ while $T^{m+1}(x) \to T(x)(\rho)$ by ρ-continuity of T. By the uniqueness of the ρ-limit, we obtain $T(x) = x$, that is, $x \in F(T)$, which completes the proof. □

Let us introduce our main iterative process for the construction of fixed points in modular function spaces.

Definition 5.20. Let $T \in \mathscr{T}_r(C)$ and let $\{n_k\}$ be an increasing sequence of natural numbers. Let $\{t_k\} \subset (0,1)$ be bounded away from 0 and 1. The *generalized Mann iteration process* generated by the mapping T, the sequence $\{t_k\}$, and the sequence $\{n_k\}$, denoted by $gM(T, \{t_k\}, \{n_k\})$ is defined by the following iterative formula:

$$x_{k+1} = t_k T^{n_k}(x_k) + (1 - t_k)x_k, \text{ where } x_1 \in C \text{ is chosen arbitrarily.}$$

Definition 5.21. We say that a generalized Mann iteration process $gM(T,\{t_k\},\{n_k\})$ is *well defined* if

$$\limsup_{k\to\infty} a_{n_k}(x_k) = 1. \tag{5.29}$$

Remark 5.17. Observe that by the definition of asymptotic pointwise nonexpansiveness, $\lim_{k\to\infty} a_k(x) = 1$ for every $x \in C$. Hence, we can always select a subsequence $\{a_{n_k}\}$ such that (5.29) holds. In other words, by a suitable choice of $\{n_k\}$ we can always make $gM(T,\{t_k\},\{n_k\})$ well defined.

Let us quote an elementary lemma about real numbers which will be used in this section, see, for example, [11].

Lemma 5.11. *Suppose $\{r_k\}$ is a bounded sequence of real numbers and $\{d_{k,n}\}$ is a doubly-index sequence of real numbers which satisfy*

$$\limsup_{k\to\infty}\limsup_{n\to\infty} d_{k,n} \le 0, \text{ and } r_{k+n} \le r_k + d_{k,n}$$

for each $k,n \ge 1$. Then $\{r_k\}$ converges to an $r \in \mathbb{R}$.

The following result provides an important technique which will be used in this paper.

Lemma 5.12. *Let $\rho \in \mathfrak{R}$ be $(UUC1)$. Let $C \subset L_\rho$ be a ρ-closed, ρ-bounded, and convex set. Let $T \in \mathscr{T}_r(C)$ and let $\{n_k\} \subset \mathbb{N}$. Assume that a sequence $\{t_k\} \subset (0,1)$ is bounded away from 0 and 1. Let w be a fixed point of T and $gM(T,\{t_k\},\{n_k\})$ be a generalized Mann process. Then, there exists $r \in \mathbb{R}$ such that $\lim_{k\to\infty} \rho(x_k - w) = r$.*

Proof. Let $w \in F(T)$. Since

$$\begin{aligned}
\rho(x_{k+1}-w) &\le t_k\rho(T^{n_k}(x_k)-w) + (1-t_k)\rho(x_k-w)\\
&= t_k\rho(T^{n_k}(x_k)-T^{n_k}(w)) + (1-t_k)\rho(x_k-w)\\
&\le t_k(1+b_{n_k}(w))\rho(x_k-w) + (1-t_k)\rho(x_k-w)\\
&= t_kb_{n_k}(w)\rho(x_k-w) + \rho(x_k-w)\\
&\le b_{n_k}(w)diam_\rho(C) + \rho(x_k-w),
\end{aligned}$$

it follows that for every $n \in \mathbb{N}$,

$$\rho(x_{k+n}-w) \le \rho(x_k-w) + diam_\rho(C)\sum_{i=k}^{k+n-1} b_{n_i}(w).$$

Denote $r_p = \rho(x_p - w)$ for every $p \in \mathbb{N}$ and $d_{k,n} = diam_\rho(C) \sum_{i=k}^{k+n-1} b_{n_i}(w)$. Observe that since $T \in \mathscr{T}_r(C)$, it follows that $\limsup_{k\to\infty}\limsup_{n\to\infty} d_{k,n} = 0$. By Lemma 5.11, there exists an $r \in \mathbb{R}$ such that $\lim_{k\to\infty} \rho(x_k - w) = r$ as claimed. □

The next result will be essential for proving the convergence theorems for iterative process.

Lemma 5.13. *Let $\rho \in \mathfrak{R}$ be $(UUC1)$. Let $C \subset L_\rho$ be a ρ-closed, ρ-bounded, and convex set, and $T \in \mathscr{T}_r(C)$. Assume that a sequence $\{t_k\} \subset (0,1)$ is bounded away from* 0 *and* 1. *Let $\{n_k\} \subset \mathbb{N}$ and $gM(T,\{t_k\},\{n_k\})$ be a generalized Mann iteration process. Then*

$$\lim_{k\to\infty} \rho(T^{n_k}(x_k) - x_k) = 0, \tag{5.30}$$

and

$$\lim_{k\to\infty} \rho(x_{k+1} - x_k) = 0. \tag{5.31}$$

Proof. By Theorem 5.12, T has at least one fixed point $w \in C$. In view of Lemma 5.12, there exists $r \in \mathbb{R}$ such that

$$\lim_{k\to\infty} \rho(x_k - w) = r. \tag{5.32}$$

Note that

$$\begin{aligned}\limsup_{k\to\infty} \rho(T^{n_k}(x_k) - w) &= \limsup_{k\to\infty} \rho(T^{n_k}(x_k) - T^{n_k}(w)) \\ &\le \limsup_{k\to\infty} a_{n_k}(w)\rho(x_k - w) \le r,\end{aligned} \tag{5.33}$$

and that

$$\lim_{k\to\infty} \rho(t_k(T^{n_k}(x_k) - w) + (1-t_k)(x_k - w)) = \lim_{k\to\infty} \rho(x_{k+1} - w) = r.$$

Set $f_k = T^{n_k}(x_k) - w$, $g_k = x_k - w$, and note that $\limsup_{k\to\infty} \rho(g_k) \le r$ by (5.32), and $\limsup_{k\to\infty} \rho(f_k) \le r$ by (5.33). Observe also that

$$\lim_{k\to\infty} \rho(t_k f_k + (1-t_k)g_k) = \lim_{k\to\infty} \rho(t_k T^{n_k}(x_k) + (1-t_k)x_k - w) = \lim_{k\to\infty} \rho(x_{k+1} - w) = r.$$

Hence, it follows from Lemma 5.2 that

$$\lim_{k\to\infty}\rho(T^{n_k}(x_k)-x_k)=\lim_{k\to\infty}\rho(f_k-g_k)=0,$$

which by the construction of the sequence $\{x_k\}$ is equivalent to

$$\lim_{k\to\infty}\rho(x_{k+1}-x_k)=0,$$

as claimed. □

In the next lemma, we prove that under suitable assumption the sequence $\{x_k\}$ becomes an approximate fixed point sequence, which will provide an important step in the proof of the generalized Mann iteration process convergence. First, we need to define the following notions.

Definition 5.22. A strictly increasing sequence $\{n_i\}\subset\mathbb{N}$ is called *quasi-periodic* if the sequence $\{n_{i+1}-n_i\}$ is bounded, or equivalently if there exists a number $p\in\mathbb{N}$ such that any block of p consecutive natural numbers must contain a term of the sequence $\{n_i\}$. The smallest of such numbers p will be called a *quasi-period* of $\{n_i\}$.

Lemma 5.14. *Let $\rho\in\Re$ be $(UUC1)$ satisfying Δ_2. Let $C\subset L_\rho$ be a ρ-closed, ρ-bounded, and convex set, and $T\in\mathcal{T}_r(C)$. Let $\{t_k\}\subset(0,1)$ be bounded away from 0 and 1. Let $\{n_k\}\subset\mathbb{N}$ be such that the generalized Mann process $gM(T,\{t_k\},\{n_k\})$ is well defined. If, in addition, the set of indices $\mathcal{J}=\{j:n_{j+1}=1+n_j\}$ is quasi-periodic, then $\{x_k\}$ is an approximate fixed point sequence, that is,*

$$\lim_{k\to\infty}\rho(T(x_k)-x_k)=0. \tag{5.34}$$

Proof. Let $p\in\mathbb{N}$ be a quasi-period of $\mathcal{J}$. Observe that it is enough to prove that $\rho(T(x_k)-x_k)\to 0$ as $k\to\infty$ through $\mathcal{J}$. Indeed, from this fact it follows that if we fix $\varepsilon>0$ then

$$\rho(T(x_k)-x_k)<\varepsilon$$

for sufficiently large $k\in\mathcal{J}$. By the quasi-periodicity of $\mathcal{J}$, to every positive integer k there exists $j_k\in\mathcal{J}$ such that $|k-j_k|\le p$. Assume that $k-p\le j_k\le k$ (the proof for the other case is identical). Since T is ρ-Lipschitzian with the Lipschitz constant $M=sup\{a_1(x):x\in C\}$, there exist a $0<\delta<\frac{\varepsilon}{3}$ such that

$$\rho(T(x)-T(y))<\varepsilon\ if\ \rho(x-y)<\delta.$$

Note that by (5.31) and by Δ_2, $\rho(p(x_{k+1}-x_k))<\frac{\delta}{p}$ for k sufficiently large. This implies that

$$\rho(x_k-x_{j_k})\le\frac{1}{p}\left(\rho(p(x_k-x_{k-1}))+\cdots+\rho(p(x_{j_k+1}-x_{j_k}))\right)\le p\frac{\delta}{p}=\delta,$$

and therefore,

$$\rho\left(\frac{x_k - T(x_k)}{3}\right) \le \frac{1}{3}\rho(x_k - x_{j_k}) + \frac{1}{3}\rho(x_{j_k} - T(x_{j_k})) + \frac{1}{3}\rho(T(x_{j_k}) - T(x_k))$$
$$\le \delta + \frac{\varepsilon}{3} + \frac{\varepsilon}{3} < \varepsilon,$$

which demonstrates that

$$\rho\left(\frac{x_k - T(x_k)}{3}\right) \to 0, \quad \text{as } k \to \infty.$$

By Δ_2 again, we get $\rho(T(x_k) - x_k) \to \infty$.

To prove that $\rho(T(x_k) - x_k) \to 0$ as $k \to \infty$ through $\mathscr{J}$, observe that, since $n_{k+1} = n_k + 1$ for such k, there holds

$$\rho\left(\frac{x_k - T(x_k)}{4}\right) \le \frac{1}{4}\rho(x_k - x_{k+1}) + \frac{1}{4}\rho(x_{k+1} - T^{n_{k+1}}(x_{k+1}))$$
$$+ \frac{1}{4}\rho(T^{n_{k+1}}(x_{k+1}) - T^{n_{k+1}}(x_k)) + \frac{1}{4}\rho(TT^{n_k}(x_k) - T(x_k))$$
$$\le \frac{1}{4}\rho(x_k - x_{k+1}) + \frac{1}{4}\rho(x_{k+1} - T^{n_{k+1}}(x_{k+1}))$$
$$+ \frac{1}{4}a_{n_{k+1}}(x_{k+1})\rho(x_k - x_{k+1}) + \frac{1}{4}M\rho(T^{n_k}(x_k) - x_k)$$

which tends to zero in view of (5.30), (5.31), and (5.29). □

We are now ready to prove our main convergence result, the proof follows [13].

Theorem 5.14. *Let $\rho \in \mathfrak{R}$. Assume that*

(a) ρ is $(UCC1)$,
(b) ρ has Strong Opial Property,
(c) ρ has Δ_2 property and is uniformly continuous.

Let $C \subset L_\rho$ be a nonempty, ρ-a.e. compact, convex, strongly ρ-bounded, and ρ-closed, and let $T \in \mathscr{T}_r(C)$. Assume that a sequence $\{t_k\} \subset (0,1)$ is bounded away from 0 and 1. Let $\{n_k\} \subset \mathbb{N}$ and $gM(T, \{t_k\}, \{n_k\})$ be a well-defined generalized Mann iteration process. Assume, in addition, that the set of indices

$$\mathscr{J} = \{j : n_{j+1} = 1 + n_j\}$$

is quasi-periodic. Then, there exists $x \in F(T)$ such that $x_n \to x$ ρ-a.e.

Proof. Observe that by Theorem 5.12 the set of fixed points $F(T)$ is nonempty and ρ-closed. Note also that by Lemma 5.9, it follows from the Strong Opial property of ρ that any ρ-type attains its minimum in C. By Lemma 5.14 the sequence $\{x_k\}$ is an approximate fixed point sequence, that is

$$\rho(T(x_k) - x_k) \to 0$$

as $k \to \infty$. Consider $y, z \in C$, two ρ-a.e. cluster points of $\{x_k\}$. There exits then $\{y_k\}$, $\{z_k\}$ subsequences of $\{x_k\}$ such that $y_k \to y$ ρ-a.e., and $z_k \to z$ ρ-a.e. By Theorem 5.13, $y \in F(T)$ and $z \in F(T)$. By Lemma 5.12, there exist $r_y, r_z \in \mathbb{R}$ such that

$$r_y = \lim_{k \to} \rho(x_k - y), \; r_z = \lim_{k \to} \rho(x_k - z).$$

We claim that $y = z$. Assume to the contrary that $y \neq z$. Then by the Strong Opial property we have

$$\begin{aligned} r_y &= \liminf_{k \to \infty} \rho(y_k - y) < \liminf_{k \to \infty} \rho(y_k - z) \\ &= \liminf_{k \to \infty} \rho(z_k - z) < \liminf_{k \to \infty} \rho(z_k - y) = r_y. \end{aligned}$$

The contradiction implies that $y = z$. Therefore, $\{x_k\}$ has at most one ρ-a.e. cluster point. Since, C is ρ-a.e. compact it follows that the sequence $\{x_k\}$ has exactly one ρ-a.e. cluster point, which means that $\rho(x_k) \to x$ ρ-a.e. Using Theorem 5.13 again, we get $x \in F(T)$ as claimed. □

Remark 5.18. It is easy to see that we can always construct a sequence $\{n_k\}$ with the quasi-periodic properties specified in the assumptions of Theorem 5.14. When constructing concrete implementations of this algorithm, the difficulty will be to ensure that the constructed sequence $\{n_k\}$ is not "too sparse" in the sense that the generalized Mann process $gM(T, \{t_k\}, \{n_k\})$ remains well defined. The similar, quasi-periodic type assumptions are common in the asymptotic fixed point theory, see, for example, [11, 48, 54].

The two-step Ishikawa iteration process is a generalization of the one-step Mann process. The Ishikawa iteration process provides more flexibility in defining the algorithm parameters which is important from the numerical implementation perspective.

Definition 5.23. Let $T \in \mathscr{T}_r(C)$ and let $\{n_k\}$ be an increasing sequence of natural numbers. Let $\{t_k\} \subset (0,1)$ be bounded away from 0 and 1, and $\{s_k\} \subset (0,1)$ be bounded away from 1. The *generalized Ishikawa iteration process* generated by the mapping T, the sequences $\{t_k\}$, $\{s_k\}$, and the sequence $\{n_k\}$, denoted by $gI(T, \{t_k\}, \{s_k\}, \{n_k\})$ is defined by the following iterative formula:

$$x_{k+1} = t_k T^{n_k}(s_k T^{n_k}(x_k) + (1 - s_k)x_k) + (1 - t_k)x_k,$$

where $x_1 \in C$ is chosen arbitrarily.

Definition 5.24. We say that a generalized Ishikawa iteration process

$$gI(T, \{t_k\}, \{s_k\}, \{n_k\})$$

is *well defined* if

$$\limsup_{k\to\infty} a_{n_k}(x_k) = 1. \tag{5.35}$$

Remark 5.19. Observe that, by the definition of asymptotic pointwise nonexpansiveness, $\lim_{k\to\infty} a_k(x) = 1$ for every $x \in C$. Hence, we can always select a subsequence $\{a_{n_k}\}$ such that (5.35) holds. In other words, by a suitable choice of $\{n_k\}$ we can always make $gI(T, \{t_k\}, \{s_k\}, \{n_k\})$ well defined.

Lemma 5.15. *Let $\rho \in \Re$ be $(UUC1)$. Let $C \subset L_\rho$ be a ρ-closed, ρ-bounded, and convex set. Let $T \in \mathscr{T}_r(C)$ and let $\{n_k\} \subset \mathbb{N}$. Let $\{t_k\} \subset (0,1)$ be bounded away from 0 and 1, and $\{s_k\} \subset (0,1)$ be bounded away from 1. Let $w \in F(T)$ and $gI(T, \{t_k\}, \{s_k\}, \{n_k\})$ be a generalized Ishikawa process. There exists then an $r \in \mathbb{R}$ such that $\lim_{k\to\infty} \rho(x_k - w) = r$.*

Proof. Define $T_k : C \to C$ by

$$T_k(x) = t_k T^{n_k}(s_k T^{n_k}(x) + (1-s_k)x) + (1-t_k)x,\ x \in C.$$

It is easy to see that $x_{k+1} = T_k(x_k)$ and that $F(T) \subset F(T_k)$. Moreover, a straight calculation shows that each T_k satisfies

$$\rho(T_k(x) - T_k(y)) \le A_k(x)\rho(x-y),$$

where

$$A_k(x) = 1 + t_k a_{n_k}(M_k(x))(1 + s_k a_{n_k}(x) - s_k) - t_k, \tag{5.36}$$

and

$$M_k(x) = s_k T^{n_k}(x) + (1-s_k)x.$$

Note that $A_k(x) \ge 1$ which follows directly from the fact that $a_{n_k}(x) \ge 1$ and from (5.36). Using (5.36) and the fact that $M_k(w) = w$ we have

$$B_k(w) = A_k(w) - 1 = t_k(1 + s_k a_{n_k}(w))(a_{n_k}(w) - 1) \le (1 + a_{n_k}(w))b_{n_k}(w).$$

Fix any $M > 1$. Since $\lim_{k\to\infty} a_{n_k}(w) = 1$, it follows that there exists a $k_0 \ge 1$ such that for $k > k_0$, $a_{n_k}(w) \le M$. Therefore, by the same argument as in the proof of Lemma 5.12, we deduce that for $k > k_0$ and $n > 1$

$$\rho(x_{k+n}-w) \le \rho(x_k-w) + diam_\rho(C) \sum_{i=k}^{k+n-1} B_i(w)$$

$$\le \rho(x_k-w) + diam_\rho(C)(1+M) \sum_{i=k}^{k+n-1} b_{n_i}(w).$$

Arguing like in the proof of Lemma 5.12, we conclude that there exists an $r \in \mathbb{R}$ such that $\lim_{k\to\infty} \rho(x_k - w) = r$. □

Lemma 5.16. *Let $\rho \in \Re$ be (UUC1). Let $C \subset L_\rho$ be a ρ-closed, ρ-bounded, and convex set. Let $T \in \mathscr{T}_r(C)$ and let $\{n_k\} \subset \mathbb{N}$. Let $\{t_k\} \subset (0,1)$ be bounded away from 0 and 1, and $\{s_k\} \subset (0,1)$ be bounded away from 1. Let $gI(T,\{t_k\},\{s_k\},\{n_k\})$ be a generalized Ishikawa process.Define $y_k = s_k T^{n_k}(x_k) + (1-s_k)x_k$. Then*

$$\lim_{k\to\infty} \rho(T^{n_k}(y_k) - x_k) = 0, \tag{5.37}$$

or equivalently

$$\lim_{k\to\infty} \rho(x_{k+1} - x_k) = 0, \tag{5.38}$$

Proof. By Theorem 5.12, $F(T) \neq \emptyset$. Let us fix $w \in F(T)$. By Lemma 5.15, the limit $\lim_{k\to\infty} \rho(x_k - w)$ exists. Let us denote it by r. Since $w \in F(T)$, $T \in \mathscr{T}_r(C)$, and $\lim_{k\to\infty} \rho(x_k - w) = r$, by Lemma 5.15 we have the following

$$\limsup_{k\to\infty} \rho(T^{n_k}(y_k) - w) = \limsup_{k\to\infty} \rho(T^{n_k}(y_k) - T^{n_k}(w))$$

$$\le \limsup_{k\to\infty} a_{n_k}(w)\rho(y_k - w) = \limsup_{k\to\infty} a_{n_k}(w)\rho(s_k T^{n_k}(x_k) + (1-s_k)x_k - w)$$

$$\le \limsup_{k\to\infty} \left(s_k a_{n_k}(w)\rho(T^{n_k}(x_k) - w) + (1-s_k)a_{n_k}(w)\rho(x_k - w)\right)$$

$$\le \limsup_{k\to\infty} \left(s_k a_{n_k}^2(w)\rho(x_k - w) + (1-s_k)a_{n_k}(w)\rho(x_k - w)\right) \le r.$$

Note that

$$\lim_{k\to\infty} \rho(t_k(T^{n_k}(y_k) - w) + (1-t_k)(x_k - w))$$

$$= \lim_{k\to\infty} \rho(t_k T^{n_k}(y_k) + (1-t_k)x_k - w) = \lim_{k\to\infty} \rho(x_{k+1} - w) = r.$$

Applying Lemma 5.2 with $u_k = T^{n_k}(y_k) - w$ and $v_k = x_k - w$, we obtain the desired equality $\lim_{k\to\infty} \rho(T^{n_k}(y_k) - x_k) = 0$, while (5.38) follows from (5.37) via the construction formulas for x_{k+1} and y_k. □

Lemma 5.17. *Let $\rho \in \mathfrak{R}$ be $(UUC1)$ satisfying Δ_2. Let $C \subset L_\rho$ be a ρ-closed, ρ-bounded, and convex set. Let $T \in \mathcal{T}_r(C)$ and let $\{n_k\} \subset \mathbb{N}$. Let $\{t_k\} \subset (0,1)$ be bounded away from 0 and 1, and $\{s_k\} \subset (0,1)$ be bounded away from 1. Let $gI(T,\{t_k\},\{s_k\},\{n_k\})$ be a well-defined generalized Ishikawa process. Then*

$$\lim_{k\to\infty} \rho(T^{n_k}(x_k) - x_k) = 0. \tag{5.39}$$

Proof. Let $y_k = s_k T^{n_k}(x_k) + (1-s_k)x_k$. Hence

$$T^{n_k}(x_k) - x_k = \frac{1}{1-s_k}(T^{n_k}(x_k) - y_k).$$

Since $\{s_k\} \subset (0,1)$ is bounded away from 1, there exists $0 < s < 1$ such that $s_k \le s$ for every $k \ge 1$. Hence,

$$\rho(T^{n_k}(x_k) - x_k) = \rho\Big(\frac{1}{1-s_k}(T^{n_k}(x_k) - y_k)\Big) \le \rho\Big(\frac{1}{1-s}(T^{n_k}(x_k) - y_k)\Big).$$

The right-hand side of this inequality tends to zero because $\rho(T^{n_k}(x_k) - y_k) \to 0$ by Lemma 5.16 and the fact that ρ satisfies Δ_2. □

Lemma 5.18. *Let $\rho \in \mathfrak{R}$ be $(UUC1)$ satisfying Δ_2. Let $C \subset L_\rho$ be a ρ-closed, ρ-bounded and convex set, and $T \in \mathcal{T}_r(C)$. Let $\{t_k\} \subset (0,1)$ be bounded away from 0 and 1 and $\{s_k\} \subset (0,1)$ be bounded away from 1. Let $\{n_k\} \subset \mathbb{N}$ be such that the generalized Ishikawa process $gI(T,\{t_k\},\{s_k\},\{n_k\})$ is well defined. If, in addition, the set $\mathcal{J} = \{j : n_{j+1} = 1 + n_j\}$ is quasi-periodic, then $\{x_k\}$ is an approximate fixed point sequence, i.e.,*

$$\lim_{k\to\infty} \rho(T(x_k) - x_k) = 0.$$

Proof. The proof is analogous to that of Lemma 5.14 with (5.38) used instead of (5.31) and (5.39) replacing (5.30). □

Theorem 5.15. *Let $\rho \in \mathfrak{R}$. Assume that*

(a) ρ is $(UCC1)$,
(b) ρ has Strong Opial Property,
(c) ρ has Δ_2 property and is uniformly continuous.

Let $C \subset L_\rho$ be a nonempty, ρ-a.e. compact, convex, strongly ρ-bounded, and ρ-closed and let $T \in \mathcal{T}_r(C)$. Let $T \in \mathcal{T}_r(C)$. Let $\{t_k\} \subset (0,1)$ be bounded away from 0 and 1, and $\{s_k\} \subset (0,1)$ be bounded away from 1. Let $\{n_k\}$ be such that the generalized Ishikawa process $gI(T,\{t_k\},\{s_k\},\{n_k\})$ is well defined. If, in addition, the set $\mathcal{J} = \{j : n_{j+1} = 1 + n_j\}$ is quasi-periodic, then $\{x_k\}$ generated by $gI(T,\{t_k\},\{s_k\},\{n_k\})$ converges ρ-a.e. to a fixed point $x \in F(T)$.

Proof. The proof is analogous to that of Theorem 5.14 with Lemma 5.14 replaced by Lemma 5.18, and Lemma 5.12 replaced by Lemma 5.15. □

It is interesting that, provided C is ρ-compact, both generalized Mann and Ishikawa processes converge strongly to a fixed point of T even without assuming the Opial property or any quasi-periodicity assumptions. The next result, following [13], proves exactly this fact.

Theorem 5.16. *Let $\rho \in \Re$ satisfy conditions $(UUC1)$ and Δ_2. Let $C \subset L_\rho$ be a ρ-compact, ρ-bounded, and convex set and let $T \in \mathscr{T}_r(C)$. Let $\{t_k\} \subset (0,1)$ be bounded away from 0 and 1, and $\{s_k\} \subset (0,1)$ be bounded away from 1. Let $\{n_k\}$ be such that the generalized Mann process $gM(T,\{t_k\},\{n_k\})$ (respectively, Ishikawa process $gI(T,\{t_k\},\{s_k\},\{n_k\})$) is well defined. Then, there exists a fixed point $x \in F(T)$ such that then $\{x_k\}$ generated by $gM(T,\{t_k\},\{n_k\})$ (respectively, $gI(T,\{t_k\},\{s_k\},\{n_k\})$) converges strongly to a fixed point of T, that is, $\lim_{k\to\infty} \rho(x_k - x) = 0$.*

Proof. By the ρ-compactness of C we can select a subsequence $\{x_{p_k}\}$ of $\{x_k\}$ such that there exists $x \in C$ with

$$\lim_{k\to\infty} \rho(T(x_{p_k}) - x) = 0. \tag{5.40}$$

Note that

$$\rho\left(\frac{x_{p_k} - x}{2}\right) \le \frac{1}{2}\rho(x_{p_k} - T(x_{p_k})) + \frac{1}{2}\rho(T(x_{p_k} - x), \tag{5.41}$$

which tends to zero by Lemma 5.13 (resp. Lemma 5.18) and by (5.40). By Δ_2 if follows from (5.41) that

$$\rho(x_{p_k} - x) \to 0 \; as \; k \to \infty. \tag{5.42}$$

Observe that by the convexity of ρ and by ρ-nonexpansiveness of T, we have

$$\begin{aligned}
\rho\left(\frac{T(x) - x}{3}\right) &\le \frac{1}{3}\rho(T(x) - T(x_{p_k})) + \frac{1}{3}\rho(T(x_{p_k}) - x_{p_k}) + \frac{1}{3}\rho(x_{p_k} - x) \\
&\le \frac{1}{3}\rho(x - x_{p_k}) + \frac{1}{3}\rho(T(x_{p_k}) - x_{p_k}) + \frac{1}{3}\rho(x_{p_k} - x),
\end{aligned}$$

which tends to zero by (5.42) and by Lemma 5.13 (resp. Lemma 5.18). From there we get $\rho(T(x) - x) = 0$ which implies that $x \in F(T)$. Applying Lemma 5.12 (resp. Lemma 5.15) we conclude that $\lim_{k\to\infty} \rho(x_k - x)$ exists. By (5.42) this limit must be equal to zero which implies that $\lim_{k\to\infty} \rho(x_k - x) = 0$, as claimed. □

Remark 5.20. Observe that in view of the Δ_2 assumption, the ρ-compactness of the set C assumed in Theorem 5.16 is equivalent to the compactness in the sense of the Luxemburg norm defined by ρ.

5.5 Semigroups of Mappings in Modular Function Spaces

Let us recall that a family $\{T_t\}_{t\geq 0}$ of mappings forms a semigroup if $T_0(x) = x$, and $T_{s+t} = T_s(T_t(x))$. Such a situation is quite typical in mathematics and applications. For instance, in the theory of dynamical systems, the modular function space L_ρ would define the state space and the mapping $(t,x) \to T_t(x)$ would represent the evolution function of a dynamical system. The question about the existence of common fixed points, and about the structure of the set of common fixed points, can be interpreted as a question whether there exist points that are fixed during the state space transformation T_t at any given point of time t, and if yes—what the structure of a set of such points may look like. In the setting of this chapter, the state space may be an infinite dimensional. Therefore, it is natural to apply these results not only to deterministic dynamical systems but also to stochastic dynamical systems.

Let us start with the modular definitions of Lipschitzian—in the modular sense—mappings, and of associated definitions of semigroups of nonlinear mappings.

Definition 5.25. Let $\rho \in \Re$ and let $C \subset L_\rho$ be nonempty and ρ-closed. A mapping $T: C \to C$ is called a *ρ-Lipschitzian* if there exists a constant $0 < L$ such that

$$\rho(T(f) - T(g)) \leq L\rho(f - g), \quad \text{for any } f,g \in L_\rho.$$

Definition 5.26. A one-parameter family $\mathscr{F} = \{T_t; t \geq 0\}$ of mappings from C into itself is said to be a *ρ-Lipschitzian (respectively, ρ-nonexpansive) semigroup* on C if $\mathscr{F}$ satisfies the following conditions:

(a) $T_0(x) = x$ for $x \in C$;
(b) $T_{t+s}(x) = T_t(T_s(x))$ for $x \in C$ and $t,s \geq 0$;
(c) for each $t \geq 0$, T_t is ρ-Lipschitzian (respectively, ρ-nonexpansive).

Definition 5.27. A one-parameter family $\mathscr{F} = \{T_t; t \geq 0\}$ of mappings from C into itself is said to be a *ρ-contractive semigroup* on C if $\mathscr{F}$ satisfies the following conditions:

(a) $T_0(x) = x$ for $x \in C$;
(b) $T_{t+s}(x) = T_t(T_s(x))$ for $x \in C$ and $t,s \geq 0$;
(c) for each $t \geq 0$, T_t is a ρ-contraction with a constant $0 < L_t < 1$ such that $\limsup_{t\to\infty} L_t < 1$.

Following [50], we will prove the existence of common fixed points for contractive and nonexpansive semigroups.

Theorem 5.17. *Let $\rho \in \Re$. Assume that L_ρ has the ρ-a.e. Strong Opial property. Let $C \subset E_\rho$ be a nonempty, ρ-a.e. compact convex subset such that $\delta_\rho(\beta\ C) = \sup\{\rho(\beta(x-y)); x,y \in C\} < \infty$, for some $\beta > 1$. Let $\mathscr{F}$ be a ρ-contractive semigroup on C. Then, $\mathscr{F}$ has a unique common fixed point $z \in C$ and for each $u \in C$, $\rho(T_t(u) - z) \to 0$ as $t \to \infty$.*

Proof. First, let us prove the uniqueness. Assume that $z, w \in \mathscr{F}$. Then we have

$$\rho(z-w) = \rho(T_t(z) - T_t(w)) \leq L_t \rho(z-w)$$

for any $t \geq 0$. If we let $t \to \infty$, we will get

$$\rho(z-w) \leq L\rho(z-w),$$

where

$$L = \limsup_{t\to\infty} L_t.$$

Since $L < 1$, we conclude that $\rho(z-w) = 0$ or equivalently that $z = w$, hence there must exist at most one common fixed point for $\mathscr{F}$.

To prove the existence of the common fixed point, let us fix any $x \in C$ and define the ρ-type τ by

$$\tau(u) = \limsup_{t\to\infty} \rho(T_t(x) - u) \tag{5.43}$$

for $u \in C$. By Lemma 5.9, there exists an $z \in C$ such that

$$\tau(z) = \inf\{\tau(y) : y \in C\}.$$

Let us prove that $\tau(z) = 0$. To this end, take $s, t \geq 0$ and observe that by the nonexpansiveness of T_t we have

$$\rho(T_{s+t}(x) - T_t(z)) \leq L_t \rho(T_s(x) - z),$$

and then letting $s \to \infty$

$$\tau(T_t(z)) \leq L_t \tau(z)$$

which implies the following

$$\tau(z) = \inf\{\tau(y) : y \in C\} \leq \tau(T_t(z)) \leq L_t \; \tau(z).$$

Passing with $t \to \infty$ we get

$$\tau(z) \leq L\tau(z).$$

Since $0 < L < 1$ it follows that $\tau(z) = 0$. Hence, by (5.43),

$$\lim_{t\to\infty} \rho(T_t(x) - z) = 0. \tag{5.44}$$

Since T_s is a ρ-contraction for any given $s \geq 0$, it follows that

$$\lim_{t\to\infty} \rho(T_t(x) - T_s(z)) = \lim_{t\to\infty} \rho(T_{s+t}(x) - T_s(z)) \leq \lim_{t\to\infty} \rho(T_t(x) - z) = 0. \qquad (5.45)$$

By the uniqueness of the ρ-limit, we conclude from (5.44) and (5.45) that $T_s(z) = z$ for any $s \geq 0$, i.e $z \in \mathscr{F}$.

To prove the convergence of orbits, let us fix any $u \in C$. We shall prove that $\{T_t(u)\}$ converges to z. Indeed we have

$$\rho(T_{t+s}(u) - z) = \rho(T_{t+s}(u) - T_t(z)) \leq L_t \rho(T_s(u) - z),$$

for any $t, s \geq 0$. Hence

$$\limsup_{s\to\infty} \rho(T_{t+s}(u) - z) \leq \limsup_{s\to\infty} L_t\ \rho(T_s(u) - z).$$

Since $\limsup_{s\to\infty} \rho(T_{t+s}(u) - z) = \limsup_{t\to\infty} \rho(T_s(u) - z)$, we get

$$\limsup_{s\to\infty} \rho(T_s(u) - z) \leq L_t \limsup_{s\to\infty} \rho(T_s(u) - z),$$

for any $t \geq 0$. If we let $t \to \infty$, we obtain

$$\limsup_{s\to\infty} \rho(T_s(u) - z) \leq L \limsup_{s\to\infty} \rho(T_s(u) - z).$$

Since $L < 1$, we get

$$\limsup_{s\to\infty} \rho(T_s(u) - z) = 0.$$

Clearly we can derive the same equality where lim sup is replaced by lim inf which implies the desired conclusion: $\lim_{s\to\infty} \rho(T_s(u) - z) = 0$. □

Theorem 5.18. *Assume $\rho \in \mathfrak{R}$ is (UUC1). Let C be a ρ-closed ρ-bounded convex nonempty subset. Let $\mathscr{F}$ be a nonexpansive semigroup on C. Then, the set $F(\mathscr{F})$ of common fixed points is nonempty, ρ-closed and convex.*

Proof. Let us fix an $x \in C$ and define the ρ-type

$$\tau(y) = \limsup_{t\to\infty} \rho(T_t(x) - y),$$

where $y \in C$. Let $\tau_0 = \inf\{\tau(y) : y \in C\}$ and let $\{z_n\}$ be a minimizing sequence for τ, i.e.

$$\lim_{n\to\infty} \tau(z_n) = \tau_0.$$

By Lemma 5.6 there exists a $z \in C$ such that

$$\lim_{n\to\infty} \rho(z_n - z) = 0.$$

We are going to prove that $z \in F(\mathscr{F})$. Noting that

$$\rho(T_{s+t}(x) - T_t(y)) \leq \rho(T_s(x) - t)$$

and passing with $s \to \infty$ we get

$$\tau(T_t(y)) \leq \tau(y).$$

In particular, for any $n \geq 1$ we have

$$\tau(T_t(z_n)) \leq \tau(z_n). \tag{5.46}$$

Let us fix any sequence $t_k \to \infty$ and note that for every $s > 0$ the sequence $\{T_{t_k+s}(z_k)\}$ is a minimizing sequence for τ. Indeed, using (5.46) we obtain

$$\tau_0 \leq \tau(T_{t_k+s}(z_k)) \leq \tau(z_k) \to \tau_0.$$

Hence, Lemma 5.6 implies that

$$\lim_{k\to\infty} \rho(T_{t_k+s}(z_k) - z) = 0, \tag{5.47}$$

and in particular for $s = 0$,

$$\lim_{k\to\infty} \rho(T_{t_k}(z_k) - z) = 0. \tag{5.48}$$

Using (5.48) we get

$$\rho(T_{t_k+s}(z_k) - T_s(z)) \leq \rho(T_{t_k}(z_k) - z) \to 0.$$

Since the ρ-limit is unique, (5.47) and (5.5) give $T_s(z) = z$, i.e. $z \in F(\mathscr{F})$ as claimed. Let us prove that $F(\mathscr{F})$ is ρ-closed. Let $x_n \in F(\mathscr{F})$ and $\rho(x_n - x) \to 0$. Observe that for every $t \geq 0$,

$$\begin{aligned}\rho\left(\frac{1}{3}(T_t(x) - x)\right) &\leq \rho(T_t(x) - T_t(x_n)) + \rho(T_t(x_n) - x_n) + \rho(x_n - x) \\ &\leq \rho(x_n - x) + \rho(x_n - x) \to 0.\end{aligned}$$

Hence, $x \in F(\mathscr{F})$ proving that $x \in F(\mathscr{F})$, and consequently, that $F(\mathscr{F})$ is ρ-closed.

To prove convexity of $F(\mathscr{F})$, we need to demonstrate that

$$w = \frac{u+v}{2} \in F(\mathscr{F})$$

provided $u, v \in F(\mathscr{F})$. Indeed, let $t > 0$. Define $x = T_t(w) - u$, $y = T_t(w) - v$. Note that

$$\frac{x+y}{2} = T_t(w) - w.$$

Define

$$r = \rho\left(\frac{v-u}{2}\right)$$

and observe that

$$\rho(x) = \rho(T_t(w) - u) = \rho(T_t(w) - T_t(u)) \leq \rho(w-u) = \rho\left(\frac{v-u}{2}\right) = r.$$

Similarly, $\rho(y) \leq r$. Hence, $x, y \in D_2(r,1)$ and therefore

$$\delta_2(r,1) \leq 1 - \frac{1}{r}\rho\left(\frac{x+y}{2}\right) = 1 - \frac{1}{r}\rho(T_t(w) - w).$$

Using the assumed $(UUC1)$ and Proposition 5.8, we conclude that ρ satisfies $(UC2)$. Hence, by Lemma 5.4, $\delta_2(r,1) = 1$, which yields

$$\frac{1}{r}\rho(T_t(w) - w) \leq 1 - \delta_2(r,1) = 0.$$

Therefore, $T_t(w) = w$, i.e. $w \in F(\mathscr{F})$, completing the proof. □

Remark 5.21. Recently, Al-Mezel, Al-Roqi, and Khamsi provided in [2] a partial generalization of the second theorem showing, under suitable assumptions, the existence of common fixed points of a commutative family of ρ-nonexpansive mappings defined on a ρ-closed ρ-bounded convex nonempty subset of L_ρ.

One can ask a legitimate question about existence of natural examples of semigroups of nonlinear mappings in modular function spaces and their applications. Let us address this issue by providing a suitable generic example.

Example 5.14. In [29] Khamsi considered the following initial value problem for an unknown function $u : [0,A] \to L_\rho$, where $C \subset L_\rho$ be ρ-closed, ρ-bounded, and convex subset of a Musielak–Orlicz space L_ρ:

$$\begin{cases} u(0) = f, \\ u'(t) + (I - T)u(t) = 0, \end{cases} \tag{5.49}$$

where $f \in C$ and $A > 0$ were fixed. If we assume in addition Δ_2, it is not difficult to show that there exists a solution u_f to (5.49), $u_f(t) \in C$ for every $t \in [0, A]$ and the solution $u_f(t)$ can be obtained as the ρ-limit of $\{u_n(t)\}$ where u_n are defined by the following recurrent sequence:

$$\begin{cases} u_0(t) = f, \\ u_{n+1} = e^{-t} f + \int_0^t e^{s-t} T(u_n(s)) ds. \end{cases}$$

Let us define

$$S_t(f) = u_f.$$

It can be proved that $\{S_t\}$ forms a ρ-nonexpansive semigroup of nonlinear mappings in the sense of Definition 5.26. Hence, if in addition ρ is $(UCC1)$, it follows from Theorem 5.18 that the set of common fixed points for $\{S_t\}$ is nonempty. To interpret this fact, observe that if f_0 is such a common fixed point and we place the initial value of our system (5.49) at f_0 then this point becomes a stationary point of the system, which means that the constant function $u_{f_0}(t) = f_0$ for every t is the solution of (5.49).

These results can be extended to systems where T is a ρ-Lipschitz operator [71] and applied to the perturbed integral equations in modular function spaces [24].

There exists an extensive literature on the question of representation of some types of semigroups of nonlinear mappings acting in Banach spaces, see, for example, [19,28,65,67,68]. It would be interesting to consider similar representation questions in modular function spaces.

Similarly, it would be interesting to discuss the modular ergodic theory for nonlinear semigroups defined in modular function spaces. For the Banach space results of this type, see, for example, [60, 69, 72].

References

1. Akimovic, B.A.: On uniformly convex and uniformly smooth Orlicz spaces. Teor. Funkc. Funkcional. Anal. i Prilozen. **15**, 114–220 (1972)
2. Al-Mezel, S.A., Al-Roqi, A., Khamsi, M.A.: One-local retract and common fixed point in modular function spaces. Fixed Point Theory Appl. **2012**(109), (2012)
3. Alspach, D.: A fixed point free nonexpansive mapping. Proc. Am. Math. Soc. **82**, 423–424 (1981)
4. Baillon, J.B.: Nonexpansive mappings and hyperconvex spaces. Contemp. Math. **72**, 11–19 (1988)

5. Beg, I.: Inequalities in metric spaces with applications. Topological Meth. Nonlinear Anal. **17**, 183–190 (2001)
6. Belluce, L.P., Kirk, W.A.: Fixed-point theorems for families of contraction mappings. Pac. J. Math. **18**, 213–217 (1966)
7. Belluce, L.P., Kirk, W.A.: Nonexpansive mappings and fixed-points in Banach spaces. Illinois J. Math. **11**, 474–479 (1967)
8. Birnbaum, Z., Orlicz, W.: Uber die Verallgemeinerung des Begriffes der zueinander konjugierten Potenzen. Studia Math. **3**, 1–67 (1931)
9. Browder, F.E.: Nonexpansive nonlinear operators in a Banach space. Proc. Natl. Acad. Sci. USA **54**, 1041–1044 (1965)
10. Bruck, R.E.: A common fixed point theorem for a commuting family of nonexpansive mappings. Pac. J. Math. **53**, 59–71 (1974)
11. Bruck, R., Kuczumow, T., Reich, S.: Convergence of iterates of asymptotically nonexpansive mappings in Banach spaces with the uniform Opial property. Colloq. Math. **65**, 169–179 (1993)
12. Chen, S.: Geometry of Orlicz spaces. Dissertation Math. **356**, 4–205 (1996)
13. Dehaish, B.A., Kozlowski, W.M.: Fixed point iterations processes for asymptotic pointwise nonexpansive mappings in modular function spaces. Fixed Point Theory Appl. **2012**(118) (2012)
14. DeMarr, R.E.: Common fixed-points for commuting contraction mappings. Pac. J. Math. **13**, 1139–1141 (1963)
15. Dhompongsa, S., Kirk, W.A., Sims, B.: Fixed points of uniformly lipschitzian mappings. Nonlinear Anal. **65**, 762–772 (2006)
16. Dominguez-Benavides, T., Khamsi, M.A., Samadi, S.: Uniformly Lipschitzian mappings in modular function spaces. Nonlinear Anal. **46**, 267–278 (2001)
17. Dominguez-Benavides, T., Khamsi, M.A., Samadi, S.: Asymptotically regular mappings in modular function spaces. Scientiae Math. Japonicae **53**, 295–304 (2001)
18. Dominguez-Benavides, T., Khamsi, M.A., Samadi, S.: Asymptotically nonexpansive mappings in modular function spaces. J. Math. Anal. Appl. **265**(2), 249–263 (2002)
19. Dorroch, J.R., Neuberger, J.W.: Linear extensions of nonlinear semigroups. In: Balakrishnan, A.V. (ed.) Semigroups and Operators: Theory and Applications, pp. 96–102. Birkhauser, Basel (2000)
20. Goebel, K., Kirk, W.A.: A fixed points theorem for asymptotically nonexpansive mappings. Proc. Am. Math. Soc. **35**, 171–174 (1972)
21. Goebel, K., Reich, S.: Uniform Convexity, Hyperbolic Geometry, and Nonexpansive Mappings. Series of Monographs and Textbooks in Pure and Applied Mathematics, vol. 83. Marcel Dekker, New York (1984)
22. Goebel, K., Sekowski, T., Stachura, A.: Uniform convexity of the hyperbolic metric and fixed points of holomorphic mappings in the Hilbert ball. Nonlinear Anal. **4**, 1011–1021 (1980)
23. Gohde, D.: Zum Prinzip der kontraktiven Abbildung. Math. Nachr. **30**, 251–258 (1965)
24. Hajji, A., Hanebaly, E.: Perturbed integral equations in modular function spaces. Electron. J. Qual. Theor. Differ. Equat. **20**, 1–7 (2003)
25. Hussain, N., Khamsi, M.A.: On asymptotic pointwise contractions in metric spaces. Nonlinear Anal. **71**, 4423–4429 (2009)
26. Ishikawa, S.: Fixed points by a new iteration method. Proc. Am. Math. Soc. **44**, 147–150 (1974)
27. Kaminska, A.: On uniform convexity of Orlicz spaces. Indag. Math. **44**, 27–36 (1982)
28. Kato, T.: Nonlinear semigroups and evolution equations. J. Math. Soc. Jpn **19**, 508–520 (1967)
29. Khamsi, M.A.: Nonlinear Semigroups in modular function spaces. Math. Japonica **37**, 1–9 (1992)
30. Khamsi, M.A.: Fixed point theory in modular function spaces. In: Proceedings of the Workshop on Recent Advances on Metric Fixed Point Theory held in Sevilla, pp. 31–35, Spain, September 1995
31. Khamsi, M.A.: A convexity property in modular function spaces. Math. Japonica **44**, 269–279 (1996)

32. Khamsi, M.A.: On asymptotically nonexpansive mappings in hyperconvex metric spaces. Proc. Am. Math. Soc. **132**, 365–373 (2004)
33. Khamsi, M.A., Kirk, W.A.: An Introduction to Metric Spaces and Fixed Point Theory. Wiley, New York (2001)
34. Khamsi, M.A., Kozlowski, W.M.: On asymptotic pointwise contractions in modular function spaces. Nonlinear Anal. **73**, 2957–2967 (2010)
35. Khamsi, M.A., Kozlowski, W.M.: On asymptotic pointwise nonexpansive mappings in modular function spaces. J. Math. Anal. Appl. **380**(2), 697–708 (2011)
36. Khamsi, M.A., Kozlowski, W.M., Reich, S.: Fixed point theory in modular function spaces. Nonlinear Anal. **14**, 935–953 (1990)
37. Khamsi, M.A., Kozlowski, W.M., Chen, S.: Some geometrical properties and fixed point theorems in Orlicz spaces. J. Math. Anal. Appl. **155**, 393–412 (1991)
38. Kilmer, S.J., Kozlowski, W.M., Lewicki, G.: Best approximants in modular function spaces. J. Approx. Theor. **63**, 338–367 (1990)
39. Kirk, W.A.: A fixed point theorem for mappings which do not increase distance. Am. Math. Mon. **72**, 1004–1006 (1965)
40. Kirk, W.A.: Mappings of generalized contractive type. J. Math. Anal. Appl. **32**, 567–572 (1970)
41. Kirk, W.A.: Fixed point theory for nonexpansive mappings, I and II. Lecture Notes in Mathematics, vol. 886, pp. 485–505. Springer, Berlin (1981)
42. Kirk, W.A.: Fixed points of asymptotic contractions. J. Math. Anal. Appl. **277**, 645–650 (2003)
43. Kirk, W.A.: Asymptotic pointwise contractions. In: Plenary Lecture, the 8th International Conference on Fixed Point Theory and Its Applications, Chiang Mai University, Thailand, 16–22 July 2007
44. Kirk, W.A., Xu, H.K.: Asymptotic pointwise contractions. Nonlinear Anal. **69**, 4706–4712 (2008)
45. Kozlowski, W.M.: Notes on modular function spaces I. Comment. Math. **28**, 91–104 (1988)
46. Kozlowski, W.M.: Notes on modular function spaces II. Comment. Math. **28**, 105–120 (1988)
47. Kozlowski, W.M.: Modular Function Spaces Series of Monographs and Textbooks in Pure and Applied Mathematics, vol. 122. Marcel Dekker, New York/Basel (1988)
48. Kozlowski, W.M.: Fixed point iteration processes for asymptotic pointwise nonexpansive mappings in Banach spaces. J. Math. Anal. Appl. **377**, 43–52 (2011)
49. Kozlowski, W.M.: Common fixed points for semigroups of pointwise Lipschitzian mappings in Banach spaces. Bull. Aust. Math. Soc. **84**, 353–361 (2011)
50. Kozlowski, W.M.: On the existence of common fixed points for semigroups of nonlinear mappings in modular function spaces. Comment. Math. **51**, 81–98 (2011)
51. Kozlowski, W.M.: Advancements in fixed point theory in modular function spaces. Arabian J. Math. doi:10.1007/s40065-012-0051-0, (2012)
52. Kozlowski, W.M.: On the construction of common fixed points for semigroups of nonlinear mappings in uniformly convex and uniformly smooth Banach spaces. Comment. Math. **52**, 113–136 (2012)
53. Kozlowski, W.M.: Pointwise Lipschitzian mappings in uniformly convex and uniformly smooth Banach spaces. Nonlinear Anal. **84**, 50–60 (2013)
54. Kozlowski, W.M., and Sims, B.: On the convergence of iteration processes for semigroups of nonlinear mappings in Banach spaces. In: Computational and Analytical Mathematics, In Honor of Jonathan Borwein's 60th Birthday. Series: Springer Proceedings in Mathematics Statistics, Vol.50 (2013)
55. Krasnosel'skii, M.A., Rutickii, Y.B.: Convex Functions and Orlicz Spaces. P. Nordhoff Ltd, Groningen (1961)
56. Lim, T.C.: A fixed point theorem for families of nonexpansive mappings. Pac. J. Math. **53**, 487–493 (1974)
57. Luxemburg, W.A.J.: Banach Function Spaces. Thesis, Delft (1955)
58. Mann, W.R.: Mean value methods in iteration. Proc. Am. Math. Soc. **4**, 506–510 (1953)
59. Milnes, H.W.: Convexity of Orlicz spaces. Pac. J. Math. **7**, 1451–1486 (1957)

60. Miyaderau, I.: Nonlinear ergodic theorems for semigroups of non-Lipschitzian mappings in Banach spaces. Nonlinear Anal. **50**, 27–39 (2002)
61. Musielak, J., Orlicz, W.: On modular spaces. Studia Math. **18**, 49–65 (1959)
62. Musielak, J., Orlicz, W.: Some remarks on modular spaces. Bull. Acad. Polon. Sci. Ser. Sci. Math. Astronom. Phys. **7**, 661–668 (1959)
63. Musielak, J.: Orlicz spaces and modular spaces. Lecture Notes in Mathematics, vol. **1034**. Springer, Berlin/Heidelberg/New York/Tokyo (1983)
64. Nakano, H.: Modulared Semi-ordered Linear Spaces. Maruzen Co., Tokyo (1950)
65. Oharu, S.: Note on the representation of semi-groups of non-linear operators. Proc. Japan. Acad. **42**, 1149–1154 (1967)
66. Opial, Z.: Weak convergence of the sequence of successive approximations for nonexpansive mappings. Bull. Am. Math. Soc. **73**, 591–597 (1967)
67. Peng, J., Chung, S-K.: Laplace transforms and generators of semigroups of operators. Proc. Am. Math. Soc. **126**, 2407–2416 (1998)
68. Peng, J., Xu, Z.: A novel approach to nonlinear semigroups of Lipschitz operators. Trans. Am. Math. Soc. **367**, 409–424 (2004)
69. Reich, S.: A note on the mean ergodic theorem for nonlinear semigroups. J. Math. Anal. Appl. **91**, 547–551 (1983)
70. Schu, J.: Weak and strong convergence to fixed points of asymptotically nonexpansive mappings. Bull. Aust. Math. Soc. **43**, 153–159 (1991)
71. Taleb, A.A., Hanebaly, E.: A fixed point theorem and its application to integral equations in modular function spaces. Proc. Am. Math. Soc. **128**, 419–427 (1999)
72. Tan, K.-K., Xu, H.-K.: An ergodic theorem for nonlinear semigroups of Lipschitzian mappings in Banach spaces. Nonlinear Anal. **19**, 805–813 (1992)
73. Xu, H.K.: Inequalities in Banach spaces with applications. Nonlinear Anal. **16**, 1127–1138 (1991)

Chapter 6
Fixed Point Theory in Ordered Sets from the Metric Point of View

M.Z. Abu-Sbeih and M.A. Khamsi

6.1 Introduction

Since early investigations of the fixed point problem, retraction played a central role in the non-discrete case, that is, topological spaces or metric spaces. But in the case of ordered sets and graphs it was neglected. Banaschewski and Bruns [5] and Hell [19] were may be the first one to consider such property in this setting. The importance of retraction in this setting took a nice turn when Duffus and Rival [15] proposed a classification scheme based upon the classes of posets closed under retracts and products, which they called *varieties*, which led to some of the most important varieties. Independently Quilliot [34, 35], while studying the pursuit games, introduced the retracts as well as posets and graphs. In his work, Quilliot offered a new bridge to understanding retraction from metric spaces. Indeed he was able to view ordered sets and graphs as "generalized" metric spaces. After his work, this approach became popular among people working in discrete sets. In recent years, Khamsi and al. [24, 25] used the same approach to offer new fixed point theorems in Logic Programming.

Very early on, mathematicians tried to generalize the concept of distance. For example, Menger [29, 30] and Blumenthal [7, 8] offered a generalized distance in the area of geometry. For logic and toposes, one may consult [20, 27] and for fuzzy sets [10, 11], for example. In this chapter, we propose to work in the framework introduced by Quilliot where the sets are equipped with a distance function $d(.,.)$

M.Z. Abu-Sbeih (✉)
Department of Mathematics & Statistics, King Fahd University of Petroleum & Minerals, Dhahran 31261, Saudi Arabia
e-mail: abusbeih@kfupm.edu.sa

M.A. Khamsi
Department of Mathematical Sciences, University of Texas at El Paso, El Paso, TX 79968-0514, USA
e-mail: mohamed@utep.edu

S. Almezel et al. (eds.), *Topics in Fixed Point Theory*, DOI 10.1007/978-3-319-01586-6_6,

whose values are not necessarily real numbers. Indeed we will assume that the values of $d(.,.)$ belong to an ordered semigroup $< \mathscr{V}, \oplus, \preceq >$, having a least element which we still denote 0. The "distance" $d(.,.)$ must satisfy the following

(1) $d(x,y) = 0$ if and only if $x = y$;
(2) $d(x,y) \preceq d(x,z) \oplus d(z,y)$, for any x, y and z.

Depending on the example at hand, one may consider the regular symmetry, that is, $d(x,y) = d(y,x)$, and in some cases an alternative symmetry is offered, that is,

(3) $d(x,y) = \tau(d(y,x))$ for any x and y,

where the function τ is order-preserving involution on the semigroup $\mathscr{V}$, that is, $\tau(\tau(v)) = v$, for any $v \in \mathscr{V}$.

For example, the authors in [23] used this generalization to show that Sine [37] and Soardi [38] fixed point theorem in hyperconvex metric setting translate for posets into the Banschewski and Bruns theorem [5]. They also proved that Isbell [22] construction of a hyperconvex hull coincides with the MacNeille completion of an ordered set [28]. Moreover the authors also developed some nice results about convergence. In particular they established a link between the theory of pre-compact spaces and the theory of partially well-ordered sets. They indeed observed that the basic characterization due to Fréchet [17] and Higman [21] are the same as the Ramsey theorem [36].

In conclusion this generalized metric approach seems to support the idea that concepts originally of infinistic nature, like those developed in the classical metric setting, can perfectly apply to the study of discrete sets (finite or infinite).

6.2 Generalized Metric Spaces

Over the years authors have tried to generalize the classical concept of a distance. They also tried to capture the topological properties of metric spaces, and some known results typical to metric spaces. For example, many attempts were offered to prove an analogue to the classical Banach Contraction Principle in generalized metric spaces. Though some attempts were successful in providing some deep understanding of these results, other attempts were created as publications producing machines. It is our belief that cone metric spaces is one of these publications producing machines. Another example is the generalization offered by Dhage [14] that produced many papers on D-metric which claimed to be a generalization of the classical distance. In [31, 32], the authors proved that this claim is false. In other words, this generalization is not a good one. An example of a good generalization which is closer to the one offered by Quilliot may be found in Grätzer's book [18], where the distance is defined on universal algebras, where $d(x,y)$ is the smallest congruence containing x and y.

In this chapter, we will discuss the generalization of a distance offered by Quilliot [34]. This generalization will bring two theories seemingly far from each other to become almost identical. It will also give us a deep understanding of how to see some concepts in a more abstract form. Throughout this chapter, $(E, \leq)$ is a partially ordered set. Before we give the formal definition of a generalized distance defined on E, let us introduce the semigroup $\mathscr{V}$. Indeed let $\mathscr{V} = \{0, \alpha, \beta, 1\}$, where

$$0 \preceq \alpha \preceq 1 \text{ and } 0 \preceq \beta \preceq 1,$$

and α and β are not comparable. It is easy to see that $(\mathscr{V}, \preceq)$ is a complete lattice. Define $\oplus$ to be the supremum, i.e. $x \oplus y = \sup(x,y) = x \vee y$, for any $x, y \in \mathscr{V}$. Finally define the involution τ on $\mathscr{V}$ to be

$$\tau(0) = 0, \ \tau(1) = 1, \ \tau(\alpha) = \beta, \ \text{ and } \tau(\beta) = \alpha.$$

Clearly we have $\tau(\tau(x)) = x$, for any $x \in \mathscr{V}$. Moreover we have

(i) if $p \preceq q$ and $p' \preceq q'$, then $p \oplus p' \preceq q \oplus q'$;
(ii) τ is order preserving, i.e., if $p \preceq q$ then $\tau(p) \preceq \tau(q)$;
(iii) $\tau(p \oplus q) = \tau(q) \oplus \tau(p)$.

Theorem 6.1. *Let $(E, \preceq)$ be partially ordered set. Set $< \mathscr{V}, \oplus, \preceq >$ as above. Define $d : E \times E \to \mathscr{V}$ by*

(i) $d(x,y) = 0$ if and only if $x = y$;
(ii) $d(x,y) = \alpha$ if and only if $x \leq y$;
(iii) $d(x,y) = \beta$ if and only if $y \leq x$;
(iv) $d(x,y) = 1$ if and only if x and y are not comparable.

Then, $d(.,.)$ satisfies the following properties

(a) $d(x,y) = 0$ if and only if $x = y$;
(b) $d(x,y) = \tau(d(y,x))$;
(c) $d(x,y) \preceq d(x,z) \oplus d(z,y) = d(x,z) \oplus \tau(d(y,z))$;

for any $x, y, z \in E$. The pair (E, d) is called a generalized metric space.

Throughout this chapter, we will only work with this distance. Of course one may consider different distances but one still has to come up with some deep results as the ones obtained by Quilliot [34] and others [24, 25].

Since the distance $d(.,.)$ is not symmetric (in the traditional sense), one has to be careful about classical concepts imported to generalized metric spaces. For example, the concept of balls will be one of these difficulties. Throughout this chapter, we will only consider closed balls and in this case, we will drop the word closed. Therefore a ball centered at $x \in E$ with radius $r \in \mathscr{V}$ will be defined as

$$B(x,r) = \{y \in E;\ d(x,y) \preceq r\} \ \text{ or } \ B(x,r) = \{y \in E;\ d(y,x) \preceq r\}.$$

The reason behind this difference is the lack of symmetry of $d(.,.)$. In this case we talk about the right ball, i.e., $B(x,r) = \{y \in E;\ d(x,y) \preceq r\}$, and the left ball, that is, $B(x,r) = \{y \in E;\ d(y,x) \preceq r\}$. Since the family of right balls is homeomorphic (through the function τ) to the family of left balls, we will only consider the right balls throughout this chapter.

Definition 6.1 (Convex Generalized Metric Space). Let $(E,\leq)$ be a partially ordered set. Consider the generalized distance $d(.,.)$ defined above. We will say that (E,d) is *convex* if and only if for any $x,y \in E$ and $p,q \in \mathscr{V}$ such that $d(x,y) \leq p \oplus q$, then there is $z \in E$ such that

$$d(x,z) \preceq p \text{ and } d(z,y) \preceq q,$$

or equivalently, we must have $B(x,p) \cap B(y,\tau(q)) \neq \emptyset$.

Another property which will be used heavily in this chapter is hyperconvexity or the 2-Helly property.

Definition 6.2 (2-Helly Property). Let $(E,\leq)$ be a partially ordered set. Consider the generalized distance $d(.,.)$ defined above. We will say that (E,d) satisfies the *2-Helly property* or the *2-ball intersection property* if and only if any family of balls which intersect pairwise, has a nonempty intersection.

This is an amazing property which the best metric space, the Hilbert space, fails. Indeed it is quite easy to find three balls in $\mathbb{R}^2$ which intersects 2-by-2 but they fail to have a nonempty intersection. It is puzzling that the best Banach space from a geometric point of view fails such property while the worst Banach space from the geometric point of view, that is, l_∞, satisfies such property. Note the real line $\mathbb{R}$ is convex and satisfies the 2-Helly property. Recall that hyperconvexity is simply convexity combined with the 2-Helly property.

Definition 6.3. Let $(E,\leq)$ be a partially ordered set. Consider the generalized distance $d(.,.)$ defined above. We will say that (E,d) is *hyperconvex* if and only if (E,d) is convex and satisfies the 2-Helly property or equivalently (E,d) is hyperconvex if and only if for any family of balls $\Big(B(x_i,r_i)\Big)_{i\in I}$ such that $d(x_i,x_j) \preceq r_i \oplus \tau(r_j)$, for any $i,j \in I$, we must have

$$\bigcap_{i\in I} B(x_i,r_i) \neq \emptyset.$$

From now on, we will consider the same distance on any partially ordered set. Next we discuss the characterization of convexity and the 2-Helly property of $d(.,.)$ [23].

Theorem 6.2. *Let $(E,\leq)$ be a partially ordered set. Consider the generalized distance $d(.,.)$ defined on E.*

(a) (E,d) is convex if and only if $(E,\leq)$ is directed up-and-down, that is, for any $x,y \in E$, there exist $z,t \in E$ such that $z \leq x \leq t$ and $z \leq y \leq t$.

(b) (E,d) *satisfies the 2-Helly property if and only if every nonempty subset A of E has a supremum and an infimum whenever every two elements of A have an upper-bound and a lower-bound.*
(c) (E,d) *is hyperconvex if and only if* $(E,\leq)$ *is a complete lattice.*

Proof. Let us, for example, show (b). Assume (E,d) satisfies the 2-Helly property. Let A be a nonempty subset of E. Assume that A has more than one element and every two elements of A have an upper-bound and a lower-bound. Let us prove that A has a supremum. The proof for the existence of the infimum is similar and will be omitted. Consider the family of balls $\Big(B(x,\alpha)\Big)_{x\in A}$. Since every two elements of A have an upper-bound, then any pair of balls will have a nonempty intersection. The 2-Helly property will then imply

$$\bigcap_{x\in A} B(x,\alpha) \neq \emptyset,$$

that is, A has an upper bound in E. Set $ub(A) = \{b\in E; x\leq b \text{ for any } x\in A\}$. Again one can easily that the family of balls $\Big(B(x,r(x))\Big)_{x\in A\cup ub(A)}$, where $r(x) = \beta$ for $x\in ub(A)\setminus A$ and $r(x) = \alpha$ for $x\in A$. It is easy to check that this family of balls do intersect pairwise. Hence, the 2-Helly property implies

$$U = \bigcap_{x\in A\cup ub(A)} B(x,r(x)) \neq \emptyset.$$

It is straightforward to see that U is reduced to one point, the least-upper bound or the supremum of A. Next we prove the converse, i.e., assume that every nonempty subset A of E has a supremum and an infimum whenever every two elements of A have an upper-bound and a lower-bound. Let us prove that E satisfies the 2-Helly property. Indeed let $\Big(B(x_i,r_i)\Big)_{i\in I}$ be a family of balls such that $B(x_i,r_i)\cap B(x_j,r_j)\neq \emptyset$, for any $i,j\in I$, we need to prove that

$$\bigcap_{i\in I} B(x_i,r_i) \neq \emptyset.$$

Without loss of generality, we may assume that $r_i\in\{\alpha,\beta\}$. Indeed if $r_{i_0} = 0$, then $x_{i_0}\in B(x_j,r_j)$, for any $j\in I$. And if $r_{i_0} = 1$, then $B(x_{i_0},r_{i_0}) = E$ which will not affect the intersection of the balls. Set $A = \{x_i;\ r_i = \alpha\}$ and $B = \{x_i;\ r_i = \beta\}$. Then, $AcupB$ is not empty. Assume that A is not empty. Since $B(x_i,r_i)\cap B(x_j,r_j)\neq\emptyset$, for any $i,j\in I$, one can easily show that any element in B is an upper bound of A. Moreover the same condition on the balls will imply that every two elements of A have a lower-bound. Therefore, the supremum of A, that is, $\sup(A)$ exists. It is easy to check that

$$\sup(A) \in \bigcap_{i\in I} B(x_i,r_i).$$

This completes the proof of (b). □

6.3 The Retraction Property

Retraction is an old concept. Initially it was developed and studied in the case of topological spaces. Since the start it was closely related to the fixed point property. Its development in the case of discrete sets, that is, ordered sets and graphs, came late through the initial work of Banaschewski and Bruns [5], Hell [19], and the wonderful contribution of Duffus and Rival [15]. Quilliot [34, 35], while studying pursuit games, pointed out a striking similarity between the properties of retractions of these discrete objects and those of metric spaces. The interested reader may found more on this in [23].

Let us first start with the metric case and the original work of Aronszajn and Panitchpakdi [1].

Definition 6.4. Let (M,d) be a metric space over $\mathbb{R}$. Let N be a nonempty subset of M. The mapping $R : M \to N$ is said to be a *retract* if and only if $R \circ R = R$, that is, $R(R(x)) = x$, for any $x \in M$.

Before we state one of the main results of [23, 34], we need the following definition:

Definition 6.5. Let (M,d) be a metric space. Let N be a nonempty subset of M. A mapping $T : N \to N$ is said to be *nonexpansive* if and only if

$$d(T(x),T(y)) \leq d(x,y),$$

for any $x,y \in N$. A point $x \in N$ is said to be a *fixed point* of T whenever $T(x) = x$. The fixed point set of T, denoted by $Fix(T)$, is defined as

$$Fix(T) = \{x \in N;\ T(x) = x\}.$$

In the next theorem, we state some of the main results of Aronszajn and Panitchpakdi [1].

Theorem 6.3. *Let (M,d) be a metric space. The following conditions are equivalent*

(a) M is an absolute nonexpansive retract (ANR in short), that is, for any metric space X which contains M isometrically, there exists a nonexpansive retract $R : X \to M$.

(b) M is injective, that is, for every metric space X, every nonempty subset $Y \subset X$, and every nonexpansive mapping $T : Y \to M$, there exists a nonexpansive mapping $\tilde{T} : X \to M$ such that $\tilde{T}(x) = T(x)$ for any $x \in Y$, that is, $\tilde{T}$ is an extension of T.

(c) M is hyperconvex, that is, for any family of balls $\left(B(x_i,r_i)\right)_{i\in I}$ such that $d(x_i,x_j) \leq r_i + r_j$, for any $i,j \in I$, we must have $\bigcap_{i\in I} B(x_i,r_i) \neq \emptyset$.

(d) M is nonexpansive retract of some $l_\infty(I)$, that is, there exists a nonexpansive retract $R : l_\infty(I) \to M$.

Note that (d) is not explicitly included in [1]. Its statement may be found in [16].

The striking similarity noticed by Quilliot is the similarity between the above theorem and the following result of Banaschewski and Bruns [5].

Theorem 6.4. *Let $(E, \leq)$ be a partially ordered set or a poset. The following conditions are equivalent*

(a) E is an absolute order-preserving retract with respect to the order embedding, that is, for any partially ordered set X which contains E (with the same order), there exists an order-preserving retract $R : X \to E$.
(b) E is injective, that is, for every partially ordered set X, every nonempty subset $Y \subset X$, and every order-preserving map $T : Y \to E$, there exists an order-preserving map $\tilde{T} : X \to E$ such that $\tilde{T}(x) = T(x)$ for any $x \in Y$, that is, $\tilde{T}$ is an extension of T.
(c) E is a complete lattice.
(d) E is an order-preserving retract of the two element chain.

Another amazing similarity is the concept of hyperconvex hull. But first let us recall one of the most beautiful results discovered by Baillon [4] in hyperconvex metric spaces.

Theorem 6.5. *Let M be a bounded metric space. Let $(H_\beta)_{\beta \in \Gamma}$ be a decreasing family of nonempty hyperconvex subsets of M, where Γ is totally ordered. Then, $\bigcap_{\beta \in \Gamma} H_\beta$ is not empty and is hyperconvex.*

One of the implications of Baillon's Theorem is the existence of hyperconvex closures or hyperconvex hulls. Indeed, let M be a metric space and consider the family $\mathscr{H}(M) = \{H; H \text{ is hyperconvex and } M \subset H\}$. Since $l_\infty(M)$ is in $\mathscr{H}(M)$, then $\mathscr{H}(M)$ is not empty. Using Baillon's result, any descending chain of elements of $\mathscr{H}(M)$ has a nonempty intersection. Therefore, one may use Zorn's lemma which will insure us of the existence of minimal elements. These minimal hyperconvex sets are called *hyperconvex hulls* of M. Isbell [22] was among the first to investigate the properties of the hyperconvex hulls. In fact he was the first one to give a concrete construction of a hyperconvex hull.

It is clear that hyperconvex hulls are not unique. But they do enjoy some kind of uniqueness. Indeed, we have:

Proposition 6.1. *Let M be a metric space. Assume that H_1 and H_2 are two hyperconvex hulls of M. Then, H_1 and H_2 are isometric.*

These metric results were thought to be unique to the metric structure and could not be similar to other results in discrete sets. In fact the authors in [23] showed that Isbell construction still extends to the case of generalized metric spaces. Though the authors did borrow most of the Isbell's ideas, they had to adapt them in the generalized case. Indeed Isbell's ideas are specific to the real line and do not extend easily to discrete sets endowed with the generalized distance. They proved the following:

Theorem 6.6. *Let $(E,\leq)$ be a partially ordered set. Consider the generalized distance $d(.,.)$ defined on E. Then, there exists a minimal injective (hyperconvex) set H which contains E isometrically. This injective envelope of E coincides with its MacNeille completion.*

The connection between the construction of the injective envelope offered by the authors in [23] and the MacNeille completion [28] may also be found in [23]. As for the metric case, the injective envelope is not unique. In fact a partially ordered set or a poset may have many injective envelopes. Despite this, we have a similar conclusion to Proposition 6.1:

Proposition 6.2. *Let $(E,\leq)$ be a partially ordered set. Assume that H_1 and H_2 are two hyperconvex hulls or injective envelopes of E. Then, H_1 and H_2 are isomorphic, that is, there exists an invertible map $T : H_1 \to H_2$ such that both T and T^{-1} are order-preserving maps.*

6.4 Fixed Point Property

While studying a paper by Brodskii and Milman [9], Kirk [26] was able to discover the following fixed point theorem:

Theorem 6.7 (Kirk Fixed Point Theorem). *Let X be a Banach space and suppose that C is a nonempty weakly compact convex subset of X which has the normal structure property. Then, any nonexpansive mapping $T : C \to C$ has a fixed point.*

Since its publication in 1965, many have tried to extend it to metric spaces. But because of its strong connection to the linear convexity structure of linear spaces, it was hard to come up with a nice and flexible extension. For example, Takahashi [39] was may be the first one to give a metric analogue to Kirk's theorem. His approach was based on defining a convexity in metric spaces extremely similar to the linear convexity also known as Menger convexity [6–8,29]. The best formulation of Kirk's fixed point theorem that had a chance to be imported to the discrete case was given by Penot [33] through the concept of convexity structures.

Definition 6.6. Let M be an abstract set. A family Σ of subsets of M is called a *convexity structure* if

(i) the empty set $\emptyset \in \Sigma$;
(ii) $M \in \Sigma$;
(iii) Σ is closed under arbitrary intersections.

The convex subsets of M are the elements of Σ. If M is a metric space, we will always assume that closed balls are convex. The smallest convexity structure which contains the closed balls is $\mathscr{A}(M)$ the family of admissible subsets of M. Recall that A is an admissible subset of M if it is an intersection of closed balls. Note that in the original Kirk's fixed point theorem, the Banach space is supposed to have the normal structure property, which means that the family of all convex sets is normal. But this family is a large one and contains the admissible sets. For example, the Banach space l^∞ is a wonderful example which illustrates the power behind Penot's formulation. Indeed, l^∞ fails to have the normal structure property but $\mathscr{A}(l^\infty)$ has the normal structure property. In fact Sine [37] and Soardi [38] fixed point theorem is just Penot's formulation of Kirk's fixed point theorem. Recall Sine and Soardi fixed point theorem:

Theorem 6.8 (Sine and Soardi Fixed Point Theorem). *Let H be a bounded hyperconvex metric space. Any nonexpansive map $T : H \to H$ has a fixed point. Moreover, the fixed point set of T, $Fix(T)$, is hyperconvex.*

In order to extend this theorem to generalized metric spaces, we will need to discuss the family of nonexpansive mappings in this setting. Recall that a mappings $T : F \to E$, where $F \subset E$, is nonexpansive if and only if

$$d(T(x),T(y)) \preceq d(x,y), \quad \text{for any } x,y \in F,$$

where $(E,\leq)$ is a partially ordered set and $d(.,.)$ is the generalized distance defined above.

Theorem 6.9. *Let $(E,\leq)$ be a partially ordered set. Consider the generalized distance $d(.,.)$ defined on E. A mapping $T : F \to F$, where $F \subset E$, is nonexpansive w. r. t. $d(.,.)$ if and only if T is order preserving w. r. t. $\leq$.*

Proof. Let $T : F \to F$, where $F \subset E$, be nonexpansive. Let us prove that T is order preserving. Indeed let $x,y \in F$ such that $x \leq y$. Assume $x \neq y$ otherwise we have nothing to prove. Then, $d(x,y) = \alpha$. Since T is nonexpansive, we get $d(T(x),T(y)) \preceq \alpha$. Hence, $d(T(x),T(y)) = 0$ or $d(T(x),T(y)) = \alpha$. In both cases, we get $T(x) \leq T(y)$.

Conversely, assume that T is order preserving. Let us show that T is nonexpansive. Let $x,y \in E$. If $d(x,y) = 1$ or $d(x,y) = 0$, then we obviously have $d(T(x),T(y)) \preceq d(x,y)$. Without loss of generality we may assume that $d(x,y) = \alpha$. In this case, we have $x \leq y$ and $x \neq y$. Since T is order preserving we get $T(x) \leq T(y)$. Hence, $d(T(x),T(y)) \leq \alpha = d(x,y)$. □

Borrowing Penot's formulation, one can define the concept of compactness in the generalized case. But one has to be careful when it comes to the definition of normal structure property. This is what the authors in [23] did by introducing the concept of accessible element in $< \mathscr{V}, \oplus, \preceq >$. Indeed an element $r \in \mathscr{V}$ is said to be *accessible* if for some $u \in \mathscr{V}$ such that $r \not\preceq u$ and $r \preceq u \oplus \tau(u)$. Otherwise we say that r is *inaccessible*. In the case under consideration in this chapter, the only inaccessible element of $\mathscr{V} = \{0, \alpha, \beta, 1\}$ is 0. Moreover, a generalized metric space (E,d) over $< \mathscr{V}, \oplus, \preceq >$ is said to be *bounded* if there are $x \in E$ and $r \in \mathscr{V}$ such that

(i) 0 is the unique inaccessible element below r,
(ii) $E \subset B(x,r)$.

The authors in [23] proved the following theorem:

Theorem 6.10. *Let (E,d) be a generalized metric space. Assume that E is bounded and hyperconvex. Then, every nonexpansive mapping $T : E \to E$ has a fixed point. Moreover, the set of fixed points $Fix(T)$ is hyperconvex.*

This generalization captures Sine and Soardi fixed point theorem. In fact in the partial ordered case, we have the following:

Theorem 6.11. *Let $(E, \leq)$ be a partially ordered set. Consider the generalized distance $d(.,.)$ defined above. Assume that (E,d) is hyperconvex. Then every nonexpansive mapping $T : E \to E$ has a fixed point. Moreover, the set of fixed points $Fix(T)$ is hyperconvex.*

Using the above characterizations, we get the following result discovered by Tarski [40]:

Theorem 6.12 (Tarski Fixed Point Theorem). *Let $(E, \leq)$ be a complete lattice. Then, every order-preserving map $T : E \to E$ has a fixed point. Moreover, the set of fixed points $Fix(T)$ is a complete sub-lattice of E.*

In fact, in the same paper, Tarski proved a more general result.

Theorem 6.13. *Let $(E, \leq)$ be a complete lattice. Then, any family of pairwise commuting order-preserving mappings $\mathscr{T} = \{T_i; i \in I\}$, $T_i : E \to E$, has a common fixed point. Moreover, the set of common fixed points is a complete sub-lattice of E.*

For quite some time, it was unknown if this conclusion is also valid in the hyperconvex metric spaces. Note how one asks questions which are known in one area to find out if it is similarly known in another area. It was Baillon [4] who proved this theorem in hyperconvex metric spaces through his intersection property. A similar intersection property of hyperconvex sets was unknown in the discrete sets. Recently the authors in [2] proved the following:

Theorem 6.14. *Let X be a partially ordered set. Let $(X_\beta)_{\beta \in \Gamma}$ be a decreasing family of nonempty complete lattice subsets of X, where Γ is a directed index set. Then, $\bigcap_{\beta \in \Gamma} X_\beta$ is not empty and is a complete lattice.*

It is worth mentioning that the proof of this theorem is inspired mostly from Baillon's original proof. This theorem will give another alternative proof of Tarski's common fixed point result. In fact this new proof allowed the authors in [2] to extend some metric concepts [12] to partially ordered sets for the first time.

Definition 6.7. Let X be a partially ordered set. The ordered pair (S,T) of two self-maps of the set X is called a *Banach operator pair* if the set $Fix(T)$ is S-invariant, namely $S(Fix(T)) \subseteq Fix(T)$.

We have the following result:

Theorem 6.15. *Let X be a complete lattice. Let $T : X \to X$ be an order preserving mapping. Let $S : X \to X$ be an order preserving mapping such that (S,T) is a Banach operator pair. Then, $Fix(S,T) = Fix(T) \cap Fix(S)$ is a nonempty complete lattice.*

In order to extend this conclusion to a family of mappings, we will need the following definition.

Definition 6.8. Let T and S be two self-maps of a partially ordered set X. The pair (S,T) is called *symmetric Banach operator pair* if both (S,T) and (T,S) are Banach operator pairs, that is, $T(Fix(S)) \subseteq Fix(S)$ and $S(Fix(T)) \subseteq Fix(T)$.

We have the following result which can be seen as an analogue to De Marr's result [13] without compactness assumption of the domain.

Theorem 6.16. *Let X be a partially ordered set. Let $\mathscr{T}$ be a family of order preserving mappings defined on X. Assume any two mappings from $\mathscr{T}$ form a symmetric Banach operator pair. Then, the family $\mathscr{T}$ has a common fixed point provided one map from $\mathscr{T}$ has a fixed point set which is a complete lattice. Moreover, the set of common fixed points is a complete lattice.*

6.5 Externally Complete Sets

Inspired by the success of the concept of the externally hyperconvex subsets introduced by Aronszajn and Panitchpakdi [1], the authors in [2] introduced a similar concept in partially ordered sets as another example of metric concepts translated to discrete sets and vice versa.

Definition 6.9. Let $(E,\leq)$ be a partially ordered set. A subset M of E is called *externally complete* if and only if for any family of points $(x_\alpha)_{\alpha\in\Gamma}$ in E such that $I(x_\alpha) \cap I(x_\beta) \neq \emptyset$ for any $\alpha, \beta \in \Gamma$, and $I(x_\alpha) \cap M \neq \emptyset$, we have

$$\left(\bigcap_{\alpha\in\Gamma} I(x_\alpha)\right) \cap M \neq \emptyset,$$

where $I(x) = (\leftarrow, x] = \{y \in E;\ y \leq x\}$ or $I(x) = [x, \rightarrow) = \{y \in E;\ x \leq y\}$.

Note that this definition given in terms of the order is similar to the metric definition using the generalized distance defined in the previous sections. The family of all nonempty externally complete subsets of E will be denoted by $\mathscr{E}\mathscr{C}(E)$.

Proposition 6.3. *Let E be a partially ordered set. Then, any $M \in \mathscr{E}\mathscr{C}(E)$ is Dedekind complete and convex.*

Recall that for $M \subset E$ is said to be *Dedekind complete* if for any nonempty subset $A \subset M$, $\sup A$ (resp. $\inf A$) exists in M provided A is bounded above (resp. bounded below) in E. And M is said to be *convex* if $[x,y] = \{z \in E;\ x \leq z \leq y\} \subset M$ for any $x, y \in M$.

Example 6.1. Let $\mathbb{N} = \{0, 1, \ldots\}$. we consider the order $0 \prec 2 \prec 4 \prec \cdots$ and $0 \prec 1 \prec 3 \prec \cdots$, and no even number (different from 0) is comparable to any odd number. Then, $(\mathbb{N}, \prec)$ is a tree. The set $M = \{0, 1, 2\}$ is in $\mathscr{E}\mathscr{C}(\mathbb{N})$. Note that M is convex and is not linearly ordered.

In [3], the authors found some interesting results of externally hyperconvex metric subsets of metric trees. The next result [2], a similar result is given for externally complete subsets of partially ordered trees. Recall that a connected partially ordered set E is called a *tree* if E has a lowest point e, and for every $m \in E$, the subset $[e,m] = \{z \in E;\ e \leq z \leq m\}$ is well ordered.

Theorem 6.17. *Let E be a partially ordered tree. A subset M of E is externally complete if and only if M is convex, Dedekind complete, and any chain $C \subset M$ has a least upper bound in M.*

The proof of the above theorem given by [2] suggests that externally complete subsets are proximinal. In fact in [1], the authors introduced externally hyperconvex subsets as an example of proximinal sets other than the admissible subsets, that is, intersection of balls. Before we state a similar result, we need the following definitions.

Definition 6.10. Let E be a partially ordered set. Let M be a nonempty subset of E. Define the lower and upper cones by

$$\mathscr{C}_l(M) = \{x \in E; \text{there exists } m \in M \text{ such that } x \prec m\}$$

and

$$\mathscr{C}_u(M) = \{x \in E; \text{there exists } m \in M \text{ such that } m \prec x\}.$$

The *cone generated by M* will be defined by $\mathscr{C}(M) = \mathscr{C}_l(M) \cup \mathscr{C}_u(M)$.

Theorem 6.18. *Let E be a partially ordered set and M a nonempty externally complete subset of E. Then, there exists an order preserving retract $P : \mathscr{C}(M) \to M$ such that*

(i) for any $x \in \mathscr{C}_l(M)$ we have $x \prec P(x)$, and
(ii) for any $x \in \mathscr{C}_u(M)$ we have $P(x) \prec x$.

We have the following result.

Theorem 6.19. *Let E be a partially ordered set and M a nonempty externally complete subset of E. Assume that E has a supremum or an infimum, then there exists an order preserving retract $P : E \to M$ which extends the retract of $\mathscr{C}(M)$ into M.*

Since a tree has an infimum, we get the following result.

Corollary 6.1. *Let E be a partially ordered tree and M a nonempty externally complete subset of E. Then, there exists an order preserving retract $P : E \to M$.*

A similar result for externally hyperconvex subsets of metric trees maybe found in [3].

References

1. Aronszajn N., Panitchpakdi, P.: Extensions of uniformly continuous transformations and hyperconvex metric spaces. Pac. J. Math. **6**, 405–439 (1956)
2. Abu-Sbeih, M.Z., Khamsi, M.A.: On externally complete subsets and common fixed points in partially ordered sets. Fixed Point Theor. Appl. **2011**, Article No. 97 (2011)
3. Aksoy, A.G., Khamsi, M.A.: A selection theorem in metric trees. Proc. Am. Math. Soc. **134**, 2957–2966 (2006)
4. Baillon, J.B.: Nonexpansive mapping and hyperconvex spaces. In: Brown, R.F. (ed.) Fixed Point Theory and Its Applications, Contemporary Math., vol. 72, pp. 11–19. Am. Math. Soc., Providence (1988)
5. Banaschewski, B., Bruns, G.: Categorical characterization of the MacNeille completion. Archiv. Math. Basel **18**, 369–377 (1967)
6. Blumenthal, L.M.: Distance Geometries: A Study of the Development of Abstract Metrics. University of Missouri Studies, vol. 13, 1938
7. Blumenthal, L.M.: Theory and Applications of Distance Geometries. Claremont Press, Oxford (1953)
8. Blumenthal, L.M., Menger, K.: Studies in Geometry. W.H.Freeman and Co., San Francisco (1970)
9. Brodskii, M.S., Milman, D.P.: On the center of a convex set. Dokl. Akad. Nauk SSSR **59**, 837–840 (1948) (Russian)
10. Caire, L., Cerruti, U.: Fuzzy relational spaces. Rendiconti Semin. della Facolta Sci. dell'Univ. Cagliari **47**, 63–87 (1977)
11. Cerruti, U., Hohle, U.: Categorical foundations of probabilistic microgeometry. Séminaire de Mathématique Floue, Lyon, 189–246 (1983–1984)
12. Chen, J., Li, Z.: Common fixed points for Banach operator pairs in best approximation. J. Math. Anal. Appl. **336**, 1466–1475 (2007)
13. De Marr, R.: Common fixed points for commuting contraction mappings. Pac. J. Math. **13**, 1139–1141 (1963)
14. Dhage, B.C.: Generalized metric spaces and mappings with fixed point. Bull. Calcutta Math. Soc. **84**, 329–336 (1992)
15. Duffus, D., Rival, I.: A structure theory for ordered sets. J. Discrete Math. **35**, 53–118 (1981)
16. Espínola, R., Khamsi, M.A.: Introduction to hyperconvex spaces. In: Kirk, W.A., Sims, B. (eds.) Handbook of Metric Fixed Point Theory, pp. 391–435. Kluwer Academic, Dordrecht (2001)
17. Fréchet, M.: Les espaces Abstraits. Gauthier-Villars, Paris (1928)
18. Grätzer, G.A.: Universal Algebra. Springer, New York (1979)

19. Hell, P.: Absolute retracts of graphs. Lect. Notes math. **406**, 291–301 (1974)
20. Higgs, D.: Injectivity in the topos of complete Heyting algebra valued sets. Canad. J. Math. **36**, 550–568 (1984)
21. Higman, G.: Ordering by divisibilty in abstract algebra. Proc. Lond. Math. Soc. **3**, 326–336 (1952)
22. Isbell, J.R.: Six theorems about injective metric spaces. Comment. Math. Helv. **39**, 439–447 (1964)
23. Jawhari, E., Misane, D., Pouzet, M.: Retracts: graphs and ordered sets from the metric point of view. Contemp. Math. **57**, 175–226 (1986)
24. Khamsi, M.A., Kreinovich, V., Misane, D.: A new method of proving the existence of Answer sets for disjunctive logic programs: A metric fixed point theorem for multivalued maps. Proc. of the Workshop on Logic Programming with Incomplete Information, Vancouver, British-Columbia, Canada, 1993
25. Khamsi, M.A., Misane, D.: Fixed point theorems in logic programming. Ann. Math. Artif. Intell. **21**, 231–243 (1997)
26. Kirk, W.A.: A fixed point theorem for mappings which do not increase distances. Am. Math. Mon. **72**, 1004–1006 (1965)
27. Lawvere, F.W.: Metric spaces, generalized logic and closed categories. Rendiconti Semin. Math. Fisico Milano **43**, 135–166 (1974)
28. MacNeille, H.: Partially ordered sets. Trans. Am. Soc. **42**, 416–460 (1937)
29. Menger, K.: Untersuchungen über allgemeine Metrik. Math. Ann. **100**, 75–163 (1928)
30. Menger, K.: Geométrie générale. Mémorial des Sciences mathématiques, No 124, Paris, Gauthiers-Villars, 1954
31. Mustafa, Z., Sims, B.: Some remarks concerning D-metric spaces. Proc. of International Conference on Fixed Point Theory and applications, pp. 189–198. Yokohama Publishers, Valencia Spain (2004)
32. Mustafa, Z., Sims, B.: A new approach to a generalized metric spaces. J. Nonlinear Convex Anal. **7**, 289–297 (2006)
33. Penot, J.P.: Fixed point theorem without convexity. Bull. Soc. Math. France Mem. **60**, 129–152 (1979)
34. Quilliot, A.: Homomorphismes, point fixes, rétractions et jeux de poursuite dans les graphes, les ensembles ordonnés et les espaces métriques. Thèse de Doctorat d'Etat, Univ. Paris VI, 1983
35. Quilliot, A.: An application of the Helly property to the partially ordered sets. J. Combin. Theor. A **35**, 185–198 (1983)
36. Ramsey, F.P.: On a problem of formal logic. Proc. Lond. Math. Soc. **30**, 264–286 (1930)
37. Sine, R.: On nonlinear contractions in sup-norm spaces. Nonlinear Anal. **3**, 885–890 (1979)
38. Soardi, P.: Existence of fixed points of nonexpansive mappings in certain Banach lattices. Proc. Am. Math. Soc. **73**, 25–29 (1979)
39. Takahashi, W.: A convexity in metric space and nonexpansive mappings. I. Kodai Math. Sem. Rep. **22**, 142–149 (1970)
40. Tarski, A.: A lattice theoretical fixpoint theorem and its applications. Pac. J. Math. **5**, 285–309 (1955)

Chapter 7
Some Fundamental Topological Fixed Point Theorems for Set-Valued Maps

Hichem Ben-El-Mechaiekh

7.1 Introduction and Preliminaries

This chapter introduces the reader to two of the most fundamental topological fixed point theorems for set-valued maps: *the Browder–Ky Fan* and *the Kakutani–Ky Fan theorems*. It provides a concise discussion including motivations, techniques, as well as some most important applications. The exposition is driven by clarity and simplicity. Generality of statements is deliberately sacrificed to the benefit of conceptual significance. Generalizations based on technicalities or artificial definitions which, with little effort, can be reduced to classical settings are set aside, unless they are motivated by convincing applications. Rather, the treatment here is reduced to the classical convex case, which is—we firmly believe—where the essence belongs. The arguments are kept elementary, as to allow the use of this chapter in a first course in topological fixed point theory and its applications.

The Browder–Ky Fan fixed point theorem asserts that:

A set-valued map A with non-empty convex values and open pre-images from a non-empty compact convex subset X of a topological vector space into itself has a fixed point, that is, an element $x_0 \in X$ *with* $x_0 \in A(x_0)$.

The statement's beautiful simplicity is only surpassed by its exceptional versatility as the geometric underpinning for a number of existence results in nonlinear functional analysis. The fixed point formulation above is due to Felix Browder [22, Theorem 1] and Ky Fan [31, Theorem 2]. It is equivalent to the earlier Ky Fan's generalization of the Knaster–Kuratowski–Mazurkiewicz finite-dimensional intersection theorem to arbitrary subsets of topological spaces [29, Lemma 1] and derives (as pointed out in [9]) from a coincidence theorem of Ky Fan for binary relations with convex/open sections in the Cartesian product of a convex compact subset of a topological vector space by itself [30, Theorem 1']. As pointed out by

H. Ben-El-Mechaiekh (✉)
Department of Mathematics, Brock University, St. Catharines, Ontario, Canada, L2S 3A1
e-mail: hmechaie@brocku.ca

S. Almezel et al. (eds.), *Topics in Fixed Point Theory*, DOI 10.1007/978-3-319-01586-6_7,

Ky Fan himself in [31], a close examination of the proof of the fundamental Lemma [29, Lemma 4] yields a strictly equivalent formulation of the fixed point result of Browder in the form of an existence result for a binary relation with convex / open sections.[1] Hence our reference, for the paternity of this fixed point theorem, to the two prolific and distinguished mathematicians. This result can be seen as a counterpart, for a class of strongly lower semicontinuous set-valued maps, to the Ky Fan generalization [28] of the celebrated Kakutani fixed point theorem, which is the second focus of this work:

An upper semicontinuous set-valued map A with non-empty convex closed values from a compact convex subset X of a locally convex topological vector space into itself has a fixed point.

The Kakutani fixed point theorem corresponds to the case where $X = B^n$, the unit ball in $\mathbb{R}^n$. It is clearly the most natural and immediate set-valued generalization of the Brouwer fixed point theorem. Let us point out that upper semicontinuity is naturally associated with the closedness of the graph (a set-valued map with values in a compact space and closed graph is upper semicontinuous); while an open graph set-valued map has open fibers and is thus lower semicontinuous. It is interesting to note that the Browder–Ky Fan fixed point theorem, unlike the Ky Fan–Kakutani theorem, does not require the underlying topological linear space to be locally convex.

The seminal papers of Ky Fan [28–31] and Browder [22] already contained a host of immediate elegant applications, in the form of existence results in diverse areas of analysis. They nurtured a large body of work which we can call now a *theory of the Browder–Ky Fan Fixed Point Theorem and its Applications*. Contributions were made along the axiomatization of the approach [2, 11–13], the weakening of compactness, the formulation of further fixed point and coincidence theorems [11–14, 37, 38, 43] as well as intersection and matching theorems (see [24] and the references there), the consideration of topological and other abstract convexity structures (see [4, 47] and the references there), the consideration of arbitrary families of mappings (see [7, 24] and the references there) and more applications to such diverse areas as fixed point theory [3, 4, 9, 16, 17], game theory [10], mathematical economics, variational inequalities [33, 39, 41, 42, 56, 58] and complementarity problems [1, 35, 41, 42] and vector equilibrium problems (see [44] and the references there).

We follow, for the treatment of the two fixed point theorems, the approach based on the reduction to the single-valued case and the use of the classical fixed point theorems of Brouwer and Schauder–Tychonoff. In the case of the Browder–Ky Fan theorem, we adopt the continuous selection approach inspired by Browder's proof

[1]Let X be a non-empty convex compact subset of a topological vector space and let A be a subset of $X \times X$ disjoint from the diagonal. If for each fixed $x \in X$, the section $\{y \in X : (x,y) \in A\}$ is convex (or empty) and for each fixed $y \in X$, the section $\{x \in X : (x,y) \in A\}$ is open in X, then $\{x_0\} \times X \not\subseteq A$ for some $x_0 \in X$. To see the equivalence with the Browder–Ky Fan fixed point theorem, set $\Phi(x) := \{y \in X : (x,y) \in A\}$.

(see [11,12,22]). In case of the Kakutani–Ky Fan theorem, we make use of a theorem of Cellina [23] on the existence of a continuous approximative selection. In both cases, convexity and continuous partitions of unity subordinated to (locally) finite open covers play a fundamental role.

The chapter's content is as follows:

- Section 7.1: Introduction and Preliminaries
- Section 7.2: Ky Fan and Kakutani Maps
- Section 7.3: Continuous Selections and Approximations
- Section 7.4: The Browder–Ky Fan and the Kakutani–Ky Fan Fixed Point Theorems and Their Consequences
- Section 7.5: Relaxing Compactness and Related Results
- Section 7.6: Systems of Nonlinear Inequalities and Applications
- Section 7.7: Concluding Remarks
- Section 7.7: References

In this chapter, all topological spaces are assumed to be Hausdorff. The *interior*, the *closure*, and the *boundary* of a subset A in a topological space are denoted by $int(A)$, $cl(A)$, and ∂A, respectively. Topological vector spaces are assumed real and Hausdorff (the complex case is treated in a similar fashion). The *convex hull* of a subset A of a linear space is written $conv(A)$.

For the reader's benefit, we include a self contained and quick exposition of an elementary proof of the Brouwer fixed point theorem that avoids the use of algebraic or combinatorial topology. We then derive the Schauder–Tychonoff theorem using a standard finite type approximation argument.

7.1.1 Elementary Proofs of the Brouwer and Schauder–Tychonoff Fixed Point Theorems

We start with a simplified account of C. A. Rogers' "less strange version" of John Milnor's analytical proof of the Brouwer fixed point theorem (see [52] and [49]).

Theorem 7.1 (Brouwer Fixed Point Theorem). *Every continuous mapping* $f : B^n \to B^n$ *of the closed unit ball* B^n *in the Euclidean space* $\mathbb{R}^n$ *has a fixed point* $x_0 = f(x_0)$.

It is well known that the Brouwer fixed point theorem is equivalent to the no retraction theorem.

Let S^{n-1} be the unit sphere (that is, the boundary of B^n) in $\mathbb{R}^n$. A continuous *retraction* of B^n into S^{n-1} is a continuous mapping $r : B^n \to S^{n-1}$ keeping S^{n-1} fixed, that is, $r(x) = x$ for all $x \in S^{n-1}$. Assume that such a retraction r exists and let $s(x) = -r(x)$, $x \in B^n$. Clearly, $s(B^n) \subseteq S^{n-1} \subset B^n$. By Brouwer's theorem, s has a fixed point $s(x_0) = x_0 \in S^{n-1}$. If follows, $x_0 = s(x_0) = -r(x_0) = -x_0$, that is, $x_0 = 0$ which is impossible. We have thus established, from Brouwer's theorem, the following:

Theorem 7.2 (No Retraction Theorem). *There is no continuous retraction of B^n into its boundary S^{n-1}.*

To show that the no retraction theorem implies Brouwer, assume that $f : B^n \to B^n$ is continuous and fixed point free, that is, $x \neq f(x)$ for all $x \in B^n$. Define $r : B^n \to S^{n-1}$ by setting, for $x \in B^n$, $r(x)$ as the point of intersection of the half-line $[f(x), x)$ with S^{n-1}. The mapping r is well defined, continuous, and keeps S^{n-1} fixed. A contradiction.

The elementary proof starts with the *smooth case*. The word *continuous* can obviously be replaced by *continuously differentiable (or class C^1)* in both Theorems 7.1 and 7.2.

The *no C^1 retraction* result reads as follows:

Lemma 7.1. *There is no C^1 retraction of B^n into S^{n-1}.*

Proof. Assume that r is a C^1 retraction of B^n into S^{n-1}. Let $L > 0$ be a Lipschitz constant for $r : \|r(x_1) - r(x_2)\| \leq L\|x_1 - x_2\|$ for all $x_1, x_2 \in B^n$.

Define $h : B^n \times [0,1] \to B^n$ by $h(x,t) = (1-t)x + tr(x)$. We show first that for small t, the mapping $h_t := h(.,t) : B^n \to B^n$ is one-to-one. Indeed, assume that for a given $t \in [0,1)$, we have $h_t(x) = h_t(x')$. Then, $\|x - x'\| = (\frac{t}{1-t})\|r(x) - r(x')\| \leq (\frac{t}{1-t})L\|\|x - x'\|$. If $x \neq x'$, then $0 < \frac{1}{1+L} \leq t < 1$. Hence, $x = x'$ whenever $0 \leq t < \frac{1}{1+L}$. It follows from the Inverse Function Theorem (see, for example, [26]), that for some $t_0 \in (0, \frac{1}{1+L})$ and all $t \in [0, t_0)$ and $x \in B^n$ the Fréchet derivative $Dh_t(x)$ is non-singular, h_t is one-to-one on B^n, and $O_t = h_t(Int(B^n))$ is an open subset of B^n.

We claim that $cl(O_t) = B^n$ for all $t \in [0, t_0)$. Obviously, $cl(O_t) \subseteq B^n$. Let $u \in B^n$ for some $t \in [0, t_0)$. Join u to an arbitrary point $v \in O_t \subset int(B^n)$ by the line segment $[u, v]$ and choose a point $w \in [u, v] \cap \partial O_t$. Since B^n is compact, $h_t(B^n)$ is also compact, hence closed and contains $cl(O_t)$ which implies that $w = h_t(x)$ for some $x \in B^n$ Since $w \notin O_t$, then $x \notin int(B^n)$, that is, $x \in S^{n-1}$. Thus $r(x) = x$ and $w = (1-t)x + tr(x) = x$. This forces $x = w = u \in S^{n-1} \cap \partial O_t$. Hence $u \in cl(O_t)$ and $B^n = cl(O_t)$. As $h_t(S^{n-1}) = S^{n-1}$, it follows that h_t maps B^n bijectively onto itself whenever $t \in [0, t_0)$.

Define a polynomial in t, $I : [0,1] \to \mathbb{R}$ as $I(t) = \int \cdots \int_{B^n} \det(\frac{\partial h_t}{\partial x_1}, \cdots, \frac{\partial h_t}{\partial x_n})$ $dx_1 \cdots dx_n$. Since, for $0 \leq t < t_0$, h_t is one-to-one and onto B^n, then $I(t) = volume(h_t(B^n)) = volume(B^n)$, a constant. A polynomial that is constant on a subinterval is constant everywhere. Thus $I(t) = volume(B^n)$ for all $t \in [0,1]$, in particular $I(1) = volume(B^n) > 0$. On the other hand, as $h_1(x) = r(x) \in S^{n-1}$ for all $x \in B^n, \|h_1(x)\|^2 = h_1(x) \cdot h_1(x) = 1$. Taking the Gateaux derivative of this equation at any given point $x \in B^n$ in the direction of any given vector $u \in \mathbb{R}^n$ yields $2Dh_1(x)u \cdot h_1(x) = 0$, i.e., the image of the linear operator $Dh_1(x)$ is contained in the orthogonal complement $\{h_1(x)\}^{\perp}$. Thus $rank(Dh_1(x)) \leq n-1$ and therefore $Dh_1(x)$ is singular, that is, $\det(Dh_1(x)) = 0$. Since x is arbitrary, $I(1) = \int_{B^n} \det(Dh_1(x))d\boldsymbol{x} = 0$ contradicting $I(1) = volume(B^n) > 0$. □

Exactly as explained above in the continuous case, Lemma 7.1 is equivalent to the:

C^1Brouwer Fixed Point Theorem: every C^1-mapping $f : B^n \to B^n$ has a fixed point.

To establish the Brouwer theorem, it suffices to invoke the density of smooth mappings in the space of continuous mappings equipped with the supnorm, that is, given a continuous mapping $f : B^n \to B^n$ and $\varepsilon > 0$, there exists a C^1-mapping $f_\varepsilon : B^n \to B^n$ such that $\max_{x\in B^n} \|f(x) - f_\varepsilon(x)\| < \varepsilon$. Indeed, by the Weierstrass approximation theorem, there exists a polynomial $p_\varepsilon : B^n \to D(B^n,\varepsilon)$ with values in the ε-disk around B^n that uniformly ε-approximate f. The mapping $f_\varepsilon : B^n \to B^n$ defined by $f_\varepsilon(x) = (\frac{1}{1+\varepsilon})p_\varepsilon(x)$ is a C^1 uniform ε-approximation of f. By the C^1 Brouwer theorem, there exists $x_\varepsilon \in B^n$ with $x_\varepsilon = f_\varepsilon(x_\varepsilon)$, an ε-almost fixed point for f, that is, $\|f(x_\varepsilon) - x_\varepsilon\| < \varepsilon$. The closed unit ball being compact, the sequence $\{x_n\}_{n=1}^\infty$ of $(1/n)$-almost fixed points has a subsequence, again denoted $\{x_n\}$, converging to a point $x_0 \in B^n$. Since the mapping f is continuous, $\|f(x_0) - x_0\| = \lim_{n\to\infty} \|f(x_n) - x_n\| \le 0$, that is, $f(x_0) = x_0$, thus establishing Theorem 7.1.

To extend the Brouwer theorem to convex compact subsets of $\mathbb{R}^n$, we make use of the three facts.

- First, the fixed point property for continuous mappings is a topological invariant. More precisely, if a topological space X has the fixed point property for continuous mappings (every continuous mapping from X into itself has a fixed point) and if a topological space Y is homeomorphic to X, then Y also has the fixed point property for continuous mappings. Indeed, given $h : X \to Y$ a homeomorphism and $f : Y \to Y$ continuous, the composite $h^{-1}fh : X \to X$ has a fixed point $x_0 \in X$; $h(x_0)$ is a fixed point for f.
- Second, the fixed point property for continuous mappings is preserved under continuous retractions; more precisely, if a topological space X has the fixed point property for continuous mappings and if Y is a retract of X (that is, $Y \subset X$ and there is a continuous mapping $r : X \to Y$ with $r(y) = y$, for all $y \in Y$), then Y also has the fixed point property for continuous mapping. The proof is self-evident: given the inclusion $i : Y \hookrightarrow X$ and a continuous $f : Y \to Y$, the composition $rfi : X \to X$ is continuous and has a fixed point $x_0 = rfi(x_0) = rf(x_0)$, thus $x_0 = f(x_0)$.
- Third, it is well known that any closed convex subset C of a Hilbert space (in particular of $\mathbb{R}^n$) is a retract of any larger subset of the space (consider the metric projection onto the nearest point of C).

Now, if C is a convex and compact subset of $\mathbb{R}^n$, then it is contained in some closed ball B_ρ with radius $\rho > 0$ and center at the origin. Thus C is a retract of B_ρ. But $B_\rho = \rho B^n$ is homeomorphic to B^n. Since B^n has the fixed point property for continuous transformations, it follows:

Corollary 7.1. *If C is a compact convex subset of a finite dimensional topological vector space, and $f : C \to C$ is a continuous mapping, then f has a fixed point.*

Remark 7.1. The first proof of the Brouwer theorem, for any n, is due to Jacques Hadamard in 1910, building on Henri Poincaré's extension (1883) to $\mathbb{R}^n$ of the Bernhard Bolzano intermediate value theorem. Luitzen E. J. Brouwer's proof appeared in 1912, although it appears that Brouwer was aware of the result in 1910 (he provided a proof for $n = 3$ in 1909). The first proof of the non-retraction theorem is due to Piers Bohl in 1904.

The Brouwer theorem was extended to continuous transformations of compact convex subsets of Banach spaces by Juliusz Schauder in 1927. He extended it in 1930 to closed convex subsets C of Banach spaces and compact mappings f (by considering the restriction of the mapping f to the compact convex set $cl(conv(f(C)))$. In 1935, Andrey Nikolayevich Tychonoff extended Schauder's result to compact convex subsets of locally convex topological vector spaces. We provide a proof of the Schauder–Tychonoff theorem based on the approximation of continuous compact[2] mappings by continuous mappings of finite type (see Singball [53]). The first step consists of showing that compact subsets of locally convex topological spaces can be approximated by convex finite polytopes.

Lemma 7.2. *Let K be a non-empty compact subset of a locally convex topological vector space E, and let U be a convex open symmetric neighborhood of the origin in E. Then, there exists a continuous mapping $\pi_U : \bigcup_{k=1}^n \{u_k + U\} \to E$ where $N_U := \{u_1, \dots, u_n\}$ is a U-net of K, satisfying the following properties:*

(i) $\pi_U(\bigcup_{k=1}^n \{u_k + U\})$ is contained in a polytope P_U.
(ii) $y - \pi_U(y) \in U$, for every $y \in \bigcup_{k=1}^n \{u_k + U\}$.

Proof. Let $N_U := \{u_1, \dots, u_n\}$ be a finite subset of K such that the collection $\{(u_k + U) \cap K : k = 1, \dots, n\}$ forms an open cover of K. Consider the so-called Schauder projection $\pi_U : \bigcup_{k=1}^n \{u_k + U\} \to P_U = conv(N_U)$ defined by:

$$\pi_U(y) := \frac{1}{\sum_{k=1}^n \mu_k(y)} \sum_{k=1}^n \mu_k(y) u_k, \text{ for all } y \in \bigcup_{k=1}^n \{u_k + U\},$$

where for $k = 1, \dots, n$, $\mu_k(y) := \max\{0, 1 - \rho_U(y - u_k)\}$ and ρ_U is the Minkowski functional (a semi-norm) associated with the neighborhood U. It follows from the convexity of U that:

$$y - \pi_U(y) \in U \text{ for all } y \in \bigcup_{k=1}^n \{u_k + U\} \text{ and } \pi_U(y) \in P_U = conv(N_U).$$

□

Proposition 7.1. *Let $f : X \to Y$ be a continuous and compact mapping from a topological spaceX into a convex subset Y of a locally convex topological vector*

[2] A mapping $f : X \to Y$ with values in a topological space Y is said to be *compact* if its image is relatively compact in Y, that is, $f(X) \subset K$ compact $\subset Y$.

space E. Then, for every convex open symmetric neighborhood of the origin U in E, there exists a continuous mapping $f_U : X \longrightarrow P_U$ where P_U is a finite convex polytope contained in Y such that $f(x) - f_U(x) \in U$, for all $x \in X$.

Proof. Consider a compact subset K of Y containing $f(X)$ and define $f_U = \pi_U \circ f : X \to K \to P_U$ as in Lemma 7.2. □

Corollary 7.2 (Schauder–Tychonoff). *Every continuous compact self-mapping f of a convex subset X in a locally convex topological vector space E has a fixed point.*

Proof. For any given convex open symmetric neighborhood of the origin U in E, the finite type approximation f_U given by Proposition 7.1 restricted to the finite polytope P_U, a convex compact finite dimensional subset contained in X, $f_U|_{P_U} : P_U \to P_U$ has a fixed point $x_U = f_U(x_U)$ by the Brouwer fixed point theorem (Theorem 7.1). Clearly, $x_U - f(x_U) \in U$. Considering a basis $\{U_i\}_{i\in I}$ of convex and symmetric open neighborhoods with $\bigcap_{i\in I} U_i = \{0_E\}$ together with the fact that $\{f(x_{U_i})\}_{i\in I} \subset K$ a compact set, and the continuity of f, we infer the existence of a subnet $\{x_{i_\nu}\}$ converging to a fixed point $\bar{x} = f(\bar{x})$. □

We refer to the excellent book by Dugundji–Granas [26], the monograph by D.R. Smart [54], the Master's thesis of T. Stuckless [55] and the papers by S. Park [50] and Park-Jeong [51] for surveys of the various proofs of the Brouwer fixed point theorem and enlightening historical remarks.

7.2 Ky Fan and Kakutani Maps

We briefly describe the types of maps under consideration in this chapter, namely:

- *Ky Fan maps,*[3] which are intimately related to the Knaster–Kuratowski–Mazurkiewicz (KMM) Principle; and
- *Kakutani maps,* which are the most natural set-valued extensions of continuous single-valued mappings.

Both types of maps have non-empty convex values but crucially depart in their "continuity" properties: while Kakutani maps are upper semicontinuous (a concept related to the closedness of the graph), Ky Fan maps have open pre-images (a strong regularity that lies between graph openess and lower semicontinuity).

Set-valued maps are point-to-set-maps (that is, their values are subsets of the range), simply called *maps.* We denote them by capital letters and double arrows, $A : X \rightrightarrows Y$, while small letters and single arrows are used for ordinary (single-valued) mappings. The map $A^{-1} : Y \rightrightarrows X$ is the *inverse* of $A : x \in A^{-1}(y) \Leftrightarrow y \in A(x)$. A map $A : X \rightrightarrows Y$ can also be seen as a relation, that is, identified with its graph $graph(A) := \{(x,y) \in X \times Y : y \in A(x)\}$.

[3]The terminology *"applications de Ky Fan"* appeared first in Ben-El-Mechaiekh et al. [11].

7.2.1 Ky Fan Maps

Definition 7.1. Let X be a topological space and Y a subset of a vector space. A map $A : X \rightrightarrows Y$ is an F-map—written $A \in \boldsymbol{F}(X,Y)$—if and only if:

(i) $A(x)$ is convex and non-empty in Y for all $x \in X$;
(ii) $A^{-1}(y)$ open in X for all $y \in Y$.

Given a class of maps $\boldsymbol{M}$, denote by $\boldsymbol{M}^* := \{F : F^{-1} \in \boldsymbol{M}\}$ the class of inverses of M-maps. Thus, the class of F^*-maps:

$$A \in \boldsymbol{F}^*(X,Y) \quad \Leftrightarrow \quad A^{-1} \in \boldsymbol{F}(Y,X).$$

We will see examples of such maps at the end of Sect. 7.4.

Observe that the non-emptiness of the values of A together with the openess of its fibers yields the hypothesis of the Heine–Borel characterization of compactness (namely, that $\{A^{-1}(y) : y \in Y\}$ forms an open cover of X). A number of authors observed that the openess of the fibers $A^{-1}(y)$ could be replaced by the seamingly weaker condition of non-empty interior. One of the first contributions along those lines was done in 1982 in [12] with the formal consideration of the class of $\boldsymbol{\Phi}$-maps defined below:

Definition 7.2. $A \in \boldsymbol{\Phi}(X,Y)$ if and only if

(i) for all $x \in Y$, $A(x)$ is a convex subset of Y,
(ii) there exists $\tilde{A} : X \rightrightarrows Y$ with
   (a) for all $x \in X, \emptyset \neq \tilde{A}(x) \subseteq A(x)$ ($\tilde{A}$ is a selection with non-empty values of A),
   (b) for all $y \in Y$, $\tilde{A}^{-1}(y)$ is an open subset of X.

Such a selection $\tilde{A}$ for a $\boldsymbol{\Phi}$-map A is called *admissible*. Clearly, $\boldsymbol{F} \subset \boldsymbol{\Phi}$.

It turns out that the consideration of the class $\boldsymbol{\Phi}$ for selectionability or fixed points in topological spaces is unnecessary as the following proposition clearly shows that every $\boldsymbol{\Phi}$-map has an F-selection. Recall that a map $\tilde{A}$ is a *selection* of a map A on a common domain X if and only if $\tilde{A}(x) \subseteq A(x), \forall x \in X$.

Proposition 7.2. *Every $\boldsymbol{\Phi}$-map has an F-map selection.*

Proof. For any given point x in a topological space X, let $\mathcal{N}(x)$ denote a local basis of open neighborhoods of x in X. Given a map $A : X \rightrightarrows Y$, define $\tilde{A} : X \rightrightarrows Y$ as

$$\tilde{A}(x) := \bigcup_{U \in \mathcal{N}(x)} \bigcap_{u \in U} A(u), \quad \forall x \in X.$$

Clearly, $\forall x \in X, \tilde{A}(x) \subseteq A(x)$. One readily verifies that

(i) $\tilde{A}^{-1}(y) = int(A^{-1}(y))$, and
(ii) $\forall x \in X, \forall y \in Y, \tilde{A}(x)$ is convex whenever $A(x)$ is convex.

Indeed, $x \in \tilde{A}^{-1}(y) \Leftrightarrow \exists U \in \mathcal{N}(x)$ with $y \in \bigcap_{u \in U} A(u) \Leftrightarrow U \subseteq A^{-1}(y) \Leftrightarrow x \in int(A^{-1}(y))$.

Moreover, $y_1, y_2 \in \tilde{A}(x) \Leftrightarrow \exists U_1, U_2 \in \mathscr{N}(x)$ with $y_i \in \bigcap_{u \in U_i} A(u), i = 1,2 \Longrightarrow$ $y_1, y_2 \in \bigcap_{u \in U} A(u)$ with $U = U_1 \cap U_2$. If A has convex values, then $\bigcap_{u \in U} A(u)$ is a convex set, thus contains all convex combinations of y_1 and y_2. Thus if A is a Φ-map, then $\tilde{A}$ is an F-map. □

In light of Proposition 7.2, the existence of a continuous selection or of a fixed point for an F-map suffices to infer the same for a Φ-map. Thus, it is wiser to restrict one's self to the simplest and most fundamental case of F-maps, at least when it comes to basic results.

Observe that given $A \in \Phi(X,Y)$, the collection $\{\tilde{A}^{-1}(y) : y \in Y\}$ covers X. Of course, if $A^{-1}(y)$ has non-empty interior, one can consider the admissible selection $\tilde{A}$ of A given by $\tilde{A}^{-1}(y) = int(A^{-1}(y))$ and have a Φ-map. The expression of the non-emptiness of $\tilde{A}(x)$ is precisely the identity $X = \bigcup\{\tilde{A}^{-1}(y) : y \in Y\} = \bigcup\{int(A^{-1}(y)) : y \in Y\}$ called by some authors *transfer openess* of the fibers $A^{-1}(y)$ and misguidedly presented as new. Transfer openess is indeed nothing new as it is covered by the consideration of Φ-maps. It is in fact unnecessary and redundant as any convex valued map with transfer open fibers contains a Ky Fan set-valued selection. A considerable amount of publications with no particularly new insight discussed at extraordinary length the so-called "generalized" but truly redundant KKM and fixed point theorems, contributing to diluting the beauty and the essence of this area of fixed point theory, which (as we mention in the Introduction) reside with the classical convex case.

7.2.2 Kakutani Maps

A map $A : X \rightrightarrows Y$ is said to be *upper semicontinuous* (*u.s.c.* for short) at the point $x_0 \in X$ if for any open neighborhood V of $A(x_0)$ in Y, there exists an open neighborhood U of x_0 in X such that $A(U) \subset V$. The map A is said to be *u.s.c.* on X if it is *u.s.c.* at every point of X.

Note that A is *u.s.c.* on X if and only if the set $A^+(V) = \{x \in X; A(x) \subset V\}$ is open in X for any open subset V of Y. Thus, upper semicontinuity amounts to continuity for single-valued mappings. Naturally, as in the single-valued case, compactness is invariant under continuous transformations.

Remark 7.2. A u.s.c. map with compact values transforms compact spaces into compact spaces.

Proof. Let $A : X \rightrightarrows Y$ be a u.s.c. map with compact values of topological spaces with X compact and let $\{V_i : i \in I\}$ be an open cover of $A(X)$. For each $x \in X$, $A(x)$ is covered by a finite collection $\{V_i : i \in I(x) \subset I\}$. By upper semicontinuity, there exists an open neighborhood U_x of x in X with $A(U_x) \subset O_x = \bigcup_{i \in I(x)} V_i$. Being compact, $X = \bigcup_{j=1}^n U_{x_j}$ and hence, $A(X) = A(\bigcup_{j=1}^n U_{x_j}) \subset \bigcup_{j=1}^n O_{x_j} = \bigcup\_j = 1^n \bigcup_{i \in I(x_j)} V_i$, a finite union. □

Given a topological space X and a set Y in a topological vector space, consider the class of *Kakutani maps*:

$$\boldsymbol{K}(X,Y) := \{A : X \rightrightarrows Y : A \text{ is u.s.c. and } \emptyset \neq A(x) \text{ is convex}, \forall x \in X\}.$$

As we mentioned earlier, upper semicontinuity, for a set-valued map of topological spaces, is closely related to the closedness of the graph.[4] Clearly, the map $A : [0,1] \rightrightarrows [0,1]$ defined by: $A(x) = \{0\}$ if $0 \leq x < \frac{1}{2}, A(x) = [0,1]$ if $x = \frac{1}{2}$, and $A(x) = \{1\}$ if $\frac{1}{2} < x \leq 1$ satisfies the hypotheses of the Kakutani theorem (it has 3 fixed points at 0, 1/2, and 1) and does not admit a continuous single-valued selection. But any open neighborhood of its graph does! This graph approximation property is the key ingredient in the proof of the Kakutani fixed point theorem. It was noted very early on by John von Neumann in the second proof (based on the Brouwer fixed point theorem) he gave of his celebrated minimax principle [57] before being formulated formally by A. Cellina [23]. This will be discussed in Sect. 7.3.2 below.

7.3 Continuous Selections and Approximations

7.3.1 Continuous Selections for Ky Fan Maps

A common approach for the solvability of multi-valued problems consists in a reduction to a single-valued case. This can be done by techniques of selections or by approximative selections (continuous, measurable, smooth, etc.). Indeed a fixed point $x_0 = s(x_0)$ of a single-valued selection $graph(s) \subset graph(A)$ is also a fixed point for the set-valued map A. Conceptually, the openess of $graph(A)$ map lends itself to the insertion of the graph of a selection s. This is the central step in the 1968 Browder's proof of the Browder–Ky Fan's fixed point theorem [22]. The existence of such a continuous selection for an F-map with paracompact domain is provided by the following result .

Theorem 7.3. *Let $A \in \boldsymbol{F}(X,Y)$ where X is paracompact and Y is convex. Then, A has a continuous selection, that is, there exists a continuous mapping $s : X \to Y$ with $s(x) \in A(x)$ for all $x \in X$.*

If X is compact, the continuous selection s has values in a finite dimensional convex polytope, that is, there exists $\{y_1, \dots, y_n\} \subset Y$ such that $s(X) \subset conv\{y_1, \dots, y_n\}$.

Proof. For each $x \in X$, there exists $y \in A(x)$, that is, $x \in A^{-1}(y)$. Hence, the collection of open sets $\omega := \{A^{-1}(y) : y \in Y\}$ covers X. Let $\mathscr{O} := \{O_i : i \in I\}$ be a locally finite open refinement of ω and let $\{\lambda_i : i \in I\}$ be a continuous partition of unity subordinated to $\mathscr{O}$. For each $x \in X$, the set of essential indices

4 We have: $[A : X \rightrightarrows Y$ is u.s.c. with closed values and Y is regular $] \Longrightarrow A$ has closed graph. Conversely, $[A$ is locally compact and has closed graph$] \Rightarrow A$ is u.s.c. (with compact values).

$I(x) := \{i \in I : \lambda_i(x) \neq 0\}$ is finite. Note that $i \in I(x) \Rightarrow x \in O_i \subset A^{-1}(y_i)$ for some $y_i \in Y$, and the finite set $\{y_i : i \in I(x)\}$ together with its convex hull is contained in $A(x)$ as the latter is convex. Define a continuous mapping $s : X \to Y$ by:

$$s(x) := \sum_{i \in I} \lambda_i(x) y_i = \sum_{i \in I(x)} \lambda_i(x) y_i, \text{ for all } x \in X.$$

Since $s(x)$ is a convex combination of $\{y_i : i \in I(x)\}$, it follows that $s(x) \in A(x)$.

If X is compact, ω admits a finite subcover $\{A^{-1}(y_i) : i = 1, \ldots, n\}$ and $s(x) = \sum_{i=1}^{n} \lambda_i(x) y_i \in A(x) \cap conv\{y_1, \ldots, y_n\} \subset Y$. □

Remark 7.3. (a) This property of F-maps, though explicitly established by Browder, was first stated as a selection theorem for a class of set-valued maps by Ben-El-Mechaiekh et al. in 1982 [11]. It was restated in 1983 by Yannelis and Prabakhar [58] and subsequently by a number of authors (see, for example, Husain and Tarafdar [34] and the references therein).

(b) The regularity condition (ii) (openess of the inverse images) in Definition 7.1 of an F-map is a very strong lower semicontinuity.[5] Thus, this selection result is to be compared with the celebrated Michael's selection theorem for lower semicontinuous maps with closed convex values in a Banach space defined on paracompact domains [48]; the more stringent regularity being compensated by a much more general codomain.

In light of Proposition 7.2, one hence has immediately that:

Corollary 7.3. *Every Φ-map from a paracompact topological space X into a convex subset Y of a topological vector space admits a continuous selection.*

Considerations of the convexification of a map with open fibers are also superfluous as they add unnecessary complications to a theorem that is otherwise simple and natural.

The paracompactness of X was weakened to X completely regular and residually paracompact (that is, the complement of every open and dense subset of X is paracompact) by Lin et al. [45].

7.3.2 Continuous Approximations of Kakutani Maps

Conceptually, a map $A : X \rightrightarrows Y$ between two topological spaces is *approachable* if for any open neighborhood W of $graph(A)$ in $X \times Y$, there exists a continuous (single-valued) mapping $s : X \to Y$ with $graph(s) \subseteq W$. Since the exposition is

---

[5] A set-valued map $A : X \rightrightarrows Y$ between two topological spaces is said to be *lower semicontinuous (l.s.c.) at* $x_0 \in X$ if for any open subset V of Y such that $V \cap A(x_0) \neq \emptyset$, the upper inverse set $A^-(V) = \{x \in X : A(x) \cap V \neq \emptyset\}$ is an open subset of X containing x_0. A is said to be l.s.c. on X if it is l.s.c. at every point of X. Also, lower semicontinuity coincides with continuity for single-valued mappings.

restricted to the classical convex case, the most general setting for the study of continuous approximative selections is that of topological vector spaces. There, approachability is expressed in terms of open neighborhoods of the origins (as in [3] and [9]). Recall that every topological vector space E is a uniform topological space if entourages of the diagonal are defined to be subsets of $E \times E$ containing sets of the form $\{(x,y) : x-y \in U\}$ where $U \in \mathscr{N}(0_E)$, a basis of open, symmetric, balanced, and radial neighborhoods of the origin in E.

When X,Y are subsets of topological vector spaces E,F, respectively,

Definition 7.3. A map $A : X \rightrightarrows Y$ is *approachable* if for every $(U,V) \in \mathscr{N}(0_E) \times \mathscr{N}(0_F)$, there exists a single-valued map $s : X \to Y$ verifying:

$$\forall x \in X, \exists x' \in (x+U) \cap X \text{ with } s(x) \in (A(x')+V) \cap Y.$$

Such a mapping s is called a *continuous (U,V)-approximative selection of A*. It is clear that a continuous approximative selection for a given map A is a continuous selection for any given open tubular neighborhood of the graph of A.

The next proposition will allow us to derive the Cellina's approximation theorem for K-maps from the selection result in Theorem 7.3.

Proposition 7.3. *Let X be a paracompact subset of a topological vector space E, Y be a subset of a topological vector space F and let $A : X \rightrightarrows Y$ be a u.s.c. map. Then, for any pair of neighborhoods $(U,V) \in \mathscr{N}(0_E) \times \mathscr{N}(0_F)$, there exists an open-graph map $B = B_{U,V} : X \rightrightarrows Y$ such that:*

$$A(x) \subseteq B(x) \subseteq (A(x+U)+V) \cap Y, \forall x \in X.$$

Proof. Let $(U,V) \in \mathscr{N}(0_E) \times \mathscr{N}(0_F)$ be given. By upper semicontinuity of A, for each $x \in X$, there is an open neighborhood $U_x \in \mathscr{N}(0_E)$ contained in U such that $A((x+U_x) \cap X) \subset (A(x)+V) \cap Y$. Let $\{O_i\}_{i \in I}$ be a point-finite open refinement of the cover $\{x+U_x\}_{x \in X}$, that is, for every $i \in I$, $O_i \subset x_i + U_{x_i}$ for some $x_i \in X$ and the set $I(x) := \{i \in I : x \in O_i\}$ is finite for each $x \in X$. Define the map $B = B_{U,V} : X \rightrightarrows Y$ by putting:

$$B(x) := \bigcap_{i \in I(x)} (A(x_i)+V) \cap Y, x \in X.$$

Clearly, $A(x) \subseteq B(x)$ for all $x \in X$. Moreover, for every $x \in X$ and every $i \in I(x)$ we have $B(x) \subseteq A(x_i)+V \subseteq A(x+U)+V$. If $x' \in \bigcap_{i \in I(x)} (x_i + U_{x_i})$, then $I(x) \subseteq I(x')$; consequently, $B(x) \subseteq B(\bigcap_{i \in I(x)} (x_i + U_{x_i})) \subseteq B(x)$. Finally, for any given $x \in X$, the open set $\bigcap_{i \in I(x)} O_i \times B(x)$ is an open set around $\{x\} \times B(x)$ in $X \times Y$, that is, B has an open graph. □

Having open graph, the majorant B of A has open pre-images. Any of its continuous selections (if it exists) is a continuous (U,V)-approximative selection of A. If Y is a convex subset of a locally convex space F, and $A \in \boldsymbol{K}(X,Y)$ is a Kakutani map, then the majorant B also has convex values (indeed, the neighborhood of the origin V can be considered convex, so that the neighborhood $(\Phi(x_i)+V)\cap Y$ of the convex set $\Phi(x_i)$ remains convex in Y, for every $i \in I(x)$. Thus, the map B is a Ky Fan map defined on a paracompact set. By Theorem 7.3, it admits a continuous selection. The Cellina's approximation theorem is thus established:

Corollary 7.4. *If $A \in \boldsymbol{K}(X,Y)$ with X paracompact and Y convex in a locally convex space, then A is approachable.*

Obviously, a partition of unity argument similar to the one used in the proof of Theorem 7.3 would directly yield the existence of a (U,V)-approximative selection for the maps A. The point here is to outline the fact that any Kakutani map is the (upper) limit of a decreasing family of selectionable maps (for example, maps with convex values and open graphs). The interested reader is referred to [3, 4, 9] for extensions of Theorem 7.3 and Corollary 7.4 to maps with non-convex values.

7.4 The Browder–Ky Fan and the Kakutani–Ky Fan Fixed Point Theorems and Their Consequences

All the ingredients are now in place for simple proofs of both fixed point theorems.

7.4.1 The Browder–Ky Fan Fixed Point Theorem

Theorem 7.4. *Let X be a non-empty compact convex subset in a topological vector space and let $A \in \boldsymbol{F}(X,X)$. Then, A has a fixed point, that is, there exists $x_0 \in X$ with $x_0 \in A(x_0)$.*

Proof. By Theorem 7.3, there exists $\{y_1,\ldots,y_n\} \subset X$ and a continuous selection $s : Xc \to X$ of A such that $s(X) \subset P := conv\{y_1,\ldots,y_n\}$. By Corollary 7.1, the restriction of s to P, $s|_P : P \to P$ has a fixed point $s(x_0) = x_0 \in A(x_0)$. □

Naturally, in view of Proposition 7.2, the result holds true for $A \in \Phi(X,X)$.

The compactness of the domain X can be replaced by the more general compactness of the map A when the underlying topological vector space is locally convex [11].

Theorem 7.5. *Let X be a non-empty convex subset in a locally convex topological vector space E and let $A \in \boldsymbol{F}(X,X)$. If A is compact, that is, $A(X) \subset K$ a compact subset of X, then it has a fixed point.*

Proof. First, we note that the convex hull C of the compact set K is a T_3 (that is, regular) σ-compact space (countable union of compact spaces), hence has the Lindelöf property (that is, is T_3 and every open cover has a countable subcover) and is therefore paracompact (see Theorem 5.1.2 in Engelking [27]). Indeed, $C=\bigcup_{n=0}^{\infty}C_n$ where $C_n=\varphi_n(\Delta^n\times K^{n+1})$ is compact, as the continuous image of the Cartesian product of the standard n-simplex Δ^n in $\mathbb{R}^{n+1}$ with the compact set $K^{n+1}=K\times\cdots\times K$, $(n+1)$-times, and $\varphi_n:\Delta^n\times K^{n+1}\to E$, is given by $\varphi_n((\alpha_i)_{i=0}^n,(x_i)_{i=0}^n)=\sum_{i=0}^n\alpha_i x_i$, a convex combination.

The set X being convex contains C. By Theorem 7.3, the restriction of A to C admits a continuous selection s with values in the compact set $K\subset C$. The Schauder–Tychonoff fixed point theorem (Corollary 7.2 above) implies the existence of a fixed point for $x_0=s(x_0)\in A(x_0)$, a fixed point for A. □

Naturally, Theorem 7.5 holds for compact Φ-maps.

7.4.2 The Kakutani–Ky Fan Fixed Point Theorem

We provide here a proof of the Himmelberg generalization of the Kakutani–Ky Fan fixed point theorem, whereby the compactness of the domain is replaced by the more general compactness of the map. A crucial step in the proof is a broad statement on the passage from almost fixed points to fixed points. Recall that given an open cover ω of topological space X, an *ω-fixed point* for a map $A:X\rightrightarrows X$ is a point $x_\omega\in X$ such that x_ω and a point $y_\omega\in A(x_\omega)$ both belong to a common member of the cover ω, that is, there exist a member $W\in\omega$ and a point $y_\omega\in A(x_\omega)$ such that $\{x_\omega,y_\omega\}\subset W$. Given a subspace K of X, denote by $Cov_X(K)$ the family of all open covers of K by open sets of X.

Lemma 7.3. *Let X be a regular topological space and $A:X\rightrightarrows X$ be u.s.c. with closed values. If A has an ω-fixed point for all $\omega\in Cov_X(cl(\Phi(X)))$, then A has a fixed point.*

Proof. Suppose that A is fixed point-free. Since X is regular, points can be separated from closed sets by open sets, that is, for each $x\in X$ there exists open sets $U_x\ni x, V_x\supset A(x)$ with $U_x\cap V_x=\emptyset$. Since A is u.s.c., U_x and V_x can be chosen so that $A(U_x)\subset V_x$. Clearly, A cannot have an ω-fixed point for the open cover $\omega=\{U_x: x\in X\}$. A contradiction. □

We derive Himmelberg's formulation of the Kakutani–Ky Fan fixed point theorem as a consequence of the Browder–Ky Fan theorem. This is, to our knowledge, novel. The proof makes use of the approximation provided by Proposition 7.3.

Theorem 7.6 (Kakutani–Ky Fan–Himmelberg). *Let X be a non-empty convex subset in a locally convex topological vector space E and let $A\in \boldsymbol{K}(X,X)$ be a compact map with closed values. Then A has a fixed point.*

Proof. As in the proof of Theorem 7.5, we have: $K=cl(A(X))$ is a compact set contained in $C=conv(K)$, a paracompact convex subset of X. The map $\tilde{A}:C\rightrightarrows C$ defined by $\tilde{A}(x)=A(x)\cap C, x\in C$, is a Kakutani map. By Proposition 7.3, given an arbitrary convex, symmetric and open neighborhood U of the origin in E, there exists a map $B_U = C \rightrightarrows C$ verifying:

$$\tilde{A}(x) \subseteq B_U(x) \subseteq (\tilde{A}((x+U)\cap C)+U)\cap C, \forall x \in C,$$

$$B_U \in \boldsymbol{F}(C,C) \text{ and } B_U(C) \subseteq K \cap C \subseteq K.$$

By Theorem 7.4, B_U has a fixed point $x_U \in B_U(x_U) \subseteq (\tilde{A}((x+U)\cap C)+U)\cap C \subseteq A(x'_U)+U$ for some $x'_U \in (x_U+U)\cap C$, that is x_U is an ω-fixed point of A for the cover $\omega = \{(x+U)\cap X : x \in X\}$. Lemma 7.3 ends the proof. □

7.4.3 Coincidence Theorems

We derive, in this section, fundamental coincidence theorems between maps of Ky Fan and Kakutani types that are keys in the study of systems of inequalities and minimax theorems. Recall that a *coincidence* between two set-valued maps $A,B : X \rightrightarrows Y$ occurs if their graphs intersect, that is, there exists a pair $(x_0,y_0) \in X \times Y$ with $y_0 \in A(x_0)\cap B(x_0)$.

We start with an elementary observation on the coincidence between a pair of "inverse" maps $(A,B) \in \boldsymbol{M} \times \boldsymbol{M}^*$ for a given abstract class of maps $\boldsymbol{M}$. Recall that $(A\times B)(x) = A(x)\times B(x)$ and $(F\circ t)(x) = F(t(x))$.

Definition 7.4. A class of maps $\boldsymbol{M}$ is said to be *coincidence stable*, if:

(i) $A,B \in \boldsymbol{M} \Rightarrow A\times B \in \boldsymbol{M}$;
(ii) $F \in \boldsymbol{M} \Rightarrow F\circ t \in \boldsymbol{M}$ for any continuous single-valued transformation t.

Note that (ii) with $t =$ inclusion, amounts to: any restriction of an $\boldsymbol{M}$-map is also an $\boldsymbol{M}$-map.

Lemma 7.4. *Let $\boldsymbol{M}$ be a coincidence stable class of maps and let X,Y be two topological spaces such that the Cartesian product $X\times Y$ has the fixed point property for the class $\boldsymbol{M}$ (that is, any map in $\boldsymbol{M}(X\times Y, X\times Y)$ has a fixed point). Then any pair of maps $(A,B) \in \boldsymbol{M}(X,Y)\times \boldsymbol{M}^*(X,Y)$ has a coincidence.*

Proof. Since $\boldsymbol{M}$ is coincidence stable, the composition $X\times Y \xrightarrow{t} Y\times X \overset{B^{-1}\times A}{\rightrightarrows} X\times Y$ where $t(x,y) = (y,x)$ is the transposition, is certainly an $\boldsymbol{M}$-map. Hence, it has a fixed point $(x_0,y_0) \in (B^{-1}\times A)(y_0,x_0)$, that is, $y_0 \in A(x_0)\cap B(x_0)$, a coincidence for the pair (A,B). □

It is easy to see that the class $\boldsymbol{F}$ of Ky Fan maps is coincidence stable, to immediately conclude:

Theorem 7.7. *If X,Y are convex compact subsets of topological vector spaces, then any pair of set-valued maps $(F,G) \in \boldsymbol{F}(X,Y)\times \boldsymbol{F}^*(X,Y)$ has a coincidence.*

This result generalizes to the case where only one of the convex sets X,Y is compact.

Corollary 7.5. *If X and Y are convex subsets of topological vector spaces and only one of them is compact, then any pair $(A,B) \in \boldsymbol{F}(X,Y) \times \boldsymbol{F}^*(X,Y)$ has a coincidence.*

Proof. Assume Y is compact. Since $B^{-1} \in \boldsymbol{F}(Y,X)$, Theorem 7.3 implies the existence of a continuous selection $s : Y \longrightarrow P$ for B^{-1} where P is a finite convex polytope contained in X. It follows that the restriction $B_P = B|_P$ to P is an F^*-map, that is, $B_P \in \boldsymbol{F}^*(P,Y)$. Indeed, $B(x)$ is open in Y for all $x \in P$, and $s(y) \in B^{-1}(y) \cap P$, a non-empty convex subset of P. Clearly, the restriction $A_P = A|_P \in \boldsymbol{F}(P,Y)$. Hence, we are back to the case of a pair (A_P, B_P) where both sets are convex compact. A coincidence $y_0 \in A_P(x_0) \cap B_P(x_0) = A(x_0) \cap B(x_0), x_0 \in P$, is provided by Theorem 7.7. The argument is similar if one supposes that X is compact. In this case, Theorem 7.3 implies the existence of a continuous selection $s : X \to P$ for A where P is a finite convex polytope contained in Y. The restriction of B^{-1} to P is an F-map, that is, $B_P(x) = B(x) \cap P$ is in $\boldsymbol{F}^*(X,P)$ and $A_P(x) = A(x) \cap P$ is in $\boldsymbol{F}(X,P)$. Again we are back to the situation of Theorem 7.7. □

To prove a coincidence for pairs of maps in $\boldsymbol{K} \times \boldsymbol{F}^*$, we require an interesting result of the author on the approachability of composites of approachable maps. Before stating this result, let us first re-frame the notion of approachability defined in Definition 7.3 in the context of uniform topological spaces as in [3, 9, 18].

Definition 7.5. Let $(X, \mathscr{U})$ and $(Y, \mathscr{V})$ be two uniform spaces and let $A : X \rightrightarrows Y$ be a set-valued map.

(i) Given $U \in \mathscr{U}$ and $V \in \mathscr{V}$, a single-valued map $s : X \to Y$ is said to be a (U,V)-approximative selection of A if and only if:

$$\forall x \in X, \exists x' \in U[x] \text{ with } s(x) \in V[A(x')].$$

(Here, $V[A(x')] = \bigcup_{y \in A(x')} V[y]$, $U[x] = \{x' \in X; (x,x') \in U\}$ and $V[y] = \{y' \in X; (y,y') \in V\}$.)

Denote by $\mathfrak{a}(A;U,V) := \{s : X \to Y : s \text{is a continuous } (U,V)-\text{approximative selection of} \Phi\}$.

(ii) The map A is said to be approachable if and only if:

$$\mathfrak{a}(A;U,V) \neq \emptyset, \forall U \in \mathscr{U}, \forall V \in \mathscr{V}.$$

Let the class of A-maps from X into Y be:

$$\boldsymbol{A}(X,Y) := \{A : X \rightrightarrows Y : A \text{ is approachable}\}.$$

Corollary 7.4 (Cellina's theorem) reads: $\boldsymbol{K}(X,Y) \subset \boldsymbol{A}(X,Y$ for a paracompact space X and a convex subset Y in a locally convex spaces. Extensions of this inclusion to non-convex maps are presented in [3, 9, 18].

Let us also specialize a little bit the concept of upper semicontinuity.

Definition 7.6. Let X, Y be topological spaces with Y equipped with a uniform structure $\mathscr{V}$ compatible with its topology. Given an entourage $V \in \mathscr{V}$, a set-valued map $A : X \rightrightarrows Y$ is said to be V*-upper semicontinuous* (V*-u.s.c.*) at a point $x_0 \in X$ if there exists an open neighborhood U of x_0 in X such that $A(U) \subset V[A(x_0)]$. The map A is V*-u.s.c. on* X if it is so at every point of X. It is said to be $\mathscr{V}$*-u.s.c.* if it is V*-u.s.c.* for every $V \in \mathscr{V}$.[6]

The next result describes the preservation of approachability under the composition product of set-valued maps. The result has an intrinsic interest. We refer the reader to [19] for a related open problem.

Proposition 7.4. *Given three uniform spaces* $(X,\mathscr{U}),(Y,\mathscr{V})$ *and* $(P,\mathscr{W})$, *let* $A \in \mathbf{A}(P,Y)$ *be* $\mathscr{V}$*-u.s.c. with non-empty values, and let* $B \in \mathbf{A}(Z,X)$ *be* $\mathscr{U}$*-u.s.c. with non-empty compact values. Then, the composition product* $A \circ B \in \mathbf{A}(P,Y)$ *and is* $\mathscr{V}$*-u.s.c. provided the first space* P *is compact.*

Proof. Let $W \in \mathscr{W}$ and $V \in \mathscr{V}$ be arbitrary but fixed. By upper semicontinuity of A, for each fixed $z \in P$, and each fixed $x_z \in B(z)$, there exists an entourage $U_{x_z} \in \mathscr{U}$ such that: $A(U_{x_z}[x_z]) \subset V'[A(x_z)]$ where V' is a member of $\mathscr{V}$ satisfying $V' \circ V' \subset V$.

For each x_z, let U'_{x_z} be a member of $\mathscr{U}$ such that $U'_{x_z} \circ U'_{x_z} \subset U_{x_z}$, let $\{U'_{x_z^i}[x_z^i]\}_{i=1}^{n_z}$ be an open cover of $B(z)$, and let $U_z = \bigcap_{i=1}^{n_z} U'_{x_z}$. Since B is upper semicontinuous, there exists an entourage $W_z \in \mathscr{W}$ contained in W such that

$$B(W_z[z]) \subset U'_z[B(z)],$$

where U'_z is a member of $\mathscr{U}$ satisfying $U'_z \circ U'_z \subset U_z$.

Let $\{W'_{z_j}[z_j]\}_{j=1}^{m}$ be an open cover of P, with $W'_{z_j} \in \mathscr{W}$ satisfying $W'_{z_j} \circ W'_{z_j} \subset W_{z_j}$ for each j. Let W' be a member of $\mathscr{W}$ contained in $\bigcap_{j=1}^{m} W'_{z_j}$, let U be a member of $\mathscr{U}$ contained in $\bigcap_{j=1}^{m} U'_{z_j}$, and let $U' \in \mathscr{U}$ be so that $U' \circ U' \subset U$.

Now let s_1 be a (W',U')-approximative selection of B, and let s_2 be a (U',V')-approximative selection of A. For any $z \in P$, we have

$$s_2 s_1(z) \in V'[A(U'[s_1(z)])] \subset V'[A(U[B(W'[z])])].$$

Since z belongs to some $W'_{z_j}[z_j]$, it follows that

$$U[B(W'[z])] \subset U'_{z_j}[B(W_{z_j})] \subset U_{z_j}[B(z_j)].$$

[6] Clearly, if the map is *u.s.c.* on X in the ordinary sense, then it is $\mathscr{V}$*-u.s.c* on X. The converse holds true in the case where Φ is compact-valued. In the case where Y is a subset of a topological vector space F, the concept of $\mathscr{V}$-upper semicontinuity ($\mathscr{V}$ being the uniformity generated by a fundamental basis of neighborhoods of the origin in F) is known as *Hausdorff upper semicontinuity*.

Hence, $s_2s_1(z) \in V'[A(U_{z_j}[B(z_j)])]$ for some j. Finally, there exists an element $x^i_{z_j} \in B(z_j)$ such that

$$s_2s_1(z) \in V[A(x^i_{z_j})] \subset V[AB(z_j)] \subset V[AB(W[z])].$$

□

We are now equipped to prove:

Theorem 7.8. *Let X be a convex subset of topological vector space E, Y be a topological space and $(A,B) \in \boldsymbol{K}(X,Y) \times \boldsymbol{F}^*(X,Y)$. Then, the pair (A,B) has a coincidence in each of the following situations:*

(a) Y is compact and A has closed values; or
(b) X is compact and A has compact values.

Proof. (a) The map $B^{-1} : Y \rightrightarrows X$ is an F-map with compact domain. By Theorem 7.3, it admits a continuous selection $s : Y \to P$ with values in a finite convex polytope P contained in X. The restriction $A|_P \in \boldsymbol{K}(P,Y) \subset \boldsymbol{A}(P,Y)$. We shall show that $A_P := s \circ A|_P : P \rightrightarrows P$ has a fixed point $x_0 = s(y_0) \in B^{-1}(y_0)$ with $y_0 \in A(x_0)$, thus establishing a coincidence $(x_0, y_0) \in graph(A) \cap graph(B)$.

Proposition 7.4 establishes the membership $A_P \in \boldsymbol{A}(P,P)$ (single-valued continuous mappings are obviously approachable). Of course, as a composition of u.s.c. maps, A_P is also u.s.c. and has closed values (A having compact values, $s \circ A$ also has compact values). Given an open neighborhood U of the origin in E, a continuous approximative selection $s_U \in \mathfrak{a}(s \circ A|_P; U, U)$ exists and has a fixed point $x_U = s_U(x_U)$ by the Brouwer fixed point theorem. Clearly $x_U = s_U(x_U) \in A_P(x'_U) + U$ for some $x'_U \in (x_U + U) \cap P$, that is x_U is an ω-fixed point of A for the cover $\omega = \{(x+U) \cap P : x \in P\}$. Lemma 7.3 ends the proof.

(b) If X is compact and A has compact values, then $A(X)$ is a compact subset of Y by Remark 7.2. We can thus use (a) with $A(X)$ in place of Y. □

Clearly, with $X = Y$ is convex and compact and $A = id_X$, we recover the Browder–Ky Fan theorem.

7.5 Relaxing Compactness and Related Results

We consider in this section the relaxation of the compactness of the domain in the Browder–Ky Fan fixed point theorem by a coercivity condition that imposes a control on the map outside of a compact set, and we briefly discuss some of its impact on related results.

7.5.1 The Coercivity Condition (κ)

In this generality (with two distinct compact sets K, C) the following general coercivity condition was first presented in [11, 12].

Definition 7.7. Let X be a topological space and let Y be a subset of a topological vector space E. A multifunction $A : X \rightrightarrows Y$ is said to satisfy the coercivity condition (κ) if there exist a compact subset K of X and a compact convex C of Y such that

$$A(x) \cap C \neq \emptyset \text{ for all } x \in X \setminus K.$$

Observe that if X is compact, then (κ) is vacuously satisfied with $K = X$.

This condition was used by Allen [1] with $K = C$ and goes back to Karamardian [35]. With K and C distinct, it has first appeared in [11]. Its equivalent counterpart in KKM theory was first used by Ky Fan [32]. Explicit instances where the condition (κ) appears naturally in more applied problems are discussed in [8].

The coincidence $(\boldsymbol{K}, \boldsymbol{F}^*)$ in Theorem 7.8 extends to the situation where Y is not compact as follows.

Theorem 7.9. *Let X be a convex subset of topological vector space E, Y be a topological space, and let $(A,B) \in \boldsymbol{K}(X,Y) \times \boldsymbol{F}^*(X,Y)$ be a pair of maps such that A has compact values and B^{-1} verifies (κ). Then, the pair (A,B) has a coincidence.*

Proof. Let $K \subseteq Y$ be compact and $C \subseteq X$ convex compact such that $B^{-1}(y) \cap C \neq \emptyset$ for all $y \in Y \setminus K$. The restriction $B^{-1}|_K$ is an F-map with compact domain. Thus, it admits a finite dimensional continuous selection $s : K \to P$ with P a finite convex polytope contained in X (Theorem 7.3). Consider the compact convex subset $\hat{C} = conv(C \cup P)$ of X and the compression map $\hat{B}^{-1} : Y \rightrightarrows \hat{C}$ given by $\hat{B}^{-1}(y) = B^{-1}(y) \cap \hat{C}$ readily seen that $\hat{B}^{-1} \in \boldsymbol{F}(Y, \hat{C})$. Indeed, on the one hand $(\hat{B}^{-1})^{-1}(x) = B(x)$ is open in Y and $\hat{B}^{-1}(y)$ is convex in $\hat{C}$; on the other hand $\hat{B}^{-1}(y)$ is non-empty as it contains $s(y) \in B^{-1}(y) \cap P \subseteq B^{-1}(y) \cap \hat{C}$ in case $y \in K$, and, if $y \in Y \setminus K$, then $\emptyset \neq B^{-1}(y) \cap C \subseteq B^{-1}(y) \cap \hat{C} = \hat{B}^{-1}(y)$. Moreover, the restriction $\hat{A} = A|_{\hat{C}} \in \boldsymbol{K}(\hat{C}, Y)$. We are now in the context of the coincidence Theorem 7.8 with a compact domain and a compact valued K-map. Thus the coincidence for $(\hat{A}, \hat{B})$ which is a coincidence for (A,B). □

As an immediate consequence, we obtain the extension of the Browder–Ky Fan fixed point theorem by considering $X = Y$ and $A = id_X$ (a fixed point for $B \in \boldsymbol{F}^*$ is a fixed point for $B^{-1} \in \boldsymbol{F}$).

Corollary 7.6. *Let X be a convex subset of a topological vector space. A map $B : X \rightrightarrows X$ has a fixed point in each of the following situation:*

(1) $B \in \boldsymbol{F}(X,X)$ satisfies (κ); or
(2) $B \in \boldsymbol{\Phi}(X,X)$ and has an admissible selection $\tilde{B}$ that satisfies (κ).

It is quite interesting to derive a Hilbert space generalization of the Brouwer fixed point theorem from this version of the Browder–Ky Fan fixed point theorem.

Definition 7.8. Let X be a subset of an inner product space $(H, \langle .,. \rangle)$, let Y be a subset of X, C a subset of H, and let $f : X \to H$ be a mapping. We say that f *strictly pulls Y towards C* if and only if

$$\forall x \in Y, \exists y \in C \text{ with } \langle f(x) - x, y - x \rangle > 0.$$

Corollary 7.7. *Let X be a convex subset of a real Hilbert space $(H,\langle .,.\rangle)$ and let $f: X \to X$ be a continuous single-valued mapping. Assume that there exist a compact subset K of X and a convex compact subset C of X such that f strictly pulls $X \setminus K$ towards C. Then, f has a fixed point in K..*

Proof. Define $\varphi : X \times X \to [-\infty, +\infty]$ by $\varphi(x,y) := \langle x - f(x), x - y\rangle$, for $(x,y) \in X \times X$, and define a map $A : X \rightrightarrows X$ by:

$$A(x) := \{y \in X : \varphi(x,y) > 0\}, x \in X.$$

The fibers $A^{-1}(y)$ of A are obviously open due to the continuity of $x \mapsto \varphi(x,y)$ and the images $A(x)$ of A are convex because of the linearity of $y \mapsto \varphi(x,y)$. If A has non-empty values, then it is an F-map and must have a fixed point $x_0 \in A(x_0)$ by the Browder–Ky Fan fixed point theorem (Theorem 7.4). But this is impossible as $\varphi(x_0,x_0) = 0$. Hence, one of the values of A must be empty:

$$\exists \bar{x} \in X \text{ with } A(\bar{x}) = \varnothing, \text{ i.e. } \langle \bar{x} - f(\bar{x}), \bar{x} - y\rangle \leq 0, \forall y \in X,$$

in particular for $y = f(\bar{x})$. This ends the proof as $0 \leq \|\bar{x} - f(\bar{x})\|^2 \leq 0$, that is, $\bar{x} = f(\bar{x})$. □

The proof of Theorem 7.9 strictly adapts to yield:

Corollary 7.8. *Let X and Y be convex subsets of topological vector spaces and $(A,B) \in \boldsymbol{F}(X,Y) \times \boldsymbol{F}^*(X,Y)$. If A verifies (κ), then A and B have a coincidence.*

7.5.2 Intersection Theorems: KKM and Matching

In 1961, Ky Fan [29] significantly extended the KKM Lemma[7] to arbitrary subsets of topological vector spaces. We refer to the Ky Fan's generalization of the KKM Lemma as the *KKMF Principle,* which we will now briefly describe.

Following Dugundji–Granas [26] let us define KKM maps as follows.

Definition 7.9. Given an arbitrary subset X of a real vector space E, a set-valued map $\Gamma : X \rightrightarrows E$ is said to be a *KKM* if for every finite subset $A := \{x_1, \ldots, x_n\} \subseteq X$ it holds:

$$conv(A) \subseteq \bigcup_{i=1}^{n} \Gamma(x_i).$$

[7] KKM stands for Knaster, Kuratowski, and Mazurkiewicz. Using the Sperner Lemma, the three famous Polish topologists established in 1929 which became known as the KKM Lemma: *if X consists of the set of vertices of a simplex in $\mathbb{R}^n$ and $\Gamma : X \rightrightarrows R^n$ is a set-valued map with non-empty compact values verifying: $\forall\{x_1,\ldots,x_k\} \subset X$, $conv(\{x_1,\ldots,x_k\}) \subset \bigcup_{i=1}^{k} \Gamma(x_i)$, then $\bigcap_{x\in X} \Gamma(x) \neq \emptyset$.*

Theorem 7.10. *Let X be a convex subset of topological vector space, let $\emptyset \neq U \subseteq X$ and let $\Gamma : U \rightrightarrows X$ be a KKM map with closed values. Assume that one of the following increasingly more general compactness conditions holds:*

(i) $\Gamma(y)$ is compact for all $y \in U$; or
(ii) there exists $y_0 \in U$ such that $\Gamma(y_0)$ is compact; or
(iii) there exists $\{y_1,\dots,y_n\} \subseteq U$ such that $\bigcap_{i=1}^n \Gamma(y_i)$ is compact; or
(iv) there exists a set U_0 contained in a compact convex C of Y such that $K = \bigcap_{y\in U_0} \Gamma(y)$ is compact or empty.

Then, $\bigcap_{y\in U} \Gamma(y) \neq \emptyset$.

Proof. Clearly (i) $\Rightarrow$ (ii) $\Rightarrow$ (iii) $\Rightarrow$ (iv). We will thus assume that (iv) holds. Consider the maps $\tilde{A}, A : X \rightrightarrows U \hookrightarrow X$ defined by:

$$\tilde{A}(x) := U \setminus \Gamma^{-1}(x) \text{ and } A(x) := conv(\tilde{A}(x)), x \in X.$$

Clearly $\tilde{A}$ is a selection of A. Also, $\tilde{A}^{-1}(y) = X \setminus \Gamma(y)$ is open in X and $A(x)$ is convex in U. Moreover, (iv) is strictly equivalent to condition (κ) for the map $\tilde{A}$. Indeed, $K = \bigcap_{y\in U_0} \Gamma(y) = \bigcap_{y\in U_0}(X \setminus \tilde{A}^{-1}(y)) = X \setminus \bigcup_{y\in U_0} \tilde{A}^{-1}(y)$. If $x \in X \setminus K$, then there exists $y_0 \in U_0 \subseteq C$ with $x \in \tilde{A}^{-1}(y_0)$, that is, $y_0 \in \tilde{A}(x) \cap C$.

Assume that K is non-empty and, for a contradiction, suppose that $\bigcap_{y\in U} \Gamma(y) = \emptyset$. This is equivalent to $X = X \setminus \bigcap_{y\in U} \Gamma(y) = \bigcup_{y\in U} \tilde{A}^{-1}(y)$, that is, every $x \in X$ belongs to $\tilde{A}^{-1}(y)$ for some $y \in U$ which means that $y \in \tilde{A}(x)$. Thus $A \in \Phi(X,X)$ and verifies (κ). Corollary 7.6 (2) would guarantee the existence of a fixed point $x_0 \in A(x_0) = conv(U \setminus \Gamma^{-1}(x_0))$, that is, x_0 is a convex combination of a finite set $\{y_1,\dots,y_n\} \subset U \setminus \Gamma^{-1}(x_0)$. Hence $x_0 \in conv(\{y_1,\dots,y_n\})$ but $x_0 \notin \bigcup_{i=1}^n \Gamma(y_i)$, contradicting the fact that Γ is a KKM map. If K is empty, then $\tilde{A}(x) \cap C \neq \emptyset$ for all $x \in X$. That is, the map $A_C(x) := A(x) \cap C$ verifies the membership $A_C \in \Phi(C,C)$. It would have a fixed point by Theorem 7.4, yielding the same contradiction as above. □

In 1984, Ky Fan rephrased the KKMF Principle (Theorem 7.10 with the most general hypothesis (iv)) as *a matching theorem* for open covers of convex sets: which guarantees the existence of a compromise between *larger* (the convex hull) and smaller (an intersection) [32].

Theorem 7.11. *Let X be a convex subset of a topological vector space E, $\emptyset \neq U \subseteq X$, and let $\{O(y) : y \in U\}$ be a open cover of X. Assume that there exists a subset U_0 of U such that $X \setminus \bigcup_{y\in U_0} O(y)$ is compact or empty and that U_0 is contained in a convex compact subset C of X, then there exists a finite subset $\{y_1,\dots,y_n\}$ of U such that:*

$$conv(\{y_1,\dots,y_n\}) \cap \left(\bigcap_{i=1}^n O(y_i) \right) \neq \varnothing.$$

Proof. The map $\Gamma : U \rightrightarrows X$ defined by $\Gamma(y) := X \setminus O(y)$ has closed values and satisfies the compactness condition (iv) of Theorem 7.10. The fact that $\{O(y) : y \in U\}$ covers X is precisely the failure of the thesis of Theorem 7.10. Consequently, Γ cannot be a KKM map. Thus, there exist a finite set $\{y_1, \ldots, y_n\} \subseteq U$ and a point $\bar{x} \in conv(\{y_1, \ldots, y_n\})$ that does not belong to $\bigcup_{i=1}^n \Gamma(y_i)$, that is, $\bar{x} \in conv(\{y_1, \ldots, y_n\}) \cap \bigcap_{i=1}^n O(y_i)$. □

Note that each step of the proofs of Theorems 7.10 and 7.11 can be reversed, resulting in the following equivalences (where BKF stands for Browder–Ky Fan fixed point theorem):

Brouwer FPT	⇔	BKF Thm for F-maps	⇔	BKF Thm for Φ – maps	⇔	KKMF Principle	⇔	Matching Open Cover

Interestingly, we can extend the Browder–Ky Fan fixed point theorem from convex to star-shaped domains.

7.5.3 Browder–Ky Fan Theorem on Star-Shaped Domains

The author extended the BrowderŰKy Fan fixed point theorem to starŰshaped domains in [5]. Recall that a subset X of a vector space E is *star-shaped at a point* $\bar{x} \in X$ if and only if for any other point $x \in X$, the line segment $[\bar{x}, x]$ remains in X; the point $\bar{x}$ is said to be a *center* of X; (X is convex if and only if it is star-shaped in each of its points.) It is readily seen that X is star-shaped at $\bar{x} \in X$ if and only if

$$X \cap C^+_{E \setminus X}(\bar{x}) = \varnothing,$$

where

$$C^+_{E \setminus X}(\bar{x}) := \{y \in E : \exists x \in E \setminus X, \exists t > 1, \text{ such that } y = \bar{x} + t(x - \bar{x})\}$$

is the truncated cone with vertex at $\bar{x}$ and based on $E \setminus X$. Indeed, assume without loss of generality that $\bar{x} = 0_E$ and let $X \cap C^+_{E \setminus X}(0_E) = \varnothing$. If X is not star-shaped at 0_E, then there exist $(x,t) \in X \times (0,1)$ with $y = tx \notin X$, that is, $y \in E \setminus X$ and $x = \dfrac{1}{t} y \in C^+_{E \setminus X}(0_E)$, a contradiction. The converse is obvious.

The main ingredient in extending the Browder–Ky Fan's theorem to star-shaped domain is the matching theorem for open covers of convex sets (Theorem 7.11 above).

Theorem 7.12. *Let X be a star-shaped compact subset of a topological vector space E and let $A \in \boldsymbol{F}(X,X)$. Then A has a fixed point.*

Proof. Since X is compact, A admits a continuous selection $s : X \to E$ with values in a convex polytope $P \subset E$ (Theorem 7.3). We can assume with no loss of generality that the center $\bar{x}$ of X belongs to P (we could otherwise replace P by the convex hull of P and $\bar{x}$, also a convex polytope).

Let $U := s(X) \cup U_0$ where $U_0 := \{\bar{x}\}$ and let $\{O(x) : x \in U\}$ be the family of open subsets of P defined by:

$$O(x) := A^{-1}(x) \cap C, \forall x \in s(X) \text{ and } O(\bar{x}) := P \setminus X.$$

Note that if $y \in P \cap X$ then $x = s(y) \in A(y) \Leftrightarrow y \in A^{-1}(x) \Rightarrow y \in O(x)$; also, $y \in P \setminus X \Leftrightarrow y \in O(\bar{x})$, that is, $\{O(x) : x \in U\}$ is an open cover of P.

Obviously, U_0 being a singleton is compact and convex and $P \setminus O(\bar{x}) = P \cap X$ is compact. Theorem 7.11 with $X = P$ implies the existence of a finite set $\{x_1, \dots, x_n\}$ of U such that $conv(\{x_1, \dots, x_n\}) \cap (\bigcap_{i=1}^n O(x_i)) \neq \varnothing$. Given an element $x_0 \in conv(\{x_1, \dots, x_n\}) \cap \bigcap_{i=1}^n O(x_i)$, two cases are possible:

Case 1: $\bar{x} \notin \{x_1, \dots, x_n\}$.
Thus, $x_0 \in O(x_i) \subseteq A^{-1}(x_i) \Rightarrow x_i \in A(x_0), i = 1, \dots, n$. Since $A(x_0)$ is a convex set, $x_0 \in A(x_0)$ is a fixed point for A.

Case 2: $\bar{x} \in \{x_1, \dots, x_n\}$.

In this case, $\bar{x} = x_{i_0}$ for some $i_0 \in \{1, \dots, n\}$. Thus, $x_0 \in P \setminus X$ and $x_0 \in A^{-1}(x_i)$ for $i \neq i_0$. Moreover, $x_0 = \sum_{i=1}^n \lambda_i x_i = \sum_{i=1, i \neq i_o}^n \lambda_i x_i + \lambda_{i_0} \bar{x}$ is a convex combination. Clearly, $\lambda = 1 - \lambda_{i_0} = \sum_{i=1, i \neq i_o}^n \lambda_i > 0$ for otherwise $\lambda_{i_0} = 1$, that is, $x_0 = \bar{x} \in P \setminus X$, contradicting the fact that $\bar{x} \in X$. Dividing by λ we obtain:

$$\frac{1}{\lambda} x_0 = \frac{1}{\lambda} \sum_{i=1, i \neq i_o}^{n} \lambda_i x_i + \frac{1-\lambda}{\lambda} \bar{x} \quad \Leftrightarrow \quad \frac{1}{\lambda} x_0 + \frac{\lambda - 1}{\lambda} \bar{x} = z$$

$$\text{where } z := \frac{1}{\lambda} \sum_{i=1, i \neq i_o}^{n} \lambda_i x_i \text{ a convex combination.}$$

Also, since $A(x_0)$ is convex and $x_i \in A(x_0)$ for all $i \neq i_0$, then $z \in A(x_0) \subset X$. On the other hand $z = \bar{x} + t(x_0 - \bar{x})$ with $t = \dfrac{1}{\lambda} > 1$, that is, $z \in C^+_{E \setminus X}(\bar{x})$ contradicting the fact that X is star-shaped. □

7.5.4 A Leray–Schauder Alternative for Kakutani Maps

We provide in this subsection an elementary proof of a nonlinear alternative of Leray–Schauder type for Kakutani maps. It is worth mentioning that the argument applies *ad verbatim* to the larger class of u.s.c. compact approachable map.

In what follows E stands for a Hausdorff locally convex topological vector space with a fundamental basis $\mathcal{N}(0_E)$ of convex symmetric neighborhoods of the origin, and X, Y are non-empty subsets of E.

The following finite-type approximation property of compact approachable maps plays a crucial role in the proof of the main theorem of this section.

Lemma 7.5. *Let $A \in \mathbf{A}(X,Y)$ be a compact approachable map. Given any $V \in \mathscr{N}(0_E)$, there exist a finite subset N_V of Y and a map $A_V \in \mathbf{A}(X, conv(N_V))$ such that $A_V(x) \subset A(x)+V$, for every $x \in X$.*

Proof. Let $V \in \mathscr{N}$ be arbitrary, and let $N_V := \{y_1,\ldots,y_n\}$ be a finite subset of Y such that the collection $\{y_i + \frac{1}{6}V : i = 1,\ldots,n\}$ forms an open cover of the compact set $cl(A(X))$. Consider the Schauder projection (cf. Lemma 7.2 above) $\pi_V : \bigcup_{i=1}^n (y_i + \frac{1}{3}V) \to conv(N_V)$ defined by:

$$\pi_V(y) := \frac{1}{\sum_{i=1}^n \mu_i(y)} \sum_{i=1}^n \mu_i(y) y_i, \text{ for all } y \in \bigcup_{i=1}^n (y_i + \frac{1}{3}V),$$

where $\mu_i(y) := \max\{0, 1 - p_{\frac{1}{3}V}(y-y_i)\}$ and $p_{\frac{1}{3}V}$ is the Minkowski functional of $\frac{1}{3}V$. One has:

$$\pi_V(y) - y \in \frac{1}{3}V, \text{ for all } y \in \bigcup_{i=1}^n (y_i + \frac{1}{3}V).$$

Let $A' : X \rightrightarrows \bigcup_{i=1}^n (y_i + \frac{1}{3}V)$ be the "compression" of A to the set $\bigcup_{i=1}^n (y_i + \frac{1}{3}V)$ defined by:

$$A'(x) := A(x) \cap \bigcup_{i=1}^n (y_i + \frac{1}{3}V), \text{ for all } x \in X.$$

(Note that truly, $A'(x) = A(x)$ for all x). Define the map $A_V : X \rightrightarrows conv(N_V)$ as the composition product $A_V := \pi_V A'$. Clearly, A_V is *u.s.c.*; it has compact values whenever A has compact values. Moreover, the following two properties hold:

$$A_V(x) \subset A(x) + V, \text{ for all } x \in X;$$

and if for a given $U \in \mathscr{N}(0_E)$, $s : X \to Y$ is a continuous approximative selection such that

$$s(x) \in [A((x+U) \cap X) + \frac{1}{6}V] \cap Y, \text{ for all } x \in X,$$

then $\pi_V s : X \to conv(N_V)$ verifies :

$$\pi_V s(x) \in [A_V((x+U) \cap X) + V] \cap co(N_V), \text{ for all } x \in X,$$

that is Φ_V is approachable. □

We shall also need the following generalization of the Fan–Kakutani fixed point theorem to approachable maps.

Proposition 7.5. *Assume that X is convex compact in a locally convex topological vector space E and that $A \in \mathbf{A}(X,X)$ has closed values. Then, A has a fixed point.*

Proof. Simply consider, for any fixed $U \in \mathscr{N}(0_E)$, a continuous (U,U)-approximative selection $s_U \in \mathfrak{a}(A;U,U)$, that is,

$$s_U(x) \in [A((x+U)\cap X)+U]\cap X, x \in X,$$

and apply the Schauder–Tychonoff fixed point theorem (Corollary 7.2 above) to obtain a fixed point $X \ni x_U = s_U(x_U) \in A(x'_U)+U, x'_U \in (x_U+U)\cap X$. A compactness argument based on Lemma 7.3 ends the proof. □

We are ready now to state and prove the main result of this subsection.

Theorem 7.13. *Assume that X is convex closed in E with boundary ∂X and that 0 is an interior point of X. Let $A \in \mathbf{K}(X,E)$ (more generally, $A \in \mathbf{A}(X,E)$) be a closed-valued compact map. Then, one of the following properties holds:*

(1) (Fixed point) $\exists x_0 \in X$, with $x_0 \in A(x_0)$;
(2) (Invariant direction) $\exists \hat{x} \in \partial X, \exists \lambda \in (0,1)$, with $\hat{x} \in \lambda A(\hat{x})$.

Proof. The proof is an adaptation of the proof of Theorem 2.2 in Granas [36] (where a similar alternative for convex-valued contractions was presented) with changes relevant to the present context. Let p_X be the Minkowski's functional of the set X, and let $r : E \longrightarrow X$ be the standard retraction of E onto X :

$$r(y) := \begin{cases} y & \text{for } y \in X, \\ \frac{y}{p_X(y)} & \text{for } y \notin X. \end{cases}$$

Let $V \in \mathscr{N}(0_E)$ be arbitrary but fixed. Consider the finite subset N_V of E and the map $A_V \in \mathscr{A}(X, conv(N_V))$ verifying $A_V(x) \subset A(x)+V$, for all $x \in X$, both provided by Lemma 7.5. Let r_V be the restriction of r to the compact convex set $conv(N_V)$. Now the composition product $B_V : conv(N_V) \overset{r_V}{\to} X \overset{A_V}{\rightrightarrows} conv(N_V)$ is an approachable map belonging to the class $\mathbf{A}(conv(N_V), conv(N_V))$ (see Proposition 7.4 above). By Proposition 7.5, it has a fixed point $x_V \in B_V(x_V)$. The point $y_V := r_V(x_V)$ of X verifies:

$$y_V \in rB_V(y_V).$$

Indeed, $y_V = r_V(x_V) \in r(B_V(x_V)) = r(A_V r_V(x_V)) = rA_V(y_V)$.
Let $z_V \in A_V(y_V)$ be such that $y_V = r(z_V)$.
Two cases are now possible:

Case 1: $z_V \in X$. In this case, $y_V = r(z_V) = z_V \in A_V(y_V) \subset A(y_V)+V$, that is, y_V is an approximative fixed point for A.

Case 2: $z_V \notin X$. Hence, $y_V = r(z_V) = \frac{z_V}{p_X(z_V)} \in \partial X$. Therefore, $y_V = \lambda_V z_V \in \lambda_V A_V(y_V) \subset \lambda_V(A(y_V)+V) \subset \lambda_V A(y_V) + V$ where $\lambda_V := 1/p_X(z_V) \in (0,1)$.

This implies the existence of a point y'_V with:

$$y_V - y'_V \in V \text{ and } y'_V \in \lambda_V A(y_V).$$

We have proven the V-alternative:

$(1)_V$ (V-fixed point) $\exists x_V \in X$, with $x_V \in A(x_V)$; or
$(2)_V$ (V-invariant direction) $\exists y_V \in \partial X, \exists y'_V \in B_V(y_V), \exists \lambda \in (0,1)$, with $y'_V \in \lambda A(y_V)$.

A standard argument based (as in [15]) on the compactness of A, its upper semicontinuity and the closedness of its values ends the proof. □

7.6 Systems of Nonlinear Inequalities and Applications

In this section, we express the fixed point and coincidence results of Sects. 7.4 and 7.5 and in terms of alternatives for systems of nonlinear inequalities. These analytical forms are much more convenient for the treatment of many problems in analysis. Convexity and continuity of functionals being two essential ingredients, let us start by recalling first the basic concepts of semicontinuity and quasiconvexity for real functions.

Definition 7.10. A real function $f : X \to \mathbb{R}$ defined on a subset X of a topological vector space is:

(i) *quasiconvex* if $\forall \lambda \in \mathbb{R}$, the level set $\{x \in X; f(x) \leq \lambda\}$ is a convex subset of X;
(ii) *quasiconcave* if $-f$ is quasiconvex;
(iii) *lower semicontinuous (l.s.c.)* if $\forall \lambda \in \mathbb{R}$, the level set $\{x \in X; f(x) \leq \lambda\}$ is a closed subset of X;
(iv) *upper semicontinuous (u.s.c.)* if $-f$ is l.s.c..

Naturally, every convex functional is quasiconvex but the converse is false. The concept of *quasiconcavity* was considered as early as 1928 by John von Neumann[8] (see [10] for references). A real function on a topological space is continuous if and only if it is both upper and lower semicontinuous.

The following result, which is an analytical formulation of the Browder–Ky Fan fixed point theorem subject to the coercivity condition (κ) in lieu of the compactness of the set X, is a general version of the celebrated *infsup inequality of Ky Fan* [31].

Theorem 7.14. *Let X be a convex subset of a topological vector space E, Y a non-empty subset of X, and $f : X \times Y \to \mathbb{R}$ a function satisfying:*

[8] He provided the characterization: f is quasiconvex on a convex set X, if and only if $f(\mu x_1 + (1-\mu)x_2) \leq \max\{f(x_1), f(x_2)\}$ for all $x_1, x_2 \in X$ and all $\mu \in [0,1]$.

(i) $x \mapsto f(x,y)$ *is l.s.c. on* X*, for each fixed* $y \in Y$.
(ii) $y \mapsto f(x,y)$ *quasiconcave on* Y*, for each fixed* $x \in X$;

Assume that for a given $\lambda \in \mathbb{R}$*, there exist a compact subset* K *of* X *and a convex compact subset* C *of* Y *such that* $\forall x \in X \setminus K, \exists y \in C$ *with* $f(x,y) > \lambda$*. Then, the following alternative holds:*

(A) *there exists* $x_0 \in X$ *such that* $f(x_0,x_0) > \lambda$*, or*
(B) *there exists* $\bar{x} \in Y$ *such that* $f(\bar{x},y) \leq \lambda$*, for all* $y \in Y$.

Consequently, when $\lambda = \sup_{x\in X} f(x,x)$*, (A) is impossible and*

$$\inf_{x\in X} \sup_{y\in Y} f(x,y) \leq \sup_{x\in X} f(x,x).$$

Proof. Let $A(x) := \{y \in Y : f(x,y) > \lambda\}$, $\forall x \in X$. All hypotheses of Corollary 7.6 (1) are satisfied except, possibly, the non-emptiness of the sections $A(x)$. Indeed, $A(x)$ is convex due to the quasiconcavity of $y \mapsto f(x,y)$ and $A^{-1}(y)$ is open because of the lower semicontinuity of $x \mapsto f(x,y)$. Thus, either $A(x) \neq \emptyset, \forall x \in X$, hence A is an F-map, and therefore has a fixed point (that is, (A) holds), or $A(\bar{x}) = \emptyset$ for some $\bar{x} \in X$, that is, (B) is satisfied. □

Landmark theorems of nonlinear functional analysis follow immediately from this nonlinear alternative (therefore, indirectly, from the Browder–Ky Fan theorem). We refer to Brézis [21] for applications of the next two fundamental results.

Corollary 7.9 (Mazur–Schauder Theorem). *Let* X *be a non-empty closed convex subset of a reflexive Banach space* E *and let* $\varphi : X \to \mathbb{R}$ *be a lower semicontinuous, quasiconvex and coercive (that is,* $\lim_{||x||\to\infty} \varphi(x) = \infty$*) functional. Then,* φ *achieves its minimum on* X.

Proof. Let $\lambda = 0$, $Y = X$, and $f(x,y) = \varphi(x) - \varphi(y)$ in Theorem 7.14. Let K be the intersection of X with a closed ball with radius $M > 0$ centered at the origin of E and such that if $x \in X$ with $\|x\| > M$, then $\varphi(x) > \varphi(y)$ for some $y \in K$. Such a non-empty set K exists due to the coercivity of φ. Since E is reflexive, K is weakly compact. One readily verifies that the hypotheses of Theorem 7.14 with $X, Y, K, C = K$, and $\lambda = 0$ all hold: f is l.s.c. in x, and quasiconcave in y. Clearly, possibility (A) of Theorem 7.14 cannot hold. Hence (B) is true: there exists $\bar{x} \in X$ such that $f(\bar{x},y) = \varphi(\bar{x}) - \varphi(y) \leq 0$, for all $y \in X$. □

We now derive from the nonlinear alternative in Theorem 7.14 the celebrated theorem of Stampacchia for variational inequalities. Recall that given a real normed space E, a form $a : E \times E \to \mathbb{R}$ is said to be:

(i) *bilinear* if it is linear in each of its arguments;
(ii) *continuous* if there exists a constant $c > 0$ with $|a(x,y)| \leq c||x||||y||$ for all $x, y \in E$; and
(iii) *coercive* if there exists a constant $\alpha > 0$ with $a(x,x) \geq \alpha||x||^2$ for all $x \in E$.

Corollary 7.10 (Stampacchia Theorem). *Let E be a reflexive Banach space, $a : E \times E \longrightarrow \mathbb{R}$ be a continuous and coercive bilinear form, and let $\ell : E \longrightarrow \mathbb{R}$ be a bounded linear functional. Given a non-empty closed and convex subset X in E, there exists a unique $\bar{x} \in X$ such that $a(\bar{x}, \bar{x} - y) \leq \ell(\bar{x}) - \ell(y)$ for all $y \in X$.*

Proof. For the existence, we apply Theorem 7.14 to $f : X \times X \to \mathbb{R}$ defined by $f(x,y) := a(x, x-y) - \ell(x-y), (x,y) \in X \times X, \lambda = 0$, $C = \{y_0\}$ with $0 \neq y_0 \in X$ arbitrary, and $K := \{x \in X : \|x\| \leq M\}$ where

$$M := \frac{1}{2}\left(\beta + \sqrt{\beta^2 + 4\gamma}\right),$$

$\beta = (c\|y_0\| + \|\ell\|)/\alpha$ and $\gamma = \|\ell\| \, \|y_0\|/\alpha$.

Indeed, first note that if E is equipped with the weak topology, then $f(x,y)$ is $l.s.c.$ in x and quasiconcave in y (it is in fact linear and continuous for the norm topology in both arguments). Since X is closed and convex, it follows that K is a closed, convex, and bounded, hence weakly compact, subset of X. C is obviously a weakly compact subset of X. Note now that if $f(x, y_0) \leq 0$ for any given $x \in X$, that is, $a(x,x) \leq a(x, y_0) + \ell(x - y_0)$, then $\|x\|$ satisfies a quadratic inequality and is bounded above by M:

$$\begin{aligned} & \alpha\|x\|^2 \leq a(x,x) \leq c\|x\|\|y_0\| + \|\ell\|\|x\| + \|\ell\|\|y_0\| \\ \Longrightarrow\ & \alpha\|x\|^2 - (c\|y_0\| + \|\ell\|)\|x\| - \|\ell\|\|y_0\| \leq 0 \\ \Longleftrightarrow\ & \|x\|^2 - \beta\|x\| - \gamma \leq 0 \\ \Longrightarrow\ & \|x\| \leq \tfrac{1}{2}(\beta + \sqrt{\beta^2 + 4\gamma}) = M. \end{aligned}$$

Consequently, if $x \in X$, $\|x\| > M$, then $f(x, y_0) > 0$ and the compactness condition in Theorem 7.14 is satisfied. Since $f(x,x) = 0$ for any $x \in X$, (A) of Theorem 7.14 is impossible, and (B) holds, that is, $f(\bar{x}, y) = a(\bar{x}, \bar{x} - y) - \ell(\bar{x}) + \ell(y) \leq 0$ for some $\bar{x} \in X$ and all $y \in X$ and the proof of the existence is complete.

The uniqueness follows at once from the bilinearity and the coercivity of the form a as follows: if $a(\bar{x}_i, \bar{x}_i - y) - \ell(\bar{x}_i) + \ell(y) \leq 0$ for two elements $\bar{x}_i \in X, i = 1,2$, and all $y \in X$, then adding $a(\bar{x}_1, \bar{x}_1 - \bar{x}_2) \leq \ell(\bar{x}_1) - \ell(\bar{x}_2)$ to $a(\bar{x}_2, \bar{x}_2 - \bar{x}_1) \leq \ell(\bar{x}_2) - \ell(\bar{x}_1)$ gives $0 \leq \alpha\|\bar{x}_1 - \bar{x}_2\|^2 \leq a(\bar{x}_1 - \bar{x}_2, \bar{x}_1 - \bar{x}_2) \leq 0$, that is, $\bar{x}_1 = \bar{x}_2$. □

As another striking application of the analytical form of the Browder–Ky Fan fixed point theorem (Theorem 7.14 above) we present a simple proof of the existence of equilibria for non-self-mapping of compact subsets of locally convex spaces. We start with a theorem on the existence of maximizable u.s.c. quasiconcave functionals.

Theorem 7.15. *Let X be a convex and compact subset of a topological vector space E, $Y \subseteq \{\varphi : X \to \mathbb{R} : \varphi$ is u.s.c. and quasiconcave, $A \in \boldsymbol{S}(X,Y)$, a class of maps having continuous selections. Then,*

$$\exists x_0 \in X, \exists \varphi_0 \in A(x_0) \textit{ with } \varphi_0(x_0) = \max_{u \in X} \varphi_0(u).$$

Proof. Let $s : X \to Y$ be a continuous selection of A and set $f(x,y) = s(x)(y) - s(x)(x)$. Clearly, $f(.,y)$ is l.s.c. and $f(x,.)$ is quasiconcave. Apply Theorem 7.14 with $\lambda = 0$: (A) is impossible; thus, (B) holds, that is, $\exists x_0 \in X$ with $s(x_0)(y) - s(x_0)(x_0) \leq 0, \forall y \in X$. Hence, $\varphi_0 = s(x_0) \in A(x_0)$ and $\varphi_0(y) \leq \varphi_0(x_0), \forall y \in X$. □

Corollary 7.11. *Let X be a convex compact subset of a topological vector space E, $Y \subseteq \{\varphi : X \to \mathbb{R} : \varphi$ is u.s.c. and quasiconcave, and $f : X \times Y \to \mathbb{R}$ verifies $x \mapsto f(x,\varphi)$ l.s.c. and $\varphi \mapsto f(x,\varphi)$ quasiconcave. Then, either*

(a) $\exists \hat{x} \in X$ with $f(\hat{x},\varphi) \leq 0$ for all $\varphi \in Y$, or
(b) $\exists x_0 \in X, \exists \varphi_0 \in Y$ with $\varphi_0(x_0) = \max_{u \in X} \varphi_0(u)$ and $f(x_0,\varphi_0) > 0$.

Proof. The map $A : X \rightrightarrows Y, A(x) := \{\varphi \in Y : f(x,\varphi) > 0\}$ has convex values and open fibers. If (a) fails, A is an $\boldsymbol{F}$-map defined on a compact set, hence selectionable (Theorem 7.3), that is, $A \in \boldsymbol{S}(X,Y)$. Theorem 7.15 applies. □

Before proving a landmark result on the solvability of nonlinear inclusions that generalizes the Poincaré–Bolzano intermediate theorem (see [6] for a detailed discussion on the subject), let us point out first that the existence of an equilibrium for a map $A : cl(X) \rightrightarrows E$, where $X \subset E$, requires adequate topological/geometric conditions on the domain, regularity on the map, and a suitable boundary condition. An *equilibrium* (also called *a zero*) for A is a point $x_0 \in X$ with $0_E \in A(x_0)$. We define the required ingredients for the statement of the theorem.

Recall that given a subset X of a topological vector space E with topological dual E', a map $A : X \rightrightarrows E$ is said to be *upper hemicontinuous on X (u.h.c.)* if for each $p \in E'$, the support functional $x \mapsto \sigma_{A(x)}(p) = \sup_{y \in A(x)} \langle p,y \rangle$ is upper semicontinuous as an extended real-valued function on X, that is, $\forall \lambda \in \mathbb{R} \cup \{\infty\}$,, the set $\{x \in X : \sigma_{A(x)}(p) < \lambda\}$ is open in X. This concept is due, as far as we can tell, to B. Cornet [25]. Every u.s.c. map is u.h.c.; conversely, a u.h.c. map with convex and weakly compact values is u.s.c..

Define next the class of maps:

$$\boldsymbol{H}(X,E) := \{A : X \rightrightarrows E : A \text{ is u.h.c. and has closed convex values}\}.$$

Clearly, $\boldsymbol{K}(X,E) \subset \boldsymbol{H}(X,E)$.

Given a convex set X in a topological vector space E and a point $x \in \partial X$, define the cones:

- $S_X(x) = \mathbb{R}_+(X - x) = \bigcup_{t>0} \frac{1}{t}(X - x)$ is the cone pointed at 0 with base $X - x$;
- $T_X(x) = cl(S_X(x))$is the *tangent cone of convex analysis* to X at x;
- $N_X(x) := S_X(x)^- = \{p \in E' : \langle p,y \rangle \leq 0, \forall y \in S_X(x)\} = T_X(x)^-$ is the negative polar cone to $T_X(x)$, known as the *normal cone of convex analysis*. It is characterized by:

$$N_X(x) = \{p \in E' : p(x) = \sup_{u \in X} \langle p,u \rangle\}.$$

Let $A : cl(X) \rightrightarrows E$, consider the following boundary conditions:

Definition 7.11. Given $x \in \partial X$:

(R) *Rothe:* $\Phi(x) \cap (X - x) \neq \emptyset$.
(H) *Halpern:* $\Phi(x) \cap S_X(x) \neq \emptyset$.
(wH) *Weak Halpern:* $\Phi(x) \cap T_X(x) \neq \emptyset$.
(KF) *Ky Fan:* $p \in N_X(x) \Rightarrow \inf_{y \in \Phi(x)} \langle p, y \rangle \leq 0$.

One easily verifies that $(R) \Rightarrow (H) \Rightarrow (wH) \Rightarrow (KF)$.

Theorem 7.16. *Assume that X is a convex compact subset in a locally convex topological vector space E and $A \in \boldsymbol{H}(X,E)$ verifies condition (KF). Then, Φ has an equilibrium in X.*

Proof. Let $Y = E', f : X \times Y \to \mathbb{R}$ be $f(x, \varphi) = \inf_{y \in \Phi(x)} \langle \varphi, y \rangle$ which is l.s.c. in x (since $x \mapsto \sigma_\Phi(x, \varphi)$ is u.s.c.) and concave in φ. Condition (KF) opposes (b) of Corollary 7.11. Hence, (a) holds: $\exists \hat{x} \in X$ with $\inf_{y \in \Phi(\hat{x})} \langle \varphi, y \rangle \leq 0$ for all $\varphi \in E'$.

If $0 \notin \Phi(\hat{x})$, by the Hahn–Banach separation theorem, $\exists \varphi \in E', \exists \lambda \in \mathbb{R}$ with $\varphi(0) = 0 < \alpha < \varphi(y), \forall y \in \Phi(\hat{x})$. This implies $0 < \alpha \leq \inf_{y \in \Phi(\hat{x})} \langle \varphi, y \rangle \leq 0$, thus $0 \in \Phi(\hat{x})$. □

Obviously, the Kakutani–Ky Fan theorem (thus Schauder–Tychonoff theorem and Brouwer theorem) follow from the equilibrium Theorem 7.16, as it corresponds to the Rothe boundary condition (R) of Definition 7.11. Hence we have the loop:

Brouwer FPT	$\Rightarrow$	BKF Thm for F-maps	$\Rightarrow$	Equil. Thm for K-maps	$\Rightarrow$	K–KF Thm Principle	$\Rightarrow$	Brouwer FPT

where Equil. Thm is Theorem 7.16, and K-KF Thm stands for the Kakutani–Ky Fan theorem.

Theorem 7.16 is crucial for many solvability theorems in analysis. We refer to the [6] for a detailed discussion.

The coincidence $(\boldsymbol{F}, \boldsymbol{F}^*)$ (Corollary 7.5) can be expressed in analytical terms as a second alternative for nonlinear systems of inequalities as follows:

Theorem 7.17. *Let X and Y be two convex subsets of topological vector spaces and let $f, g : X \times Y \to \mathbb{R}$ be two functions satisfying:*

(i) $f(x,y) \leq g(x,y)$ *for all* $(x,y) \in X \times Y$;
(ii) $x \mapsto f(x,y)$ *is quasiconcave on* X, *for each fixed* $y \in Y$;
(iii) $y \mapsto f(x,y)$ *is l.s.c. on* Y, *for each fixed* $x \in X$;
(iv) $x \mapsto g(x,y)$ *is u.s.c. on* X, *for each fixed* $x \in X$;
(v) $y \mapsto g(x,y)$ *is quasiconvex on* Y, *for each fixed* $x \in X$.
(vi) Given $\lambda \in \mathbb{R}$ arbitrary, assume that either Y is compact, or X is compact, or there exist a compact subset K of X and a convex compact subset C of Y such that for any $x \in X \setminus K$ there exists $y \in C$ with $g(x,y) < \lambda$.

Then, one of the following statements holds:

(A) there exists $\bar{x} \in X$ such that $g(\bar{x},y) \geq \lambda$, for all $y \in Y$; or
(B) there exists $\bar{y} \in Y$ such that $f(x,\bar{y}) \leq \lambda$, for all $x \in X$.

Proof. Simply apply Corollary 7.5 to $A, B \subset X \times Y$ defined as:

$$A := \{(x,y) : g(x,y) < \lambda\} \text{ and } B := \{(x,y) : f(x,y) > \lambda\}.$$

Note that in view of (i) a coincidence between A and B is impossible as it yields $\lambda < \lambda$. Since all hypotheses of Corollary 7.5 are satisfied save for $A(x) \neq \emptyset$ for all $x \in X$ and $B^{-1}(y) \neq \emptyset$ for all $y \in Y$, it follows that either $A(\bar{x}) = \emptyset$ for some $\bar{x} \in X$ (thesis (A)) or $B^{-1}(\bar{y}) = \emptyset$ for some $\bar{y} \in Y$ (thesis (B)). □

Remark 7.4. Theorem 7.17 implies $\alpha = \sup_X \inf_Y g(x,y) \geq \inf_Y \sup_X f(x,y) = \beta$.

Indeed, assuming that $\alpha < \beta < \infty$, let λ be an arbitrary but fixed real number strictly between α and β. By Theorem 7.17, either there exists $\bar{y} \in Y$ such that $f(x,\bar{y}) \leq \lambda$, for all $x \in X$ thus $\beta \leq \lambda < \beta$ which is impossible, or there exists $\bar{x} \in X$ such that $g(\bar{x},y) \geq \lambda$, for all $y \in Y$ thus $\alpha \geq \lambda > \alpha$ which is absurd. Hence $\alpha \geq \beta$.

Maurice Sion's formulation of the von Neumann Minimax Theorem follows immediately with $f = g$:

Theorem 7.18 (Sion–von Neumann Minimax Theorem). *Let X and Y be convex subsets of topological vector spaces and let f be a real function on $X \times Y$ such that*

(i) $x \mapsto f(x,y)$ is quasiconcave and u.s.c. on X, for each fixed $y \in Y$;
(ii) $y \mapsto f(x,y)$ is quasiconvex and l.s.c. on Y, for each fixed $x \in X$;

Assume that either X is compact or Y is compact. Then

$$\alpha = \sup_X \inf_Y f(x,y) = \inf_Y \sup_X f(x,y) = \beta.$$

Proof. The inequality $\alpha \leq \beta$ is always true and $\alpha \geq \beta$ follows from Remark 7.4. □

Remark 7.5. If both X and Y are compact, the infsup equality in Theorem 7.18 is a minmax equality and is equivalent to the existence of a saddle point (x_0, y_0) for the function $f(x,y)$, that is, $f(x,y_0) \leq f(x_0,y), \forall (x,y) \in X \times Y$.

We proceed with a short proof of the Markov–Kakutani fixed point theorem for abelian families of continuous affine mappings in linear topological spaces having separating duals.[9]

---

[9] A topological vector space E *has separating dual* if for each $x \in E, x \neq 0$, there exists a bounded linear form $\ell \in E'$, the topological dual of E, such that $\ell(x) \neq 0$. Locally convex topological vector spaces have separating duals. Sequence spaces $\ell^p, 0 < p < 1$, and Hardy spaces $H^p, 0 < p < 1$, are instances of non-locally convex spaces with separating duals.

Recall that a mapping ϕ from a convex set X into a vector space is said to be *affine* if and only if $\phi(\sum_{i=1}^n \lambda_i x_i) = \sum_{i=1}^n \lambda_i \phi(x_i)$ for any convex combination $\sum_{i=1}^n \lambda_i x_i, \lambda_i \geq 0, \sum_{i=1}^n \lambda_i = 1$, in X. The key ingredient is the following fixed point property for continuous affine transformations of a compact convex set.

Corollary 7.12. *Let X be a non-empty compact convex subset of a topological vector space E with separating dual E' and let $\phi : X \to X$ be a continuous affine mapping. Then, ϕ has a fixed point.*

Proof. Define $f : X \times E' \to \mathbb{R}$ by $f(x,\ell) = \ell(\phi(x) - x), (x,\ell) \in X \times E'$. It suffices to prove the existence of $x_0 \in X$ such that $f(x_0,\ell) \leq 0, \forall \ell \in E'$, for this would imply $\ell(\phi(x_0) - x_0) = 0, \forall \ell \in E'$, that is, $\phi(x_0) - x_0 = 0$ and the proof is complete.

This amounts to showing that $\bigcap_{\ell \in E'} A(\ell) \neq \emptyset$ for the relation $A := \{(\ell,x) \in E' \times X : f(x,\ell) \leq 0\}$.

Since for each fixed $\ell \in E'$, the function $f(x,\ell)$ is l.s.c. in x, then each $A(\ell)$ is a closed, hence compact, subset of X. It suffices, therefore, to show that the collection $\{A(\ell) : \ell \in E'\}$ has the finite intersection property. Consider, to this aim, a finite collection of bounded linear functionals $L := \{\ell_1, \ldots, \ell_n\} \subset E'$, and let $Y = conv(L)$, a convex compact subset of E'. The restriction of $f(x,\ell)$ to $X \times Y$ is obviously u.s.c. and quasiconcave in x and l.s.c. and quasiconvex in ℓ. Since both X and Y are compact and convex, it follows from Remark 7.5 that there exists $(x_0,\ell_0) \in X \times Y$ with $f(x_0,\ell) \leq f(x,\ell_0)$ for all $(x,\ell) \in X \times Y$, i.e., $\ell(\phi(x_0) - x_0) \leq \ell_0(\phi(x) - x)$ for all $(x,\ell) \in X \times Y$. Let $\hat{x}$ be such that $\ell_0(\hat{x}) = \max_{x \in X} \ell_0(x)$. Since $\phi(\hat{x}) \in X$, it follows that $\ell_0(\phi(\hat{x}) - \hat{x}) \leq 0$ and, consequently, $\ell(\phi(x_0) - x_0) \leq 0$, for all $\ell \in Y$, in particular, $\ell_i(\phi(x_0) - x_0) = f(x_0,\ell_i) \leq 0$ for all $\ell_i \in L$, and the proof is complete. □

The Markov–Kakutani follows by a standard compactness argument. Recall that a family $\mathscr{F} = \{\phi\}$ of mappings is said to be *abelian* if $\phi_1\phi_2 = \phi_2\phi_1$ for all $\phi_1, \phi_2 \in \mathscr{F}$.

Corollary 7.13 (Theorem of Markov–Kakutani). *Let X be a non-empty compact convex subset of a topological vector space E with separating dual E' and let $\mathscr{F}$ be an abelian family of continuous affine transformations from X into itself. Then, there exists $x_0 \in X$ such that $\phi(x_0) = x_0$ for every $\phi \in \mathscr{F}$.*

Proof. For any given $\phi \in \mathscr{F}$, let $Fix(\phi)$ be the set of its fixed points. We show that $\bigcap_{\phi \in \mathscr{F}} Fix(\phi) \neq \emptyset$. Clearly, for each $\phi \in \mathscr{F}$, $Fix(\phi)$ is non-empty (by Corollary 7.12), convex (as ϕ is affine), and closed hence compact in X. It suffices to show that the family $\{Fix(\phi) : \phi \in \mathscr{F}\}$ has the finite intersection property, i.e., $\bigcap_{i=1}^n Fix(\phi_i) \neq \emptyset$ for any $\{\phi_1, \ldots, \phi_n\} \subset \mathscr{F}$. The proof is by induction on n. For $n = 1$, clearly $Fix(\phi_1) \neq \emptyset$ (again by Corollary 7.12). Assume that the statement is true for any family $\{\phi_1, \ldots, \phi_k\} \subset \mathscr{F}$ with $k = n-1$ and let $\{\phi_1, \ldots, \phi_n\} \subset \mathscr{F}$ be arbitrary. For any $x \in \bigcap_{i=1}^{n-1} Fix(\phi_i)$, $\phi_n(x) = \phi_n(\phi_i(x)) = \phi_i(\phi_n(x))$ for all $i = 1, \ldots, n-1$, that is, $\phi_n(x) \in \bigcap_{i=1}^{n-1} Fix(\phi_i)$. Thus ϕ_n maps the non-empty compact convex set $\bigcap_{i=1}^{n-1} Fix(\phi_i)$ into itself. By Corollary 7.12 $N_X(x)$ again, it has a fixed point $\bar{x} = \phi_n(\bar{x}) \cap \bigcap_{i=1}^{n-1} Fix(\phi_i)$, that is, $\bar{x} \in \bigcap_{i=1}^n Fix(\phi_i)$. □

The coincidence result (Φ, Φ^*) in Corollary 7.5 can be expressed as an alternative for systems of nonlinear inequalities with four functionals.

Theorem 7.19 (Nonlinear Alternative with Four Functions). *Let X and Y be two convex subsets of topological vector spaces and let $f : X \times Y \to \mathbb{R}$ be a function satisfying:*

(i) $f_1(x,y) \le f_2(x,y) \le f_3(x,y) \le f_4(x,y)$, *for all* $(x,y) \in X \times Y$;
(ii) $y \mapsto f_1(x,y)$ *is l.s.c. on* Y, *for each fixed* $x \in X$;
(iii $x \mapsto f_2(x,y)$ *is quasiconcave on* X, *for each fixed* $y \in Y$;
(iv) $y \mapsto f_3(x,y)$ *is quasiconvex on* Y, *for each fixed* $x \in X$;
(v) $x \mapsto f_4(x,y)$ *is u.s.c. on* X, *for each fixed* $y \in Y$.

If either X or Y is compact, then for any $\lambda \in \mathbb{R}$, the following alternative holds:

(A) there exists $\bar{x} \in X$ *such that* $f_4(\bar{x},y) \ge \lambda$, *for all* $y \in Y$; *or*
(B) there exists $\bar{y} \in Y$ *such that* $f_1(x,\bar{y}) \le \lambda$, *for all* $x \in X$.

Proof. Given $\lambda \in \mathbb{R}$, define $A, \tilde{A}, B, \tilde{B} : X \rightrightarrows Y$ for any given $x \in X$ as:

$$A(x) := \{y \in Y : f_3(x,y) < \lambda\} \text{ and } \tilde{A}(x) := \{y \in Y : f_4(x,y) < \lambda\}.$$
$$B(x) := \{y \in Y : f_2(x,y) > \lambda\} \text{ and } \tilde{B}(x) := \{y \in Y : f_1(x,y) > \lambda\}.$$

Clearly, if $\tilde{A}(x) \neq \emptyset$ for all $x \in X$ and $\tilde{B}^{-1}(y) \neq \emptyset$ for all $y \in Y$, then $(A,B) \in \Phi \times \Phi^*$ and would have a coincidence point $\lambda < f_2(x_0,y_0) \le f_3(x_0,y_0) < \lambda$, which is impossible. Thus, $\exists \bar{x} \in X$ with $\tilde{A}(\bar{x}) = \emptyset$, that is, (A) holds, or $\exists \bar{y} \in Y$ with $\tilde{B}^{-1}(\bar{y}) = \emptyset$, that is, (B) is true. □

7.7 Concluding Remarks

The Browder–Ky Fan and the Kakutani–Ky Fan fixed point theorems are central existence results in nonlinear functional analysis as they lie at the foundation of solvability problems in a wide array of areas, from dynamical systems to game theory, including variational inequalities and optimization. As suggested in this chapter, there is a unity between the most fundamental solvability theorems of nonlinear and convex analysis. This unity is best expressed by the many equivalences between the results of Brouwer, Schauder–Tychonoff, Browder–Ky Fan, Kakutani–Ky Fan, the KKMF Principle, Sperner's lemma, the Ky Fan inf-sup inequality, the von-Neumann minimax principle, etc. (see, for example, [10, 39, 50]). Significant developments on theoretical aspects took place in the last 40 years. The lack of convexity on the spaces and the values of the maps were tackled by topological methods from homotopy or homology theories (see, for example, [3, 4, 6, 26] and reference therein) for the case where contractibility, proximal contractibility, or trivial homology are the natural substitutes for convexity. The motivations there stem from the methods used in the classical convex setting were adapted to situation where convexity is "metric" or "topological" and usually defined by means of simplicial transformations or set-valued maps (see, for example, [4, 20, 24, 40, 46] and references therein). In the absence of convincing applications, the consideration of "generalized convexities" has only limited theoretical significance. In this latter case, a lot is yet to be done by way of meaningful and striking applications, particularly in promising areas of computer sciences and game theory, where convexity is not linear.

References

1. Allen, G.: Variational inequalities, complementarity problems and duality theorems. J. math. Anal. Appl. **58**, 1–10 (1977)
2. Ben-El-Mechaiekh, H.: Note on a class of set-valued maps having continuous selections. In: Théra, M.A., Baillon, J.B. (eds.) Fixed Point Theory and Applications, Proceedings of the "Colloque International sur la Théorie du Point Fixe et Applications", France, 1989. Pitman Research Notes in Math. Series, vol. 252, 1991, pp. 33–43
3. Ben-El-Mechaiekh, H.: Continuous approximations of multifunctions, fixed points and coincidences. In: Florenzano, M., et al. (eds.) Approximation and Optimization in the Carribean II, Proceedings of the Second International Conference on Approximation and Optimization in the Carribean, pp. 69–97. Peter Lang Verlag, Frankfurt (1995)
4. Ben-El-Mechaiekh, H.: Spaces and maps approximation and fixed points. J. Comp. Appl. Math. **113**, 283–308 (2000)
5. Ben-El-Mechaiekh, H.: The Ky Fan fixed point theorem on star-shaped domains. Math. Rep. Acad. Sci. Can. **27**, 97–100 (2005)
6. Ben-El-Mechaiekh, H.: On nonlinear inclusions in non-smooth domains. Arabian J. Math. **1**, 395–416 (2012)
7. Ben-El-Mechaiekh, H., Chebbi, S., Florenzano, M.: A generalized KKMF principle. J. Math. Anal. Appl. **309**, 583–590 (2005)
8. Ben-El-Mechaiekh, H., Chebbi, S., Florenzano, M., Llinares, J.V.: Abstract convexity and fixed points. J. Math. Anal. Appl. **222**, 138–150 (1998)
9. Ben-El-Mechaiekh, H., Deguire, P.: Approachability and fixed points for non-convex set-valued maps. J. Math. Anal. Appl. **170**, 477–500 (1992)
10. Ben-El-Mechaiekh, H., Dimand, R.: The von Neumann minimax principle revisited. In: Fixed Point Theory and Its Applications, Banach Center Publications, vol. 77, pp. 23–34. Inst. of Mathematics, Polish Academy of Sciences, Warszawa, Poland (2007)
11. Ben-El-Mechaiekh, H., Deguire, P., Granas, A.: Points fixes et coincidences pour les fonctions multivoques I (applications de Ky Fan). C. R. Acad. Sci. Paris **295**, 337–340 (1982)
12. Ben-El-Mechaiekh, H., Deguire, P., Granas, A.: Points fixes et coincidences pour les fonctions multivoques II (applications de type Φ et Φ^*). C. R. Acad. Sci. Paris **295**, 381–384 (1982)
13. Ben-El-Mechaiekh, H., Deguire, P., Granas, A.: Points fixes et coincidences pour les fonctions multivoques III (applications de type M et M^*). C. R. Acad. Sci. Paris **305**, 381–384 (1987)
14. Ben-El-Mechaiekh, H., Idzik, A.: Ky Fan type coincidence theorems I. Bull. Pol. Acad. Sci. **122**, 105–109 (1994)
15. Ben-El-Mechaiekh, H., Idzik, A.: A Leray–Schauder type theorem for approximable maps. Proc. Am. Math. Soc. **122**, 105–109 (1994)
16. Ben-El-Mechaiekh, H., Isac, G.: Some geometric solvability theorems in topological vector spaces. Bull. Kor. Math. Soc. **34**, 273–285 (1997)
17. Ben-El-Mechaiekh, H., Kryszewski, W.: Equilibria for perturbations of upper semicontinuous set-valued maps by convex processes. Georgian J. Math. **3**, 201–215 (1996)

18. Ben-El-Mechaiekh, H., Oudadess, M., Tounkara, J.: Approximation of multifunctions on uniform spaces and fixed points. In: Lau, A.T-L., Tweddle, I. (eds.) Topological Vector Spaces, Algebras and Related Areas. Pitman Research Notes in Mathematics Series, vol. 316, pp. 239–250. Longman, New York (1994)
19. Ben-El-Mechaiekh, H., Saidi, F.B.: On the continuous approximation of upper semicontinuous set-valued maps. Q & A in General Topology, in print
20. Bielwaski, R.: Simplicial convexity and its applications. J. Math. Anal. Appl. **127**, 155–171 (1987)
21. Brézis, H.: Analyse fonctionelle, théorie et applications. Masson, Paris (1983)
22. Browder, F.E.: The fixed point theory of multi-valued mappings in topological vector spaces. Math. Ann. **177**, 283–301 (1968)
23. Cellina, A.: A theorem on the approximation of compact multivalued mappings. Atti Accad. Naz. Lincei Rend. **8**, 149–153 (1969)
24. Chebbi, S., Gourdel, P., Hammami, H.: A generalization of Fan's matching theorem. J. Fixed Point Theor. Appl. **9**, 117–124 (2011)
25. Cornet, B.: Paris avec handicap et théorème de surjectivité de correspondances. C. R. Acad. Sci. Paris **281**, 479–482 (1975)
26. Dugundji, J., Granas, A.: Fixed Point Theory, vol. 1. PWN—Polish Scientific Publishers, Warszawa (1982)
27. Engelking, R.: General Topology. PWN—-Polish Scientific Publishers, Warszawa (1977)
28. Fan, K.: Fixed point and minimax theorems in locally convex topological linear spaces. Proc. Natl. Acad. Sci. USA **38**, 121–126 (1952)
29. Fan, K.: A generalization of Tychonoff's fixed point theorem. Math. Ann. **142**, 305–310 (1961)
30. Fan, K.: Applications of a theorem concerning sets with convex sections. Math. Ann. **163**, 189–203 (1966)
31. Fan, K.: A minimax inequality and applications. In: Shisha, O. (ed.) Inequalities III, pp. 103–113. Academic Press, New York, London (1972)
32. Fan, K.: Some properties of convex sets related to fixed point theorems. Math. Ann. **266**, 519–537 (1984)
33. Florenzano, M.: L'équilibre économique général transitif et intransitif: problème d'existence. Monographie du séminaire d'économétrie. C. N. R. S., Paris (1981)
34. Husain, T., Tarafdar, E.: A selection and a fixed point theorem and an equilibrium point of an abstract economy. Int. J. Math. Math. Sci. **18**, 179–184 (1995)
35. Karamardian, S.: Generalized complementarity problems. J. Optim. Theor. Appl. **8**, 161–168 (1971)
36. Granas, A.: On the Leray-Schauder alternative. Top. Meth. Nonlinear Anal. **2**, 225–230 (1993)
37. Granas, A.: Méthodes topologiques en analyse convexe, Séminaire de mathématiques supérieures. Les Presses de l'Université de Montréal **110**, 11–77 (1990)
38. Granas, A., Liu, F.C.: Coincidences for set-valued maps and minimax inequalities. J. Math. Pure. Appl. **65**, 119–148 (1986)
39. Gwinner, J.: On fixed points and variational inequalities. A circular tour. Nonlinear Anal. **5**, 565–583 (1981)
40. Horvath, C.D.: Contractibility and generalized convexity. J. Math. Anal. Appl. **156**, 341–357 (1991)
41. Isac, G.: On some generalization of Karamardian's theorem on the complementarity problem. Boll. Unione Math. Ital. **7**(2)-B, 323–332 (1988)
42. Isac, G.: Complementarity Problems. Lecture Notes in Mathematics, vol. 1528. Springer, Berlin (1992)
43. Lassonde, M.: On the use of KKM multifunctions in fixed point theory and related topics. J. Math. Anal. Appl. **97**, 151–201 (1983)
44. Lin, L.-J., Ansari, Q.H., Wu, C.Y.: Geometric properties and coincidence theorems with applications to, generalized vector equilibrium problems. J. Optim. Theor. Appl. **117**, 121–137 (2003)

45. Lin, L.-J., Wong, N.-C., Yu, Z.-T.: Continuous selections and fixed points of multi-valued mappings on noncompact or nonmetrizable spaces. Proc. Am. Math. Soc. **133**, 3421–3427 (2005)
46. Liu, F.C.: On a form of KKM principle and SupInfSup inequalities of von Neumann and of Ky Fan type. J. Math. Anal. Appl. **155**, 420–436 (1991)
47. Llinares, J.V.: Abstract convexity, some relations and applications. Optimization **51**, 799–818 (2003)
48. Michael, E.: Continuous selections I. Ann. Math. **63**, 361–382 (1956)
49. Milnor, J.: Analytic proofs of the "hairy ball theorem" and the Brouwer fixed point theorem. Am. Math. Mon. **85**, 525–527 (1978)
50. Park, S.: Ninety years of the Brouwer fixed point theorem. Vietnam J. Math. **27**, 187–222 (1999)
51. Park, S., Jeong, K.S.: Fixed point and non-retract theorems: classical circular tours. Taiwanese J. Math. **5**, 97–108 (2001)
52. Rogers, C.A.: A less strange version of Milnor's proof of Brouwer fixed point theorem. Am. Math. Mon. **8**, 521–524 (1980)
53. Singbal, B.V.: Generalized form of Schauder-Tychnoff fixed-point principle. In: Bonsall, F.F. (ed.) Lectures on Some Fixed-point Theorems of Functional Analysis. Mimeographed Notes. TFIR, Bombay (1962)
54. Smart, D.R.: Fixed Point Theorems. Cambridge University Press, Cambridge (1974)
55. Stuckless, T.: Brouwer's fixed point theorem: methods of proof and generalizations. M.Sc. Thesis, Simon Fraser University (2003)
56. Toussaint, S.: On the existence of equilibria in economies with infinitely many commodities and without ordered preferences. J. Econ. Theor. **33**, 98–115 (1984)
57. von Neumann, J.: Über ein Ökonomisches Gleichungssytem und eine Verallgemeinerung des Brouwerwschen Fixpunktsatzes. In: Menger, K. (ed.) Ergebnisse eines Mathematischen Kolloquium. Vienna. Translated by Morton G. as *A Model of General Economic Equilibrium*, Rev. Econ. Stud. **13**, 1–9 (1945)
58. Yannelis, N.C., Prabakhar, N.D.: Existence of maximal elements and equilibria in linear topological spaces. J. Math. Econ. **12**, 233–245 (1983)

Chapter 8
Some Iterative Methods for Fixed Point Problems

Q.H. Ansari and D.R. Sahu

8.1 Introduction

Many problems arise in different areas of mathematics, such as optimization, variational analysis, and differential equations, can be modeled by the following fixed point problem:

$$\text{find } x \in C \quad \text{such that} \quad x = Tx, \tag{8.1}$$

where T is an operator (possibly nonlinear) defined on a subset C of a suitable space X (see [4]). The solutions to this problem are called *fixed points* of the mapping T. We denote by $\mathrm{Fix}(T)$ the set of all fixed points of T.

There is a classical general existence theory of fixed points for mappings satisfying compactness conditions associated with the names of Browder, Schauder, Leray, etc., as well as an abundant literature of metrical fixed point theorems. We concentrate on approximation of fixed points of nonexpansive type mappings in non-compact setting.

An iteration method for solving problem (8.1) consists of the construction of a sequence $\{x_n\}$ in C which converges to x^*, given a suitable starting value x_0, and a procedure for calculating the value of x_{n+1} once x_n, is known. Picard, Mann, Ishikawa are well-known fixed point iteration methods.

It is well known [1, 3] that if T is a contraction mapping defined on a complete metric space X, then the Banach contraction principle establishes that T has a unique fixed point x^* and for any $x \in X$, the sequence of Picard iterates $\{T^n x\}$ converges to

Q.H. Ansari (✉)
Department of Mathematics, Aligarh Muslim University, Aligarh, India
e-mail: qhansari@gmail.com

D.R. Sahu
Department of Mathematics, Banaras Hindu University, Varanasi, India
e-mail: drsahudr@gmail.com

S. Almezel et al. (eds.), *Topics in Fixed Point Theory*, DOI 10.1007/978-3-319-01586-6_8,

x^* (see Sect. 8.3). However, if the mapping T is nonexpansive, then we must assume additional conditions on T and/or the underlying space to ensure the existence of fixed points.

The interest and importance of construction of fixed points of nonexpansive mappings stem mainly from the fact that it may be applied in many areas, such as image recovery and signal processing, see, for example, [8, 40].

Iterative techniques for approximating fixed points of nonexpansive mappings have been studied by using famous. Some variations of Mann and Ishikawa iterations for approximation of several problems can be found in [2, 9, 15, 17, 18, 22, 28, 29, 45, 47, 50].

In this chapter, we focus on some iterative methods for approximation of fixed points of nonlinear operators of nonexpansive type with non-compact convex domain in Banach/Hilbert spaces. We include some knowledge of geometry of Banach/Hilbert spaces as well as some fundamental properties of iterative methods for approximating fixed points of nonexpansive type mappings. Subsequently, we introduce iterative methods and discuss their convergence analysis for approximation of fixed points of nonexpansive type mappings in suitable spaces.

8.2 Preliminaries

We present some fundamental results which will be used in the sequel.

Lemma 8.1 ([38]). *Let $\{\delta_n\}$ be a sequence of nonnegative numbers satisfying:*

$$\delta_{n+1} \le \beta_n \delta_n + \gamma_n, \quad \textit{for all } n \in \mathbb{N},$$

where $\{\beta_n\}$ and $\{\gamma_n\}$ are sequences of nonnegative numbers such that

$$\{\beta_n\} \subseteq [1,\infty), \ \sum_{n=1}^{\infty} (\beta_n - 1) < \infty, \tag{8.2}$$

$$\sum_{n=1}^{\infty} \gamma_n < \infty. \tag{8.3}$$

Then, $\lim_{n\to\infty} \delta_n$ exists. If $\liminf_{n\to\infty} \delta_n = 0$, then $\lim_{n\to\infty} \delta_n = 0$.

Proof. For $m, n \in \mathbb{N}$, we have

$$\begin{aligned}
\delta_{n+m+1} &\le \beta_{n+m}\delta_{n+m} + \gamma_{n+m} \\
&\le \beta_{n+m}(\delta_{n+m} + \gamma_{n+m}) \\
&\le \beta_{n+m}(\beta_{n+m-1}(\delta_{n+m-1} + \gamma_{n+m-1}) + \gamma_{n+m}) \\
&\cdots \\
&\le \left(\prod_{i=n}^{n+m} \beta_i\right)\left(\delta_n + \sum_{i=n}^{n+m} \gamma_i\right).
\end{aligned}$$

Hence, $\limsup_{m\to\infty} \delta_m \le \left(\prod_{i=n}^{\infty} \beta_i\right)\left(\delta_n + \sum_{i=n}^{\infty} \gamma_i\right)$. By the conditions (8.2) and (8.3), we have $\lim_{n\to\infty} \left(\prod_{i=n}^{\infty} \beta_i\right) = 1$ and $\lim_{n\to\infty} \sum_{i=n}^{\infty} \gamma_i = 0$. It follows that $\limsup_{n\to\infty} \delta_n \le \liminf_{n\to\infty} \delta_n$. Therefore, $\lim_{n\to\infty} \delta_n$ exists. Suppose $\liminf_{n\to\infty} \delta_n = 0$. Then, $\lim_{n\to\infty} \delta_n = \liminf_{n\to\infty} \delta_n = 0$. □

Lemma 8.2 ([55]). *Let $\{\alpha_n\}$ be a sequence of nonnegative real numbers such that*

$$\alpha_{n+1} \le (1-\gamma_n)\alpha_n + \gamma_n \sigma_n, \quad \text{for all } n \ge 0, \tag{8.4}$$

where $\{\gamma_n\}$ is a sequence in $(0,1)$ and $\{\sigma_n\}$ is a sequence of real numbers such that

(i) $\sum_{n=0}^{\infty} \gamma_n = \infty$,

(ii) either $\limsup_{n\to\infty} \sigma_n \le 0$ *or* $\sum_{n=0}^{\infty} |\gamma_n \sigma_n| < \infty$.

Then, $\{\alpha_n\}_{n=0}^{\infty}$ converges to zero.

Proof. For any $\varepsilon > 0$, let N be an integer large enough so that

$$\sigma_n < \varepsilon, \quad \text{for } n \ge N.$$

Using (8.4) and by induction, we obtain, for $n > N$,

$$\alpha_{n+1} \le \left(\prod_{k=N}^{n}(1-\gamma_k)\right)\alpha_N + \left((1-\prod_{k=N}^{n}(1-\gamma_k)\right)\varepsilon.$$

Then, condition (ii) implies that $\limsup_{n\to\infty} \alpha_n \le 2\varepsilon$. □

Lemma 8.3 ([20]). *Suppose $\{a_n\}$ and $\{b_n\}$ are two sequences in a normed space X. Let $\{t_n\}$ be a sequence of real numbers satisfying the following conditions:*

(a) $0 \le t_n \le t < 1$ *and* $\sum_{n=1}^{\infty} t_n = \infty$,
(b) $a_{n+1} = (1-t_n)a_n + t_n b_n$ *for all* $n \in \mathbb{N}$,
(c) $\lim_{n\to\infty} \|a_n\| = a$,
(d) $\limsup_{n\to\infty} \|b_n\| \le a$ *and* $\{\sum_{i=1}^{n} t_i b_i\}$ *is bounded.*

Then, $a = 0$.

Let X be a normed space. We denote by $x_n \to x$ and $x_n \rightharpoonup x$, the strong convergence and the weak convergence, respectively, of the sequence $\{x_n\}$ to x. We also adopt the following notation:

$$w_\omega(\{x_n\}) = \left\{x : \exists x_{n_j} \rightharpoonup x \text{ denotes the weak w-limit set of } \{x_n\}\right\}.$$

A normed space X is said to satisfy the *Opial condition* [37] if for each sequence $\{x_n\}$ in X converges weakly to a point $x \in X$, we have

$$\liminf_{n\to\infty} \|x_n - x\| < \liminf_{n\to\infty} \|x_n - y\|, \quad \text{for all } y \ (\neq x) \in X.$$

Let H be a real Hilbert space. It is well known that

$$\|x-y\|^2 = \|x\|^2 - \|y\|^2 - 2\langle x-y, y\rangle, \quad \text{for all } x,y \in H. \tag{8.5}$$

Also,

(a) $\|(1-t)x+ty\|^2 = (1-t)\|x\|^2 + t\|y\|^2 - t(1-t)\|x-y\|^2$, for all $t \in [0,1]$ and all $x,y \in H$;
(b) If $\{x_n\}$ is a sequence in H such that $x_n \rightharpoonup x$, it follows that

$$\limsup_{n\to\infty} \|x_n - y\|^2 = \limsup_{n\to\infty} \|x_n - x\|^2 + \|x-y\|^2, \quad \text{for all } y \in H. \tag{8.6}$$

We conclude from identity (8.6) that every Hilbert space enjoys the Opial condition.

Lemma 8.4. *Let H be a real Hilbert space and C be a closed convex subset of H. Then forall $x,y,z \in H$ and a real number $a \in \mathbb{R}$, the set*

$$\left\{v \in C : \|y-v\|^2 \le \|x-v\|^2 + \langle z,v\rangle + a\right\}$$

is convex (and closed).

A subset C of a Hilbert space H is called a *retract* of H if there is exists a continuous mapping P from H onto C such that $Px = x$ for all $x \in C$. Such P is a called a *retraction* of H onto C. It follows that if P is a retraction, then $Py = y$ for all y in the range of P.

A retraction P is said to be *sunny* if $P(Px + t(x - Px)) = Px$ for all $x \in H$ and $t \ge 0$. If a sunny retraction P is also nonexpansive, then C is said to be a *sunny nonexpansive retract* of H.

Let C be a nonempty subset of H and $x \in H$. An element $\bar{x} \in C$ is said to be a *best approximation* to x if $\|x-\bar{x}\| = d(x,C)$, where $d(x,C) = \inf_{y\in C} \|x-y\|$. The set of all best approximations from x to C is denoted by

$$P_C(x) = \{y \in C : \|x-y\| = d(x,C)\}.$$

This defines a mapping P_C from X into 2^C and is called the *nearest point projection mapping* (or *metric projection mapping*) onto C.

It is well known that if C is a nonempty closed convex subset of a real Hilbert space H, then the nearest point projection P_C from H onto C is the unique sunny nonexpansive retraction of H onto C (see [1]). It is also known that $P_C x \in C$ and,

$$\langle x - P_C x, P_C x - y\rangle \ge 0, \quad \text{for all } x \in H \text{ and } y \in C. \tag{8.7}$$

8.2.1 The LE Property and the AF Point Property for Nonlinear Operators

Definition 8.1 ([41]). Let X be a normed space, C a nonempty convex subset of X, $T : C \to C$ an operator with $\mathrm{Fix}(T) \neq \emptyset$ and $\{x_n\}$ a sequence in C. We say that $\{x_n\}$ has:

(D1) *the limit existence property* (for short, the *LE property*) for T if $\lim_{n\to\infty} \|x_n - p\|$ exists for all $p \in \mathrm{Fix}(T)$;
(D2) *the approximate fixed point property* (for short, the *AF point property*) for T if $\lim_{n\to\infty} \|x_n - Tx_n\| = 0$;
(D3) *the LEAF point property* if $\{x_n\}$ has both LE property and AF point property.

Both LE property and AF point property play fundamental role in fixed point theory of nonlinear operators. These basic properties are initially studied in [1].

Let X be a Banach space and T be a mapping with domain $D(T)$ and range $R(T)$ in X. Then, T is said to be *demiclosed at a point* $p \in R(T)$ if whenever $\{x_n\}$ is a sequence in $D(T)$ converges weakly to a point $z \in D(T)$ and $\{Tx_n\}$ converges strongly to p, then $Tz = p$.

The following demiclosed principle for nonexpansive mappings is one of the most important results in fixed point theory.

Proposition 8.1 (Demiclosed Principle). *Let X be a Banach space satisfying the Opial condition. Let C be a nonempty weakly closed subset of X and $T : C \to C$ be a nonexpansive mapping. Then, $I - T$ is demiclosed at zero.*

Proof. Suppose that $\{x_n\}$ is a sequence in C converges weakly to x and $\{x_n\}$ has the AF point property for T. Suppose, for contradiction, that $x \neq Tx$. Then, by the Opial condition, we have

$$\begin{aligned}\limsup_{n\to\infty} \|x_n - x\| &< \limsup_{n\to\infty} \|x_n - Tx\| \\ &\leq \limsup_{n\to\infty} (\|x_n - Tx_n\| + \|Tx_n - Tx\|) \\ &\leq \limsup_{n\to\infty} \|x_n - x\|,\end{aligned}$$

a contradiction. This proves that $(I - T)x = 0$. □

An important feature of a sequence with *AF* point property for any mapping T is that any weak subsequential limit of the sequence is a fixed point of T when $I - T$ is demiclosed at zero.

Proposition 8.2. *Let X be a Banach space satisfying the Opial condition, C a nonempty weakly compact subset of X and $T : C \to C$ a mapping such that*

(a) $\mathrm{Fix}(T) \neq \emptyset$,
(b) $I - T$ is demiclosed at zero.

Let $\{x_n\}$ be a sequence in C such that $\{x_n\}$ has the LEAF point property. Then, $\{x_n\}$ converges weakly to a fixed point of T.

Proof. Since C is weakly compact, it follows that $\{x_n\}$ has a weakly convergent subsequence $\{x_{n_j}\}$. Suppose $\{x_{n_j}\}$ converges weakly to p. Since $\{x_{n_j}\} \subset C$ and C is weakly closed, then $p \in C$. Since $\lim_{n\to\infty} ||x_n - Tx_n|| = 0$ and $I - T$ is demiclosed at zero, we have $(I-T)p = 0$, so that $p \in \text{Fix}(T)$. To complete the proof, we show that $\{x_n\}$ converges weakly to a fixed point of T. It suffices to show that $w_\omega(\{x_n\})$ consists exactly one point, namely, p. Suppose that there exists another subsequence $\{x_{n_k}\}$ of $\{x_n\}$ which converges weakly to some $q \neq p$. As in the case of p, we must have $q \in C$ and $q \in \text{Fix}(T)$. It follows from the LE property that $\lim_{n\to\infty} ||x_n - p||$ and $\lim_{n\to\infty} ||x_n - q||$ exist. Since X satisfies the Opial condition, we have

$$\lim_{n\to\infty} ||x_n - p|| = \lim_{j\to\infty} ||x_{n_j} - p|| < \lim_{j\to\infty} ||x_{n_j} - q|| = \lim_{n\to\infty} ||x_n - q||,$$

$$\lim_{n\to\infty} ||x_n - q|| = \lim_{k\to\infty} ||x_{n_k} - q|| < \lim_{k\to\infty} ||x_{n_k} - p|| = \lim_{n\to\infty} ||x_n - p||,$$

a contradiction. Hence, $p = q$ and $\{x_n\}$ converges weakly to p. □

8.2.2 Nearly Lipschitzian Mappings

Definition 8.2 ([42]). Let C be a nonempty subset of a Banach space X and fix a sequence $\{a_n\}$ in $[0,\infty)$ with $a_n \to 0$. A mapping $T : C \to C$ is said to be *nearly Lipschitzian* with respect to $\{a_n\}$ if for each $n \in \mathbb{N}$, there exists a constant $k_n \geq 0$ such that

$$\|T^n x - T^n y\| \leq k_n(\|x-y\| + a_n), \quad \text{for all } x, y \in C. \tag{8.8}$$

The infimum of constants k_n for which (8.8) holds is denoted by $\eta(T^n)$ and called the *nearly Lipschitz constant* of T^n.

Definition 8.3. A nearly Lipschitzian mapping T with sequence $\{(a_n, \eta(T^n)\}$ is said to be

(a) *nearly nonexpansive* if $\eta(T^n) \leq 1$ for all $n \in \mathbb{N}$;
(b) *nearly asymptotically nonexpansive* if $\eta(T^n) \geq 1$ for all $n \in \mathbb{N}$ and $\lim_{n\to\infty} \eta(T^n) = 1$;
(c) *nearly uniformly k-Lipschitzian* if $\eta(T^n) \leq k$ for all $n \in \mathbb{N}$;
(d) *nearly uniform k-contraction* if $\eta(T^n) \leq k < 1$ for all $n \in \mathbb{N}$.

Remark 8.1. A nearly asymptotically nonexpansive with sequence $\{(a_n, \eta(T^n))\}$ is asymptotically nonexpansive if $a_n = 0$ for all $n \in \mathbb{N}$.

Example 8.1. Let $X = \mathbb{R}$, $C = [0,1]$ and $T : C \to C$ be a mapping defined by

$$Tx = \begin{cases} \frac{1}{2}, & \text{if } x \in [0, \frac{1}{2}], \\ 0, & \text{if } x \in (\frac{1}{2}, 1]. \end{cases}$$

Clearly, T is discontinuous and non-Lipschitzian mapping. However, it is nearly nonexpansive mapping, and hence, nearly asymptotically nonexpansive. Indeed, for a sequence $\{a_n\}$ with $a_1 = \frac{1}{2}$ and $a_n \to 0$, we have

$$\|Tx - Ty\| \le \|x - y\| + a_1, \quad \text{for all } x, y \in C,$$

and

$$\|T^n x - T^n y\| \le \|x - y\| + a_n, \quad \text{for all } x, y \in C \text{ and } n \ge 2,$$

since

$$T^n x = \frac{1}{2}, \quad \text{for all } x \in [0,1] \text{ and } n \ge 2.$$

Sahu [42] developed a nearly contraction mapping principle for the existence and uniqueness of fixed points of demicontinuous mappings, more general than contraction mappings, in Banach spaces. The details of the fixed point theory of nearly Lipschitzian mappings can be found in [1].

The following result is about AF point property for sequence of nearly asymptotically nonexpansive mappings in general normed spaces.

Proposition 8.3. *Let C be a nonempty convex subset of a normed space X and for each $n \in \mathbb{N}$, let $T_n : C \to C$ be a mapping satisfying the following condition:*

$$\|T_n x - T_n y\| \le L_n(\|x - y\| + \rho_n), \quad \textit{for all } x, y \in C \textit{ and } n \in \mathbb{N},$$

where $\{(\rho_n, L_n)\}$ is a sequence in $[0,\infty) \times [1,\infty)$ such that $\sum \rho_n < \infty$ and $\sum_{n=1}^{\infty}(L_n - 1) < \infty$. Let $\{\alpha_n\}$ be a sequence in $(0,1)$ such that $0 < a \le \alpha_n \le b < 1$ for all $n \in \mathbb{N}$. For arbitrary $x_1 \in C$, define a sequence $\{x_n\}$ in C by

$$x_{n+1} = (1 - \alpha_n)x_n + \alpha_n T_n x_n, \quad \textit{for } n \in \mathbb{N}. \tag{8.9}$$

Suppose that $\{x_n\}$ is bounded and $\sum_{n=1}^{\infty} \|T_n x_n - T_{n+1} x_n\| < \infty$. Then,

(a) $(I - T_n)x_n \to 0$ as $n \to \infty$, and
(b) $x_n - x_{n+1} \to 0$ as $n \to \infty$.

Proof. (a) Set $d_n := x_n - T_n x_n$ and $\varepsilon_n := \|T_n x_n - T_{n+1} x_n\|$ for all $n \in \mathbb{N}$. By (8.9), we have

$$\begin{aligned}
\|d_{n+1}\| &= \|x_{n+1} - T_{n+1}x_{n+1}\| \\
&\le (1-\alpha_n)\|x_n - T_{n+1}x_{n+1}\| + \alpha_n\|T_nx_n - T_{n+1}x_{n+1}\| \\
&\le (1-\alpha_n)(\|x_n - x_{n+1}\| + \|x_{n+1} - T_{n+1}x_{n+1}\|) \\
&\quad + \alpha_n(\|T_nx_n - T_{n+1}x_n\| + \|T_{n+1}x_n - T_{n+1}x_{n+1}\|) \\
&\le (1-\alpha_n)(\alpha_n\|d_n\| + \|d_{n+1}\|) + \alpha_n(\varepsilon_n + L_{n+1}(\|x_n - x_{n+1}\| + \rho_n)),
\end{aligned}$$

which implies that

$$\begin{aligned}
\|d_{n+1}\| &\le (1-\alpha_n)\|d_n\| + \varepsilon_n + L_{n+1}(\alpha_n\|d_n\| + \rho_n) \\
&\le L_{n+1}\|d_n\| + \varepsilon_n + K\rho_n,
\end{aligned}$$

where $K = \sup_{n\in\mathbb{N}} L_n$. Since $\sum_{n=1}^{\infty}(\varepsilon_n + K\rho_n) < \infty$, it follows from Lemma 8.1 that $\lim_{n\to\infty}\|d_n\|$ exists.

Let $\lim_{n\to\infty}\|d_n\| = d$ and set $b_n := \alpha_n^{-1}(T_nx_n - T_{n+1}x_{n+1})$. Then, $d_{n+1} = (1-\alpha_n)d_n + \alpha_nb_n$. Observe that

$$\begin{aligned}
\|b_n\| &= \alpha_n^{-1}\|T_nx_n - T_{n+1}x_{n+1}\| \\
&\le \alpha_n^{-1}(\|T_nx_n - T_{n+1}x_n\| + \|T_{n+1}x_n - T_{n+1}x_{n+1}\|) \\
&\le \alpha_n^{-1}(\varepsilon_n + (L_{n+1}\|x_n - x_{n+1}\| + \rho_n)) \\
&\le \alpha_n^{-1}(\varepsilon_n + L_{n+1}(\alpha_n\|d_n\| + \rho_n)) \\
&\le L_{n+1}\|d_n\| + a^{-1}(\varepsilon_n + L_{n+1}\rho_n),
\end{aligned}$$

which implies that $\limsup_{n\to\infty}\|b_n\| \le d$. Moreover, for $n \in \mathbb{N}$, we have

$$\begin{aligned}
\left\|\sum_{i=1}^{n}\alpha_ib_i\right\| &= \left\|\sum_{i=1}^{n}(T_ix_i - T_{i+1}x_{i+1})\right\| \\
&= \|T_1x_1 - T_{n+1}x_{n+1}\| \\
&\le \|T_1x_1 - x_n\| + \|x_n - T_nx_n\| + \|T_nx_n - T_{n+1}x_n\| \\
&\quad + \|T_{n+1}x_n - T_{n+1}x_{n+1}\| \\
&\le \|T_1x_1 - x_n\| + \|d_n\| + \varepsilon_n + L_{n+1}(\alpha_n\|d_n\| + \rho_n).
\end{aligned}$$

Note that $\{\|d_n\|\}$, $\{L_n\}$, $\{\varepsilon_n\}$, $\{\rho_n\}$ are convergent sequences and $\{x_n\}$ is bounded. So $\{\sum_{i=1}^{n}\alpha_ib_i\}$ is bounded. Since $\{\alpha_n\}$ is bounded away from 0 and 1, it follows that $\sum_{n=1}^{\infty}\alpha_n = \infty$. Applying Lemma 8.3, we conclude that $(I - T_n)x_n \to 0$ as $n \to \infty$.

(b) From (8.9), we have

$$\|x_{n+1} - x_n\| \leq \alpha_n \|x_n - T_n x_n\| \leq b\|x_n - T_n x_n\|, \quad \text{for all } n \in \mathbb{N}. \tag{8.10}$$

By Part (a), we have $\|x_n - T_n x_n\| \to 0$ as $n \to \infty$. It follows from (8.10) that $x_{n+1} - x_n \to 0$ as $n \to \infty$. □

Proposition 8.3 extends Theorem 1 of Deng [20] and the corresponding result of Ishikawa [27] for a sequence of nearly asymptotically nonexpansive mappings.

8.2.3 *Asymptotically κ-strict Pseudocontractive Mappings in the Intermediate Sense*

As a generalization of class of nonexpansive mappings, the class of κ-strict pseudocontractive mapping was introduced by Browder and Petryshyn [7]. Some iterative methods for fixed points of κ-strict pseudocontractive mappings are initiated in [7, 11]. Kim and Xu [30] introduced the concept of asymptotically κ-strict pseudocontractive mappings in Hilbert space as a generalization of asymptotically nonexpansive mapping [23] and κ-strict pseudocontractive mapping. Recently, the notion of an asymptotically κ-strict pseudocontractive mapping in the intermediate sense was introduced by Sahu et al. [44] in a Hilbert space.

Definition 8.4. Let C be a nonempty subset of a Hilbert space H. A mapping $T : C \to C$ is called *asymptotically κ-strict pseudocontractive in the intermediate sense* with sequence $\{\gamma_n\}$ if there exists a constant $\kappa \in [0,1)$ and a sequence $\{\gamma_n\}$ in $[0,\infty)$ with $\lim_{n\to\infty} \gamma_n = 0$ such that

$$\limsup_{n\to\infty} \sup_{x,y\in C} \Big(\|T^n x - T^n y\|^2 - (1+\gamma_n)\|x-y\|^2 - \kappa\|x - T^n x - (y - T^n y)\|^2 \Big) \leq 0. \tag{8.11}$$

We assume that

$$c_n := \max\left\{0, \sup_{x,y\in C} \left(\|T^n x - T^n y\|^2 - (1+\gamma_n)\|x-y\|^2 - \kappa\|x - T^n x - (y - T^n y)\|^2\right)\right\}.$$

Then, $c_n \geq 0$ for all $n \in \mathbb{N}$, $c_n \to 0$ as $n \to \infty$ and (8.11) reduces to the relation

$$\|T^n x - T^n y\|^2 \leq (1+\gamma_n)\|x-y\|^2 + \kappa\|x - T^n x - (y - T^n y)\|^2 + c_n, \tag{8.12}$$

for all $x, y \in C$ and $n \in \mathbb{N}$.

We give basic properties of asymptotically κ-strict pseudocontractive mappings in the intermediate sense.

Proposition 8.4. *Let C be a nonempty subset of a Hilbert space H and $T : C \to C$ an asymptotically κ-strict pseudocontractive mapping in the intermediate sense with sequence $\{\gamma_n\}$. Then,*

$$\|T^n x - T^n y\| \leq \frac{1}{1-\kappa}\left(\kappa\|x-y\| + \sqrt{(1+(1-\kappa)\gamma_n)\|x-y\|^2 + (1-\kappa)c_n}\right),$$

for all $x, y \in C$ and $n \in \mathbb{N}$.

Proof. For $x, y \in C$, we have

$$\begin{aligned}\|T^n x - T^n y\|^2 &\leq (1+\gamma_n)\|x-y\|^2 + \kappa\|x - T^n x - (y - T^n y)\|^2 + c_n \\ &\leq (1+\gamma_n)\|x-y\|^2 + \kappa(\|x-y\| + \|T^n x - T^n y\|)^2 + c_n \\ &\leq (1+\kappa+\gamma_n)\|x-y\|^2 + \kappa(2\|x-y\|\ \|T^n x - T^n y\| \\ &\quad + \|T^n x - T^n y\|^2) + c_n.\end{aligned}$$

It gives us that

$$(1-\kappa)\|T^n x - T^n y\|^2 - 2\kappa\|x-y\|\ \|T^n x - T^n y\| - (1+\kappa+\gamma_n)\|x-y\|^2 - c_n \leq 0,$$

which is a quadratic inequality in $\|T^n x - T^n y\|$. Hence, the result follows. □

Proposition 8.5. *Let C be a nonempty subset of a Hilbert space H and $T : C \to C$ a uniformly continuous asymptotically κ-strict pseudocontractive mapping in the intermediate sense with sequence $\{\gamma_n\}$. Let $\{x_n\}$ be a sequence in C such that $\|x_n - x_{n+1}\| \to 0$ and $\|x_n - T^n x_n\| \to 0$ as $n \to \infty$. Then, $\|x_n - Tx_n\| \to 0$ as $n \to \infty$.*

Proof. Since T is an asymptotically κ-strict pseudocontractive mapping in the intermediate sense, we obtain from Proposition 8.4 that

$$\begin{aligned}\|T^{n+1}x_n - T^{n+1}x_{n+1}\| \leq \frac{1}{1-\kappa}\Bigg(&\kappa\ \|x_n - x_{n+1}\| \\ &+ \sqrt{(1+(1-\kappa)\gamma_{n+1})\|x_n - x_{n+1}\|^2 + (1-\kappa)c_{n+1}}\Bigg).\end{aligned}$$

Note that $\|x_n - x_{n+1}\| \to 0$ which implies that $\|T^{n+1}x_n - T^{n+1}x_{n+1}\| \to 0$. Observe that

$$\begin{aligned}\|x_n - Tx_n\| &\leq \|x_n - x_{n+1}\| + \|x_{n+1} - T^{n+1}x_{n+1}\| \\ &\quad + \|T^{n+1}x_{n+1} - T^{n+1}x_n\| + \|T^{n+1}x_n - Tx_n\|. \qquad (8.13)\end{aligned}$$

By the uniform continuity of T, we have

$$\|Tx_n - T^{n+1}x_n\| \to 0, \quad \text{as } n \to \infty. \tag{8.14}$$

Since $x_n - T^n x_n \to 0$ and $x_n - x_{n+1} \to 0$, it follows from (8.13) and (8.14) that $\lim_{n\to\infty} \|x_n - Tx_n\| = 0$. □

Remark 8.2. It is shown in [44] that if C is a nonempty subset of a Hilbert space H and $T : C \to C$ is a uniformly continuous asymptotically κ-strict pseudocontractive mapping in the intermediate sense, then $I - T$ is demiclosed at zero in the sense that if $\{x_n\}$ is a sequence in C such that $x_n \rightharpoonup x \in C$ and $\limsup_{m\to\infty} \limsup_{n\to\infty} \|x_n - T^m x_n\| = 0$, then $(I - T)x = 0$.

8.3 Picard Iterative Method

The form of (8.1) suggests immediately the classical method of successive substitutions, in which one chooses some x_0, and uses the relationship

$$x_{n+1} = Tx_n, \quad \text{for } n \geq 0,$$

to construct the remainder of the sequence $\{x_n\}$. This successive approximation technique is also known as *Picard iteration method* [39]. More precisely, we have

Definition 8.5. Let C be a nonempty subset of a metric space (X, d) and $T : C \to C$ a mapping. Then, the Picard iteration sequence $\{x_n\}$ is given by

$$\begin{cases} x_0 \in C, \\ x_{n+1} = Tx_n, \quad n \in \mathbb{N}_0, \end{cases} \tag{8.15}$$

where $\mathbb{N}_0$ denotes the set of whole numbers.

We prove the well-known Banach contraction principle and discuss the role of Picard iteration sequence in sense of the AF point property.

Theorem 8.1. *Let C be a closed subset of a complete metric space (X, d) and $T : C \to C$ a contraction mapping. Then,*

(a) The Picard iteration sequence $\{x_n\}$ is defined by (8.15) has the AF point property for T;
(b) T has a unique fixed point or (8.1) has a unique solution (in C).

Proof. (a) Note that

$$d(x_2, x_1) = d(Tx_1, Tx_0) \leq kd(x_1, x_0),$$
$$d(x_3, x_2) = d(Tx_2, Tx_1) \leq kd(x_2, x_1) \leq k^2 d(x_1, x_0).$$

Inductively, we have

$$d(x_{n+1},x_n) \leq k^n(x_1,x_0)\ n \in \mathbb{N}.$$

It follows that $\{x_n\}$ has the AF point property for T.

(b) We show that $\{x_n\}$ is a Cauchy sequence. For $m > n$, we have

$$\begin{aligned} d(x_n,x_m) &\leq d(x_n,x_{n+1}) + d(x_{n+1},x_{n+2}) + \cdots + d(x_{m-1},x_m) \\ &\leq k^n d(x_1,x_0)[1+k+k^2+\cdots+k^{m-n-1}] \\ &\leq k^n d(x_1,x_0)[1+k+k^2+\cdots+k^{m-n-1}+\cdots] \\ &\leq k^n d(x_1,x_0)[1/(1-k)] \to 0y. \end{aligned} \tag{8.16}$$

Since $k < 1$, it follows that $\{x_n\}$ is a Cauchy sequence. Since X is a complete metric space, $x_n \to y$ for some $y \in C$. From (8.16), we have $d(x_{n+1},x_n) \to 0$ as $n \to \infty$ and hence $\lim_{n\to\infty} x_{n+1} = y$. Since T is continuous, $y = \lim_{n\to\infty} x_{n+1} = \lim_{n\to\infty} Tx_n = Ty$. It is easy to see the uniqueness. If $Tx = x$ and $Ty = y$ with $x \neq y$, then we get

$$d(x,y) = d(Tx,Ty) \leq kd(x,y),$$

a contradiction. Hence, fixed point of T is unique.

□

Remark 8.3. If X is not a complete metric space, then T may not have a fixed point. For example, if $X = C = (0,1]$ and $Tx = x/2$, then T has no fixed point in $(0,1]$.

Since the 1960s, the study of the class of nonexpansive mappings is one of the major and most active research areas of nonlinear analysis. This is due to the firm connection with the geometry of Banach spaces along with the relevance of these mappings in the theory of monotone and accretive operators.

Theorem 8.1 shows for each $x \in C$, $\{T^n x\}$ has the *AF* point property for contraction mappings. We can see that Picard iteration does not have AF point property for the case of nonexpansive mappings, in general. For example, if $T : [0,1] \to [0,1]$ is defined by $Tx = 1-x$ for all $x \in [0,1]$, then $\{T^n x\}$ does not have the *AF* point property for T, when $x \neq 1/2$. Proposition 8.2 guarantees the weak convergence of any sequence $\{x_n\}$ under the AF point property. Asymptotic regularity assumption of arbitrary nonlinear mapping T at $x \in C$, that is, $\lim_{n\to\infty} \|T^n x - T^{n+1}x\| = 0$ provides the facility that $\{T^n x\}$ has the *AF* point property for T. Using this fact, we deal with weak convergence of Picard iterates under asymptotic regularity.

Theorem 8.2. *Let X be a Banach space satisfying the Opial condition, C a nonempty weakly compact subset of X and $T : C \to C$ a nonexpansive mapping. If for $x \in C$, $\|T^n x - T^{n+1}x\| \to 0$ as $n \to \infty$, then $\{T^n x\}$ converges weakly to an element of $Fix(T)$.*

Proof. For $x \in C$, define a sequence $\{x_n\}$ in C by $x_n = T^n x$ for $n \in \mathbb{N}_0$. Then, for $v \in \text{Fix}(T)$,

$$\|x_{n+1} - v\| = \|T^{n+1}x - v\| \leq \|T^n x - v\| \leq \|x_n - v\|, \quad \text{for all } n \in \mathbb{N}_0,$$

and it follows that $\lim_{n\to\infty} \|x_n - v\|$ exists. By assumption $\|x_n - Tx_n\| \to 0$ as $n \to \infty$. It follows that $\{x_n\}$ has the AF Point property for T. By Proposition 8.1, $I - T$ is demiclosed at zero. Applying Proposition 8.2, we conclude that $\{T^n x\}$ converges weakly to some $z \in \text{Fix}(T)$. □

One can derive the following:

Theorem 8.3. *Let X be a Banach space satisfying the Opial condition, C a nonempty weakly compact convex subset of X and $T : C \to C$ a quasi-nonexpansive mapping such that $I - T$ is demiclosed at zero. If for $x \in C$, $\|T^n x - T^{n+1}x\| \to 0$ as $n \to \infty$, then $\{T^n x\}$ converges weakly to an element of Fix(T).*

8.4 Mann Iterative Method

Theorem 8.2 shows that the weak convergence of Picard iterates of nonexpansive mappings is guaranteed under asymptotic regularity. In [35], Mann introduced mean value method by which asymptotic regularity can be relaxed.

Definition 8.6 (Mann Iteration Process [35]). Let C be a convex subset of a linear space X and $T : C \to C$ a mapping. Then, the *Mann iteration sequence* $\{x_n\}$ is given by

$$\begin{cases} x_1 \in C, \\ x_{n+1} = (1 - \alpha_n)x_n + \alpha_n T x_n, \quad \text{for } n \in \mathbb{N}, \end{cases} \tag{8.17}$$

where $\{\alpha_n\}$ is a sequence in $[0, 1]$ satisfying appropriate conditions.

We denote $(1 - \alpha_n)x_n + \alpha_n T x_n$ by $M(x_n, \alpha_n, T)$.

It is well known, for continuous mapping T, that if the Mann iterative process converges, then it must converge to a fixed point of T. But if T is not continuous, there is no guarantee that, even if the Mann process converges, it will converge to a fixed point of T. It is shown in the following example.

Example 8.2. Let $X = [0, 1]$ and $T : X \to X$ a mapping defined by $T0 = T1 = 0$ and $Tx = 1$, for $x \in (0, 1)$. Then, T is a self-mapping of [0, 1], which has the unique fixed point $x = 0$. However, the Mann iteration, with $\alpha_n = 1/(n+1)$ and $x_1 \in (0, 1)$ converges to 1, which is not a fixed point of T.

Let $T : C \to C$ be a nonexpansive mapping and $T_\alpha = (1 - \alpha)I + \alpha T$ for $\alpha \in (0, 1)$, where I denotes the identity mapping. In 1955, Krasnoseleskii [32] proved that if X

is uniformly convex, then for each $x \in C$, $\left\{T^n_{\frac{1}{2}}x\right\}$ has the AF point property for T. Schaefer [46] proved that if X is uniformly convex, then for each $x \in C$, $\{T^n_\alpha x\}$ has the AF point property for T. Edelstein [21] observed that for Scheafer's result strict convexity of X suffices. The removal of strict convexity remained an open problem for many years. In 1976, this problem was solved by Ishikawa [27] in a normed space setting. We now derive the following result of Ishikawa [27] from Proposition 8.3.

Theorem 8.4 (Ishikawa [27]). *Let C be a nonempty subset of a normed space X and $T : C \to X$ a nonexpansive mapping. For $x_1 \in C$, define the sequence $\{x_n\}$ by (8.17), where $\{\alpha_n\}$ is a sequence satisfying the following:*

(a) $0 \leq \alpha_n \leq 1$ *for all* $n \in \mathbb{N}$, $\limsup_{n\to\infty} \alpha_n < 1$ *and* $\sum_{n=1}^{\infty} \alpha_n = \infty$,

(b) $\{x_n\}$ *is contained in* C.

If $\{x_n\}$ is bounded, then $\{x_n\}$ has the AF point property.

Proof. Set $T_n := T$ for all $n \in \mathbb{N}$. Then, $L_n = 1$ and $\rho_n = 0$ for all $n \in \mathbb{N}$. Therefore, the result follows from Theorem 8.3 (a). □

Now we are able to apply Proposition 8.2 for weak convergence of sequence $\{x_n\}$ defined (8.17) for nonexpansive mappings in Banach spaces.

Theorem 8.5. *Let X be a Banach space satisfying the Opial condition, C a nonempty weakly compact convex subset of X and $T : C \to C$ a nonexpansive mapping. For $x_1 \in C$, define the sequence $\{x_n\}$ by (8.17), where $\{\alpha_n\}$ is a sequence satisfying the following condition:*

$$0 \leq \alpha_n \leq 1, \textit{ for all } n \in \mathbb{N}, \quad \limsup_{n\to\infty} \alpha_n < 1 \textit{ and } \sum_{n=1}^{\infty} \alpha_n = \infty.$$

Then, $\{x_n\}$ converges weakly to an element of $Fix(T)$.

Proof. Proposition 8.1 implies that $I - T$ is demiclosed at 0. Theorem 8.4 shows that the iterative sequence $\{x_n\}$ defined by (8.17) has the AF point property for T. Applying Proposition 8.2, we conclude that $\{x_n\}$ converges weakly to some $z \in \mathrm{Fix}(T)$. □

Bose [5] initiated the study of approximation of fixed points of asymptotically nonexpansive mappings in a uniformly convex Banach space satisfying the Opial condition. In [48], Schu proved that the Mann iteration sequence enjoys the AF point property for asymptotically nonexpansive mappings on a uniformly convex Banach space.

The following result shows that the sequence $\{x_n\}$ generated by *modified Mann iteration sequence* enjoys the AF point property for the class of nearly asymptotically nonexpansive mappings on an arbitrary normed space.

Theorem 8.6. *Let C be a nonempty convex subset of a normed space X and $T : C \to C$ be a nearly asymptotically nonexpansive mappings with sequence $\{(a_n, \eta(T^n))\}$ such that $\sum_{n=1}^{\infty} a_n < \infty$ and $\sum_{n=1}^{\infty}(\eta(T^n) - 1) < \infty$. Let $\{\alpha_n\}$ be a sequence in $(0,1)$ such that $0 < a \leq \alpha_n \leq b < 1$ for all $n \in \mathbb{N}$. For arbitrary $x_1 \in C$, let $\{x_n\}$ be sequence defined by the* modified Mann iteration process:

$$x_{n+1} = M(x_n, \alpha_n, T^n), \quad \text{for } n \in \mathbb{N}. \tag{8.18}$$

Suppose that $\{x_n\}$ is bounded and $\sum_{n=1}^{\infty} \|(I - T)T^n x_n\| < \infty$. Then, the following conclusions hold.

(a) $\lim_{n\to\infty} \|x_n - T^n x_n\| = 0$,
(b) If T is uniformly continuous, then the sequence $\{x_n\}$ has the AF point property for T.

Proof. (a) Set $T_n := T^n$ for all $n \in \mathbb{N}$. Then, the result follows from Theorem 8.3 (a).
(b) Since T is uniformly continuous and $\|x_n - T^n x_n\| \to 0$ as $n \to \infty$, it follows that $\|T^{n+1}x_n - Tx_n\| \to 0$ as $n \to \infty$. So,

$$\|x_n - Tx_n\| \leq \|x_n - T^n x_n\| + \|T^n x_n - T^{n+1}x_n\| + \|T^{n+1}x_n - Tx_n\| \to 0, \quad \text{as } n \to \infty.$$

□

The following result extends [20, Theorem 2] from the class of nonexpansive mappings to the class of nearly asymptotically nonexpansive mappings.

Theorem 8.7. *Let X be a Banach space satisfying the Opial condition, C a nonempty weakly compact convex subset of X and $T : C \to C$ a uniformly continuous nearly asymptotically nonexpansive mapping with sequence $\{(a_n, \eta(T^n))\}$ such that*

(i) $\sum_{n=1}^{\infty} a_n < \infty$ *and* $\sum_{n=1}^{\infty} (\eta(T^n) - 1) < \infty$,
(ii) $I - T$ is demiclosed at zero.

Let $\{\alpha_n\}$ be a sequence in $(0,1)$ such that $0 < a \leq \alpha_n \leq b < 1$ for all $n \in \mathbb{N}$. For any $x_1 \in C$, define the sequence $\{x_n\}$ by (8.18) and assume that $\sum_{n=1}^{\infty} \|(I - T)T^n x_n\| < \infty$. Then, $\{x_n\}$ converges weakly to an element of $Fix(T)$.

Proof. From (8.18), we have

$$\begin{aligned} \|x_{n+1} - v\| &\leq (1 - \alpha_n)\|x_n - v\| + \alpha_n \|T^n x_n - v\| \\ &\leq (1 - \alpha_n)\|x_n - v\| + \alpha_n(\|x_n - v\| + a_n) \\ &\leq \|x_n - v\| + a_n, \quad \text{for all } n \in \mathbb{N}. \end{aligned} \tag{8.19}$$

By Lemma 8.1, $\lim_{n\to\infty}\|x_n - v\|$ exists. By Theorem 8.6 (b), $\lim_{n\to\infty}\|x_n - Tx_n\| = 0$, that is, $\{x_n\}$ has the AF Point property for T. Applying Proposition 8.2, we conclude that $\{x_n\}$ converges weakly to some $z \in \mathrm{Fix}(T)$. □

Corollary 8.1. *Let X be a Banach space satisfying the Opial condition, C a nonempty weakly compact convex subset of X and $T : C \to C$ an asymptotically nonexpansive mapping with sequence $\{k_n\}$ such that*

(a) $\sum_{n=1}^{\infty}(k_n - 1) < \infty$,
(b) *$I - T$ is demiclosed at zero.*

Let $\{\alpha_n\}$ be a sequence in $(0,1)$ such that $0 < a \le \alpha_n \le b < 1$ for all $n \in \mathbb{N}$. For any $x_1 \in C$, define the sequence $\{x_n\}$ by (8.18) and assume that $\sum_{n=1}^{\infty}\|(I-T)T^n x_n\| < \infty$. Then, $\{x_n\}$ converges weakly to an element of $Fix(T)$.

8.5 Ishikawa Iterative Method

The Ishikawa iteration process was introduced by Ishikawa [26] in 1974 for approximation of pseudo-contractive mappings in Hilbert space setting.

Definition 8.7. Let C be a nonempty convex subset of a Hilbert space X and $T : C \to C$ a mapping. For arbitrary $x_1 \in C$, define a sequence $\{x_n\}$ in C by

$$\begin{cases} x_{n+1} = (1-\alpha_n)x_n + \alpha_n T y_n, \\ y_n \quad = (1-\beta_n)x_n + \beta_n T x_n, \quad \text{for } n \in \mathbb{N}, \end{cases} \tag{8.20}$$

where $\{\alpha_n\}$ and $\{\beta_n\}$ are two sequences in $[0,1]$. Then, $\{x_n\}$ is called the *Ishikawa iteration*.

Ishikawa iteration process (8.20) is indeed more general than Mann iteration process (8.17).

Ishikawa [26] considered this iteration scheme by imposing the following different restrictions:

(a) $0 \le \alpha_n \le \beta_n \le 1$ and $\lim_{n\to\infty}\beta_n = 0$,
(b) $\sum_{n=1}^{\infty}\alpha_n\beta_n = \infty$.

The class of Lipschitzian pseudo-contractive operators has been studied extensively by various authors, see [18] and references therein. We deal with Ishikawa iteration for nonexpansive mappings in Banach spaces.

We show that the sequence $\{x_n\}$ defined by Ishikawa iteration process (8.20) has the *AF* point property for nonexpansive mappings in normed space.

Theorem 8.8. *Let C be a nonempty subset of a normed space X and $T : C \to X$ a nonexpansive mapping. For $x_1 \in C$, define the sequence $\{x_n\}$ by (8.20), where $\{\alpha_n\}$ and $\{\beta_n\}$ are sequences satisfying the following conditions:*

(a) $0 \le \alpha_n \le 1$ *for all* $n \in \mathbb{N}$, $\limsup_{n\to\infty} \alpha_n < 1$ *and* $\sum_{n=1}^{\infty} \alpha_n = \infty$,
(b) $0 \le \beta_n \le 1$ *for all* $n \in \mathbb{N}$, *and* $\sum_{n=1}^{\infty} \beta_n < \infty$,
(c) $\{x_n\}$ *is contained in* C.

If $\{x_n\}$ is bounded, then $\{x_n\}$ has the AF point property.

Proof. From (8.20), we have

$$\begin{aligned}\|x_{n+1} - x_n\| &= \alpha_n \|x_n - Ty_n\| \\ &\le \alpha_n(\|x_n - Tx_n\| + \|Tx_n - Ty_n\|) \\ &\le \alpha_n(1+\beta_n)\|x_n - Tx_n\|,\end{aligned}$$

$$\begin{aligned}\|x_{n+1} - y_n\| &\le (1-\alpha_n)\|x_n - y_n\| + \alpha_n\|Ty_n - y_n\| \\ &\le (1-\alpha_n)\|x_n - y_n\| + \alpha_n(\|Ty_n - Tx_n\| + \|Tx_n - y_n\|) \\ &\le (1-\alpha_n)\|x_n - y_n\| + \alpha_n(\|y_n - x_n\| + \|Tx_n - y_n\|) \\ &= \|y_n - x_n\| + \alpha_n(1-\beta_n)\|x_n - Tx_n\| \\ &= (\beta_n + \alpha_n(1-\beta_n))\|x_n - Tx_n\|,\end{aligned}$$

and

$$\begin{aligned}\|x_{n+1} - Tx_{n+1}\| &\le (1-\alpha_n)\|x_n - Tx_{n+1}\| + \alpha_n\|Tx_{n+1} - Ty_n\| \\ &\le (1-\alpha_n)\|x_n - Tx_{n+1}\| + \alpha_n\|x_{n+1} - y_n\| \\ &\le (1-\alpha_n)(\|x_n - x_{n+1}\| + \|x_{n+1} - Tx_{n+1}\|) + \alpha_n\|x_{n+1} - y_n\| \\ &\le (1-\alpha_n)(\alpha_n(1+\beta_n)\|x_n - Tx_n\| + \|x_{n+1} - Tx_{n+1}\|) \\ &\quad + \alpha_n\|x_{n+1} - y_n\|.\end{aligned}$$

Thus,

$$\begin{aligned}\|x_{n+1} - Tx_{n+1}\| &\le (1-\alpha_n)(1+\beta_n)\|x_n - Tx_n\| + \|x_{n+1} - y_n\|\| \\ &\le (1-\alpha_n)(1+\beta_n)\|x_n - Tx_n\| + (\beta_n + \alpha_n(1-\beta_n))\|x_n - Tx_n\| \\ &= [1 + 2\beta_n(1-\alpha_n)]\|x_n - Tx_n\|.\end{aligned}$$

Since $\sum_{n=1}^{\infty} \beta_n < \infty$, it follows that $\lim_{n\to\infty}\|x_n - Tx_n\|$ exists. One can assume that $\alpha_n > 0$ for all $n \ge 0$. Set $d := \lim_{n\to\infty}\|x_n - Tx_n\|$, $a_n := Tx_n - x_n$ and $b_n := \alpha_n^{-1}(Tx_{n+1} - Tx_n) + (Tx_n - Ty_n)$. Then, $d_{n+1} = (1-\alpha_n)d_n + \alpha_n b_n$. Observe that

$$\begin{aligned}\|b_n\| &\leq \alpha_n^{-1}\|Tx_n - Tx_{n+1}\| + \|Tx_n - Ty_n\| \\ &\leq \alpha_n^{-1}\|x_n - x_{n+1}\| + \|x_n - y_n\| \\ &\leq (1+2\beta_n)\|x_n - Tx_n\|,\end{aligned}$$

which implies that $\limsup_{n\to\infty} \|b_n\| \leq d$. Moreover, for $n \in \mathbb{N}$, we have

$$\begin{aligned}\left\|\sum_{i=1}^{n} \alpha_i b_i\right\| &= \left\|\sum_{i=1}^{n}((Tx_{i+1} - Tx_i) + \alpha_i(Tx_i - Ty_i))\right\| \\ &= \left\|(Tx_{n+1} - Tx_1) + \sum_{i=1}^{n} \alpha_i(Tx_i - Ty_i)\right\| \\ &\leq \|Tx_{n+1} - Tx_1\| + \sum_{i=1}^{n} \alpha_i \|Tx_i - Ty_i\| \\ &\leq \|Tx_{n+1} - Tx_1\| + \sum_{i=1}^{n} \alpha_i\beta_i \|x_i - Tx_i\|.\end{aligned}$$

Note that $\{\|a_n\|\}$ is a convergent sequence and $\{x_n\}$ is bounded. So, $\{\sum_{i=1}^{n} \alpha_i b_i\}$ is bounded. Note that $\sum_{n=1}^{\infty} \alpha_n = \infty$. Applying Lemma 8.3, we conclude that $(I-T)x_n \to 0$ as $n \to \infty$. □

Theorem 8.9 ([20]). *Let X be a Banach space satisfying the Opial condition, C a nonempty weakly compact convex subset of X and $T : C \to C$ a nonexpansive mapping. For $x_1 \in C$, define the sequence $\{x_n\}$ by (8.20), where $\{\alpha_n\}$ and $\{\beta_n\}$ are sequences satisfying the following conditions:*

(a) $0 \leq \alpha_n \leq 1$ for all $n \in \mathbb{N}$, $\limsup_{n\to\infty} \alpha_n < 1$ and $\sum_{n=1}^{\infty} \alpha_n = \infty$;

(b) $0 \leq \beta_n \leq 1$ for all $n \in \mathbb{N}$, and $\sum_{n=1}^{\infty} \beta_n < \infty$.

Then, $\{x_n\}$ converges weakly to an element of $Fix(T)$.

Proof. For $v \in \mathrm{Fix}(T)$, we have $\|x_{n+1} - v\| \leq \|x_n - v\|$ for all $n \in \mathbb{N}$. It follows that $\lim_{n\to\infty} \|x_n - v\|$ exists. From Theorem 8.8, we have that $\{x_n\}$ has the AF Point property for T. By Proposition 8.1, $I-T$ is demiclosed at zero. Applying Proposition 8.2, we conclude that $\{x_n\}$ converges weakly to some $z \in \mathrm{Fix}(T)$. □

Theorem 8.5 follows now as a corollary of Theorem 8.9.

8.6 Helpern Iterative Method

As we have seen that processes (8.15), (8.17) and (8.20) have only weak convergence, in general. Halpern iterative method [24] is one of the most effective methods to find a fixed point of a nonexpansive mapping, which guarantees strong convergence of the approximating sequence.

Definition 8.8. Let C be a nonempty convex subset of a linear space X and $T : C \to C$ a mapping. Let $u \in C$ and $\{\alpha_n\}$ a sequence in $[0,1]$. Then, a sequence $\{x_n\}$ in C defined by

$$\begin{cases} x_0 \in C, \\ x_{n+1} = \alpha_n u + (1-\alpha_n)Tx_n, \quad \text{for } n \in \mathbb{N}_0, \end{cases} \tag{8.21}$$

is called the *Halpern iteration*.

For numerical solutions of fixed point problems of nonexpansive mappings, Halpern [24] gave the following:

Let C be a closed convex bounded subset of a real Hilbert space H and $T : C \to C$ a nonexpansive mapping. Let $\{\alpha_n\}$ be a sequence in $(0,1)$, and define a sequence $\{x_n\}$ in C by (8.21). Then, the sequence $\{x_n\}$ converges strongly to z, a fixed point of T, closest to u if $\lim\limits_{n\to\infty} \alpha_n = 0$, $\sum_{n=1}^{\infty} \alpha_n = \infty$, and either $\sum_{n=1}^{\infty} |\alpha_{n+1} - \alpha_n| < \infty$ or $\lim\limits_{n\to\infty} \alpha_n/\alpha_{n+1} = 1$ (for example, if $\alpha_n = 1/(n+1)$).

The following result is more general than the result of Halpern [24] in the sense that u is not necessarily an element of C.

Theorem 8.10. *Let C be a closed convex subset of a real Hilbert space H, $u \in H$ and $T : C \to C$ a nonexpansive mapping. Let $\{\alpha_n\}$ be a sequence in $(0,1)$, and define a sequence $\{x_n\}$ in C by*

$$\begin{cases} x_0 \in C, \\ x_{n+1} = P_C[\alpha_n u + (1-\alpha_n)Tx_n], \quad \text{for } n \in \mathbb{N}_0. \end{cases} \tag{8.22}$$

Then, the sequence $\{x_n\}$ converges strongly to $\tilde{x}$, a fixed point of T, where $\tilde{x} = P_{Fix(T)}u$ if $\lim\limits_{n\to\infty} \alpha_n = 0$, $\sum_{n=1}^{\infty} \alpha_n = \infty$, and either $\sum_{n=1}^{\infty} |\alpha_{n+1} - \alpha_n| < \infty$ or $\lim\limits_{n\to\infty} \alpha_n/\alpha_{n+1} = 1$.

Proof. We proceed with the following steps.

STEP 1. $\{x_n\}$ *is bounded.*

Let $z \in \mathrm{Fix}(T)$. From (8.22), we have

$$\begin{aligned} \|x_{n+1} - z\| &= \|P_C[\alpha_n u + (1-\alpha_n)Tx_n] - P_C(z)\| \\ &\le \|\alpha_n u + (1-\alpha_n)Tx_n - z\| \\ &\le (1-\alpha_n)\|x_n - z\| + \alpha_n \|u - z\| \\ &\le \max\{\|x_n - z\|, \|u - z\|\} \\ &\le \max\{\|x_0 - z\|, \|u - z\|\}, \quad \text{for all } n \in \mathbb{N}_0. \end{aligned}$$

Hence, $\{x_n\}$ is bounded.

STEP 2. $\|x_{n+1} - x_n\| \to 0$ *as* $n \to \infty$.
From (8.22), we have

$$\begin{aligned}\|x_{n+1} - x_n\| &= \|P_C[\alpha_n u + (1-\alpha_n)Tx_n] - P_C[\alpha_{n-1}u + (1-\alpha_{n-1})Tx_{n-1}]\| \\ &\le \|\alpha_n u + (1-\alpha_n)Tx_n - [\alpha_{n-1}u + (1-\alpha_{n-1})Tx_{n-1}]\| \\ &= \|(\alpha_n - \alpha_{n-1})u \ +(1-\alpha_n)Tx_n - (1-\alpha_n)Tx_{n-1} + Tx_{n-1} - Tx_{n-1} \\ &\quad +\alpha_{n-1}Tx_{n-1} - \alpha_n Tx_{n-1}\| \\ &\le (1-\alpha_n)\|Tx_n - Tx_{n-1}\| + |\alpha_n - \alpha_{n-1}|(\|u\| + \|Tx_{n-1}\|) \\ &\le (1-\alpha_n)\|x_n - x_{n-1}\| + |\alpha_n - \alpha_{n-1}|(\|u\| + \|Tx_{n-1}\|).\end{aligned}$$

Using Lemma 8.2, we get $\|x_{n+1} - x_n\| \to 0$ as $n \to \infty$.
STEP 3. $\{x_n\}$ *has the AF Point property for* T.
Note that

$$\begin{aligned}\|x_n - Tx_n\| &\le \|x_n - x_{n+1}\| + \|x_{n+1} - Tx_n\| \\ &= \|x_n - x_{n+1}\| + \|P_C[\alpha_n u + (1-\alpha_n)Tx_n] - P_C(Tx_n)\| \\ &\le \|x_n - x_{n+1}\| + \|\alpha_n u + (1-\alpha_n)Tx_n - Tx_n\| \\ &= \|x_n - x_{n+1}\| + \alpha_n\|u - Tx_n\| \to 0, \quad \text{as } n \to \infty.\end{aligned}$$

STEP 4. $\limsup_{n\to\infty}\langle x_n - \tilde{x}, u - \tilde{x}\rangle \le 0.$
Let us choose a subsequence $\{x_{n_k}\}$ of $\{x_n\}$ such that

$$\limsup_{n\to\infty}\langle x_n - \tilde{x}, u - \tilde{x}\rangle = \lim_{k\to\infty}\langle x_{n_k} - \tilde{x}, u - \tilde{x}\rangle.$$

Without loss of generality, we may assume that $x_{n_k} \rightharpoonup z \in C$. Since $\{x_n\}$ has the AF point property for T, it follows form Proposition 8.1 that $z \in \mathrm{Fix}(T)$. Hence, from (8.7), we get

$$\limsup_{n\to\infty}\langle x_n - \tilde{x}, u - \tilde{x}\rangle = \langle z - \tilde{x}, u - \tilde{x}\rangle \le 0.$$

STEP 5. $x_n \to \tilde{x}$ *as* $n \to \infty$.

Set $y_n = \alpha_n u + (1-\alpha_n u)Tx_n$. Noticing that $x_{n+1} = P_C(y_n)$. From (8.22), we have

$$\begin{aligned}\|x_{n+1} - \tilde{x}\|^2 &= \langle y_n - \tilde{x}, x_{n+1} - \tilde{x}\rangle + \langle P_C(y_n) - y_n, P_C(y_n) - \tilde{x}\rangle \\ &\le \langle y_n - \tilde{x}, x_{n+1} - \tilde{x}\rangle \\ &= \alpha_n\langle u - \tilde{x}, x_{n+1} - \tilde{x}\rangle + \langle (1-\alpha_n)Tx_n - (1-\alpha_n)T\tilde{x}, x_{n+1} - \tilde{x}\rangle \\ &\le \alpha_n\langle u - \tilde{x}, x_{n+1} - \tilde{x}\rangle + (1-\alpha_n)\|x_n - \tilde{x}\|\,\|x_{n+1} - \tilde{x}\| \\ &\le \frac{1-\alpha_n}{2}\left(\|x_n - \tilde{x}\|^2 + \|x_{n+1} - \tilde{x}\|^2\right) + \alpha_n\langle u - \tilde{x}, x_{n+1} - \tilde{x}\rangle.\end{aligned}$$

Hence, we have

$$\|x_{n+1}-\tilde{x}\|^2 \le \frac{1-\alpha_n}{1+\alpha_n}\|x_n-\tilde{x}\|^2 + \frac{2\alpha_n}{1+\alpha_n}\langle u-\tilde{x}, x_{n+1}-\tilde{x}\rangle$$
$$= (1-\gamma_n)\|x_n-\tilde{x}\|^2 + \alpha_n\delta_n,$$

where $\gamma_n = \frac{2\alpha_n}{(1+\alpha_n)}$ and $\delta_n = \frac{2}{(1+\alpha_n)}\langle u-\tilde{x}, x_{n+1}-\tilde{x}\rangle$. It follows from Lemma 8.2 that $\lim_{n\to\infty} x_n = \tilde{x}$. □

In 1992, Wittmann [51] obtained a remarkable result for nonlinear mappings in the setting of Hilbert spaces. Since then, it has been investigated by a large number of researchers, and obtained different types of strong convergence theorems for nonexpansive mappings and their variations. Some further work on this topic in the framework of Banach spaces can be found in [19,41,43,49,52,53].

8.7 CQ Iteration Method

It is well-known that the modified Mann's iteration method (8.18) is, in general, not strongly convergent. In this section we deal with approximation of fixed points of uniformly continuous asymptotically κ-strict pseudocontractive mappings in the intermediate sense. We also observe the role of LEAF point property of CQ iteration method.

Theorem 8.11 ([44]). *Let C be a nonempty closed convex subset of a real Hilbert space H and $T : C \to C$ a uniformly continuous asymptotically κ-strict pseudocontractive mapping in the intermediate sense with sequence $\{\gamma_n\}$ such that $Fix(T)$ is nonempty and bounded. Let $\{\alpha_n\}$ be a sequence in $[0,1]$ such that $0 < \delta \le \alpha_n \le 1-\kappa$ for all $n \in \mathbb{N}$. Let $\{x_n\}$ be a sequence in C generated by the following CQ algorithm:*

$$\begin{cases} u = x_1 \in C \text{ chosen arbitrary}, \\ y_n = (1-\alpha_n)x_n + \alpha_n T^n x_n, \\ C_n = \{z \in C : \|y_n - z\|^2 \le \|x_n - z\|^2 + \theta_n\}, \\ Q_n = \{z \in C : \langle x_n - z, u - x_n\rangle \ge 0\}, \\ x_{n+1} = P_{C_n \cap Q_n}(u), \quad \text{for all } n \in \mathbb{N}, \end{cases} \tag{8.23}$$

where $\theta_n = c_n + \gamma_n\Delta_n$ and $\Delta_n = sup\{\|x_n - z\| : z \in Fix(T)\} < \infty$. Then, $\{x_n\}$ converges strongly to $P_{Fix(T)}(u)$.

Proof. We break the proof into following six steps:

STEP 1. *C_n is convex.*
Indeed, the defining inequality in C_n is equivalent to the inequality

$$\langle 2(x_n - y_n), v\rangle \le \|x_n\|^2 - \|y_n\|^2 + \theta_n,$$

it follows from Lemma 8.4 that C_n is convex.

STEP 2. *$Fix(T) \subset C_n$.*

Let $p \in \text{Fix}(T)$. From (8.23), we have

$$\begin{aligned}
\|y_n - p\|^2 &= \|(1-\alpha_n)(x_n - p) + \alpha_n(T^n x_n - p)\|^2 \\
&\le (1-\alpha_n)\|x_n - p\|^2 + \alpha_n\|T^n x_n - p\|^2 - \alpha_n(1-\alpha_n)\|x_n - T^n x_n\|^2 \\
&\le (1-\alpha_n)\|x_n - p\|^2 + \alpha_n\left((1+\gamma_n)\|x_n - p\|^2 + \kappa\|x_n - T^n x_n\|^2 + c_n\right) \\
&\quad -\alpha_n(1-\alpha_n)\|x_n - T^n x_n\|^2 \\
&\le \|x_n - p\|^2 + \alpha_n(\kappa - (1-\alpha_n))\|x_n - T^n x_n\|^2 + c_n + \gamma_n\Delta_n \\
&\le \|x_n - p\|^2 + c_n + \gamma_n\Delta_n.
\end{aligned}$$

Hence, $p \in C_n$.

STEP 3. *$Fix(T) \subset C_n \cap Q_n$, for all $n \in \mathbb{N}$.*

It suffices to show that $\text{Fix}(T) \subset Q_n$. We prove this by induction.

For $n = 1$, we have $\text{Fix}(T) \subset C = Q_1$. Assume that $\text{Fix}(T) \subset Q_n$. Since x_{n+1} is the projection of u onto $C_n \cap Q_n$, it follows that

$$\langle x_{n+1} - z, u - x_{n+1}\rangle \ge 0, \quad \text{for all } z \in C_n \cap Q_n.$$

As $\text{Fix}(T) \subset C_n \cap Q_n$, the last inequality holds, in particular for all $z \in \text{Fix}(T)$. By definition,

$$Q_{n+1} = \{z \in C : \langle x_{n+1} - z, u - x_{n+1}\rangle \ge 0\},$$

it follows that $\text{Fix}(T) \subset Q_{n+1}$. By the principle of mathematical induction, we have $\text{Fix}(T) \subset Q_n$ for all $n \in \mathbb{N}$.

STEP 4. $\|x_n - x_{n+1}\| \to 0$.

By the definition of Q_n, we have $x_n = P_{Q_n}(u)$ and

$$\|u - x_n\| \le \|u - y\|, \quad \text{for all } y \in \text{Fix}(T) \subset Q_n.$$

Note that boundedness of $\text{Fix}(T)$ implies that $\{\|x_n - u\|\}$ is bounded. Since $x_n = P_{Q_n}(u)$ which together with the fact that $x_{n+1} \in C_n \cap Q_n \subseteq Q_n$ implies that $\|u - x_n\| \le \|u - x_{n+1}\|$. Thus, $\{\|x_n - u\|\}$ is increasing. Since $\{\|x_n - u\|\}$ is bounded, we obtain that $\lim_{n\to\infty}\|x_n - u\|$ exists.

Observe that $x_n = P_{Q_n}(u)$ and $x_{n+1} \in Q_n$ which imply that

$$\langle x_{n+1} - x_n, x_n - u\rangle \ge 0.$$

Using (8.5), we obtain

$$\begin{aligned}
\|x_{n+1} - x_n\|^2 &= \|x_{n+1} - u - (x_n - u)\|^2 \\
&= \|x_{n+1} - u\|^2 - \|x_n - u\|^2 - 2\langle x_{n+1} - x_n, x_n - u\rangle \\
&\le \|x_{n+1} - u\|^2 - \|x_n - u\|^2 \to 0, \text{ as } n \to \infty.
\end{aligned}$$

STEP 5. *$\{x_n\}$ has AF point property for T.*
By the definition of y_n, we have

$$\begin{aligned}\|x_n - T^n x_n\| &= \alpha_n^{-1}\|x_n - y_n\| \\ &\le \alpha_n^{-1}(\|x_n - x_{n+1}\| + \|x_{n+1} - y_n\|) \\ &\le \delta^{-1}(\|x_n - x_{n+1}\| + \|x_{n+1} - y_n\|). \end{aligned} \tag{8.24}$$

Since $x_{n+1} \in C_n$, we have

$$\|y_n - x_{n+1}\|^2 \le \|x_n - x_{n+1}\|^2 + c_n + \gamma_n \Delta_n \to 0.$$

It follows from (8.24) that

$$\|x_n - T^n x_n\| \to 0 \text{ as } n \to \infty. \tag{8.25}$$

By Step 4 and (8.25), we obtain from Proposition 8.5 that $\|x_n - Tx_n\| \to 0$ as $n \to \infty$.

STEP 6. *$x_n \to v \in Fix(T)$.*
Since H is reflexive and $\{x_n\}$ is bounded, we get $w_\omega(\{x_n\})$ is nonempty. First, we show that $w_\omega(\{x_n\})$ is a singleton. Assume that $\{x_{n_i}\}$ is a subsequence of $\{x_n\}$ such that $x_{n_i} \rightharpoonup v \in C$. Since $x_n - Tx_n \to 0$ by step 4, it follows from uniform continuity of T that $x_n - T^m x_n \to 0$ for all $m \in \mathbb{N}$. By Remark 8.2, $v \in w_\omega(\{x_n\}) \subset \operatorname{Fix}(T)$.

Since $x_{n+1} = P_{C_n \cap Q_n}(u)$, we obtain

$$\|u - x_{n+1}\| \le \|u - P_{\operatorname{Fix}(T)}(u)\|, \quad \text{for all } n \in \mathbb{N}.$$

Observe that $u - x_{n_i} \rightharpoonup u - v$. By the weak lower semicontinuity of norm, we have

$$\begin{aligned}\|u - P_{\operatorname{Fix}(T)(u)}\| \le \|u - v\| &\le \liminf_{i\to\infty} \|u - x_{n_i}\| \\ &\le \limsup_{i\to\infty} \|u - x_{n_i}\| \le \|u - P_{\operatorname{Fix}(T)}(u)\|,\end{aligned}$$

which yields $\|u - P_{\operatorname{Fix}(T)}(u)\| = \|u - v\|$ and

$$\lim_{i\to\infty} \|u - x_{n_i}\| = \|u - P_{\operatorname{Fix}(T)}(u)\|. \tag{8.26}$$

Hence, $v = P_{\operatorname{Fix}(T)}(u)$ by the uniqueness of the nearest point projection of u onto $\operatorname{Fix}(T)$. Thus, $\|x_{n_i} - u\| \to \|v - u\|$. It shows that $x_{n_i} - u \to v - u$, that is, $x_{n_i} \to v$. Since $\{x_{n_i}\}$ is an arbitrary weakly convergent subsequence, it follows that $w_\omega(\{x_n\}) = \{v\}$. This shows that $x_n \rightharpoonup v$. It is easy to see, as (8.26), that $\|x_n - u\| \to \|v - u\|$. Therefore, $x_n \to v$. □

Corollary 8.2 ([31, Theorem 2.2]). *Let C be a nonempty closed convex bounded subset of a real Hilbert space H and $T : C \to C$ an asymptotically nonexpansive mapping with sequence $\{k_n\}$ in $[1,\infty)$. Let $\{\alpha_n\}$ be a sequence in $[0,1]$ such that $0 < \delta \leq \alpha_n \leq 1$. Define a sequence $\{x_n\}_{n=1}^{\infty}$ in C by the following algorithm:*

$$\begin{cases} u = x_1 \in C \text{ chosen arbitrary}, \\ y_n = (1-\alpha_n)x_n + \alpha_n T^n x_n, \\ C_n = \left\{z \in C : \|y_n - z\|^2 \leq \|x_n - z\|^2 + \theta_n\right\}, \\ Q_n = \{z \in C : \langle x_n - z, u - x_n\rangle \geq 0\}, \\ x_{n+1} = P_{C_n \cap Q_n}(u), \end{cases} \tag{8.27}$$

where $\theta_n = (k_n^2 - 1)\ diam(C)^2$ for all $n \in \mathbb{N}$. Then, $\{x_n\}$ converges strongly to $P_{Fix(T)}(u)$.

Corollary 8.3 ([36, Theorem 3.4]). *Let C be a nonempty closed convex bounded subset of a real Hilbert space H and $T : C \to C$ a nonexpansive mapping with $Fix(T) \neq \emptyset$. Let $\{\alpha_n\}$ be a sequence in $[0,1]$ such that $0 < \delta \leq \alpha_n \leq 1$. Define a sequence $\{x_n\}_{n=1}^{\infty}$ in C by the following algorithm:*

$$\begin{cases} u = x_1 \in C \text{ chosen arbitrary}, \\ y_n = (1-\alpha_n)x_n + \alpha_n T^n x_n, \\ C_n = \{z \in C : \|y_n - z\| \leq \|x_n - z\|\}, \\ Q_n = \{z \in C : \langle x_n - z, u - x_n\rangle \geq 0\}, \\ x_{n+1} = P_{C_n \cap Q_n}(u) \text{ for all } n \in \mathbb{N}. \end{cases} \tag{8.28}$$

Then, $\{x_n\}$ converges strongly to $P_{Fix(T)}(u)$.

Iterative methods for approximation of fixed points of asymptotically κ-strict pseudocontractive mappings in the intermediate sense are also studied in [12,16,25] and the references therein.

8.8 Browder Iterative Method

Nonexpansive mappings can be approximated by contractions. The details are given below:

Let C be a closed convex subset of a Banach space X, and let $f : C \to C$ be a nonexpansive mapping. For any $r_i \in [0,1)$ and $x_0 \in C$, the mapping defined by

$$f_{r_i}x = r_i fx + (1-r_i)x_0, \quad \text{for all } x \in C$$

maps C into itself and is a contraction map with Lipschitz constant r_i. If r_i sufficiently close to 1, then f_{r_i} is a contractive approximation of f. By Banach contraction principle each f_{r_i} has a unique fixed point, say x_{r_i}. Thus,

$$x_{r_i} = f_{r_i}x_{r_i} = r_i f x_{r_i} + (1-r_i)x_0.$$

As the mapping f is a contractive approximation of f_{r_i}, so does the fixed point of f is the limit of the sequence of fixed points of f_{r_i}? The answer was given by Browder [6] as follows:

Theorem 8.12 (Browder's Convergence Theorem). *Let C be a nonempty closed convex bounded subset of a Hilbert space H. Let u be an element in C and $G_t : C \to C$, $t \in (0,1)$ the family of mappings defined by $G_t x = (1-t)u + tTx$ for all $x \in C$. Then, the following statements hold:*

(a) There is exactly one fixed point x_t of G_t, that is,

$$x_t = (1-t)u + tTx_t. \tag{8.29}$$

(b) The path $\{x_t\}$ converges strongly to $P_{Fix(T)}(u)$ as $t \to 1$.

Proof. (a) Note that for each $t \in (0,1), G_t$ is a contraction mapping from C into itself. Hence, G_t has a unique fixed point x_t in C.

(b) Since $\mathrm{Fix}(T)$ is a nonempty closed convex subset of C, there exists an element $u_0 \in \mathrm{Fix}(T)$ which is the nearest point of u. By boundedness of $\{x_t\}$, there exists a subsequence $\{x_{t_n}\}$ of $\{x_t\}$ such that $x_{t_n} \rightharpoonup z \in C$. Write $x_{t_n} = x_n$. Since H has the Opial condition and $\|x_n - Tx_n\| \to 0$, we obtain from Proposition 8.1 that $z = Tz$. Set $U := I - T$. Observe that

$$(1-t_n)x_n + t_n(x_n - Tx_n) = (1-t_n)u,$$

and

$$(1-t_n)u_0 + t_n(u_0 - Tu_0) = (1-t_n)u_0.$$

Subtracting and taking the inner product of the difference with $x_n - u_0$, we get

$$\begin{aligned}(1-t_n)\langle x_n - u_0, x_n - u_0\rangle + t_n\langle Ux_n - Uu_0, x_n - u_0\rangle \\ = (1-t_n)\langle u - u_0, x_n - u_0\rangle,\end{aligned}$$

where $U = I - T$. Since U is monotone, that is, $\langle Ux_n - Uu_0, x_n - u_0\rangle \geq 0$, it follows that

$$\|x_n - u_0\|^2 \leq \langle u - u_0, x_n - u_0\rangle, \quad \text{for all } n \in \mathbb{N}.$$

Since $u_0 \in \mathrm{Fix}(T)$ is the nearest point to u, we have

$$\langle u - u_0, z - u_0\rangle \leq 0,$$

which gives

$$\begin{aligned}\|x_n - u_0\|^2 &\leq \langle u - u_0, x_n - u_0\rangle \\ &= \langle u - u_0, x_n - z\rangle + \langle u - u_0, z - u_0\rangle \\ &\leq \langle u - u_0, x_n - z\rangle.\end{aligned}$$

Thus, from $x_n \rightharpoonup z$, we obtain $x_n \to u_0$ as $n \to \infty$. We show that $x_t \to u_0$ as $t \to 1$, that is, u_0 is the only strong cluster point of $\{x_t\}$. Suppose, for contradiction, that $\{x_{t_{n'}}\}$ is another subsequence of $\{x_t\}$ such that $x_{t_{n'}} \to v \neq u_0$ as $n' \to \infty$. Set $x_{n'} := x_{t_{n'}}$. Since $x_{n'} - Tx_{n'} \to 0$, it follows that $v \in \mathrm{Fix}(T)$. From (8.29), we have

$$x_t - Tx_t = (1-t)(u - Tx_t). \tag{8.30}$$

Since for $y \in \mathrm{Fix}(T)$, we have

$$\begin{aligned}\langle x_t - Tx_t, x_t - y\rangle &= \langle x_t - Ty + Ty - Tx_t, x_t - y\rangle \\ &= \|x_t - y\|^2 - \langle Tx_t - Ty, x_t - y\rangle \geq 0.\end{aligned}$$

From this and (8.30), we have $\langle u - Tx_t, x_t - y\rangle \geq 0$. Thus, $\langle x_t - u, x_t - y\rangle \leq 0$ for all $t \in (0,1)$ and $y \in \mathrm{Fix}(T)$. It follows that

$$\langle u_0 - u, u_0 - v\rangle \leq 0 \quad \text{and} \quad \langle v - u, v - u_0\rangle \leq 0,$$

which imply that $u_0 = v$, a contradiction. Therefore, $\{x_t\}$ converges strongly to $P_{\mathrm{Fix}(T)}(u)$.

□

Browder's strong convergence theorem has been widely studied for more general class of mappings and semigroups in Hilbert and Banach spaces. Details can be found in [1, 41, 43, 52]. Recently, strong convergence theorems for the class of nonexpansive mappings have been established in Banach space X under one of the following assumptions:

(a) X has weakly continuous duality mapping.
(b) X has a uniformly Gâteaux differentiable norm and C has the fixed point property for nonexpansive mappings.

For further details on approximation of fixed points of asymptotically nonexpansive mappings by Browder method, we refer [33].

References

1. Agarwal, R.P., O'Regan, D., Sahu, D.R.: Fixed Point Theory for Lipschitzian-type Mappings with Applications. Springer, New York (2009)
2. Agarwal, R.P., O'Regan, D., Sahu, D.R.: Iterative construction of fixed points of nearly asymptotically nonexpansive mappings. J. Nonlinear Convex Anal. **8**, 61–79 (2007)
3. Ansari, Q.H.: Metric Spaces: Including Fixed Point Theory and Set-Valued Maps. Narosa Publishing House, New Delhi (2010)
4. Ansari, Q.H. (ed.): Topics in Nonlinear Analysis and Optimization. World Education, Delhi (2012)

5. Bose, S.C.: Weak convergence to the fixed of an asymptotically nonexpansive map. Proc. Am. Math. Soc. **68**, 305–308 (1978)
6. Browder, F.E.: Convergence of approximants to fixed point of nonexpansive nonlinear mappings in Banach spaces. Arch. Ration Mech. Anal. **21**, 259–269 (1967)
7. Browder, F.E., Petryshyn, W.V.: Construction of fixed points of nonlinear mappings. J. Math. Anal. Appl. **20**, 197–228 (1967)
8. Byrne, C.: A unified treatment of some iterative algorithms in signal processing and imagine reconstruction. Inverse Probl. **20**, 103–120 (2004)
9. Ceng, L.C., Al-Homidan, S., Ansari, Q.H., Yao, J.-C.: An iterative scheme for equilibrium problems and fixed points problems of strict pseudo-contraction mappings. J. Computat. Appl. Math. **223**, 967–974 (2009)
10. Ceng, L.C., Ansari, Q.H., Schaible, S., Yao, J.C.: Iterative methods for generalized equilibrium problems, systems of general generalized equilibrium problems and fixed point problems for nonexpansive mappings in Hilbert spaces. Fixed Point Theor. **12**, 293–308 (2011)
11. Ceng, L.C., Ansari, Q.H., Wen, C.-F.: Implicit relaxed and hybrid methods with regularization for minimization problems and asymptotically strict pseudocontractive mappings in the intermediate sense. Abstr. Appl. Anal. **2013**, Article ID 854297 (2013)
12. Ceng, L.C., Ansari, Q.H., Yao, J.C.: Strong and weak convergence theorems for asymptotically strict pseudocontractive mappings in intermediate sense. J. Nonlinear Convex Anal. **11**, 283–308 (2010)
13. Ceng, L.C., Ansari, Q.H., Yao, J.C.: Hybrid pseudoviscosity approximation schemes for equilibrium problems and fixed point problems of infinitely many nonexpansive mappings. Nonlinear Anal. Hybrid Syst. **4**, 743–754 (2010)
14. Ceng, L.C., Ansari, Q.H., Yao, J.C.: Hybrid proximal-type and hybrid shrinking projection algorithms for equilibrium problems, maximal monotone operators and relatively nonexpansive mappings. Numer. Funct. Anal. Optim. **31**, 763–797 (2010)
15. Ceng, L.C., Ansari, Q.H., Yao, J.C.: Mann type iterative methods for finding a common solution of split feasibility and fixed point problems. Positivity **16**, 471–495 (2012)
16. Ceng, L.C., Yao, J.C.: Strong convergence theorems for variational inequalities and fixed point problems of asymptotically strict pseudocontractive mappings in the intermediate sense. Acta Appl. Math. **115**, 167–191 (2011)
17. Chang, S.S.: On the approximation problem of fixed points for asymptotically nonexpansive mappings. Indian J. Pure Appl. Math. **32**, 1297–1307 (2001)
18. Chidume, C.E.: Global iteration schemes for strongly pseudocontractive maps. Proc. Am. Math. Soc. **126**, 2641–2649 (1998)
19. Cui, Y.-L., Liu, X.: Notes on Browder's and Halpern's methods for nonexpansive mappings. Fixed Point Theor. **10**, 89–98 (2009)
20. Deng, L.: Convergence of the Ishikawa iteration process for nonexpansive mappings. J. Math. Anal. Appl. **199**, 769–775 (1996)
21. Edelstein, M.: A remark on a theorem of Krasnoselskii. Am. Math. Mon. **73**, 509–510 (1966)
22. Fukhar-ud-din, H., Khan, A.R.: Approximating common fixed points of asymptotically nonexpansive maps in uniformaly convex Banach spaces. Comput. Math. Appl. **53**, 1349–1360 (2007)
23. Goebel, K., Kirk, W.A.: A fixed point theorem of asymptotically nonexpansive mappings. Proc. Am. Math. Soc. **35**, 171–174 (1972)
24. Halpern, B.: Fixed points of nonexpansive maps. Bull. Am. Math. Soc. **73**, 957–961 (1967)
25. Huang, S.: Hybrid extragradient methods for asymptotically strict pseudocontractions in the intermediate sense and variational inequality problems. Optimization **60**, 739–754 (2011)
26. Ishikawa, S.: Fixed points by a new iteration method. Proc. Am. Math. Soc. **44**, 147–150 (1974)
27. Ishikawa, S.: Fixed points and iteration of a nonexpansive mapping in a Banach space. Proc. Am. Math. Soc. **59**, 65–71 (1976)
28. Khan, A.R., Domlo, A.A., Fukhar-ud-din, H.: Common fixed points Noor iteration for a finite family of asymptotically quasi-nonexpansive mappings in Banach spaces. J. Math. Anal. Appl. **341**, 1–11 (2008)

29. Khan, A.R., Fukhar ud-din, H., Domlo, A.A.: Approximating fixed points of some maps in uniformly convex metric spaces. Fixed Point Theor. Appl. **2010**, Article ID 385986 (2010)
30. Kim, T.H., Xu, H.K.: Convergence of the modified Mann's iteration method for asymptotically strict pseudocontractions. Nonlinear Anal. **68**, 2828–2836 (2008)
31. Kim, T.H., Xu, H.K.: Strong convergence of modified Mann iterations for asymptotically nonexpansive mappings and semigroups. Nonlinear Anal. **64**, 1140–1152 (2006)
32. Krasnoselskii, M.A.: Two observations about the method of successive approximations. Uspehi Math. Nauk **10**, 123–127 (1955)
33. Lim, T.C., Xu, H.: Fixed point theorems for asymptotically nonexpansive mappings. Nonlinear Anal. **22**, 1345–1355 (1994)
34. Liu, Z.Q., Kang, S.M.: Weak and strong convergence for fixed points of asymptotically nonexpansive mappings. Acta Math. Sinica **20**, 1009–1018 (2004)
35. Mann, W.R.: Mean value methods in iteration. Proc. Am. Math. Soc. **4**, 506–610 (1953)
36. Nakajo, K., Takahashi, W.: Strong convergence theorems for nonexpansive mappings and nonexpansive semigroups. J. Math. Anal. Appl. **279**, 372–379 (2003)
37. Opial, Z.: Weak convergence of successive approximations for nonexpansive mappings. Bull. Am. Math. Soc. **73**, 591–597 (1967)
38. Osilike, M.O., Aniagbosor, S.C.: Weak and strong convergence theorems for fixed points of asymptotically nonexpansive mappings. Math. Comput. Model. **32**, 1181–1191 (2000)
39. Picard, E.: Mémoire sur la théorie des equations aux derivée partielles et la metode des approximations successives. J. Math. Pures. Appl. **6**, 145–210 (1890)
40. Podilchuk, C.I., Mammone, R.J.: Imagine recovery by convex projections using a least-squares constraint. J. Opt. Soc. Am. A **7**, 517–521 (1990)
41. Sahu, D.R.: Applications of the S-iteration process to constrained minimization problems and split feasibility problems. Fixed Point Theor. **12**, 187–204 (2011)
42. Sahu, D.R.: Fixed points of demicontinuous nearly Lipschitzian mappings in Banach spaces. Comment. Math. Univ. Carol. **46**, 653–666 (2005)
43. Sahu, D.R., Wong, N.C., Yao, J.C.: A unified hybrid iterative method for solving variational inequalities involving generalized pseudo-contractive mappings. SIAM J. Contr. Optim. **50**, 2335–2354 (2012)
44. Sahu, D.R., Xu, H.K., Yao, J.C.: Asymptotically strict pseudocontractive mappings in the intermediate sense. Nonlinear Anal. **70**, 3502–3511 (2009)
45. Sahu, D.R., Beg, I.: Weak and strong convergence for fixed points of nearly asymptotically non-expansive mappings. Int. J. Mod. Math. **3**(2), 135–151 (2008)
46. Schaefer, H.: Uber die methode sukzessiver Approximationen. Jber. Deutch. Math. Verein **59**, 131–140 (1957)
47. Schu, J.: Iterative construction of fixed points of asymptotically nonexpansive mapping. J. Math. Anal. Appl. **159**, 407–413 (1991)
48. Schu, J.: Weak and strong convergence of fixed points of asymptotically nonexpansive maps. Bull. Aust. Math. Soc. **43**, 153–159 (1991)
49. Takahashi, W.: Viscosity approximation methods for resolvents of accretive operators in Banach spaces. J. Fixed Point Theor. Appl. **1**, 135–147 (2007)
50. Tan, K.K., Xu, H.K.: Fixed point iteration process for asymptotically nonexpansive mappings. Proc. Am. Math. Soc. **122**, 733–739 (1994)
51. Wittmann, R.: Approximation of fixed points of nonexpansive mappings. Arch. Math. **58**, 486–491 (1992)
52. Wong, N.C., Sahu, D.R., Yao, J.C.: Solving variational inequalities involving nonexpansive type mappings. Nonlinear Anal. **69**, 4732–4753 (2008)
53. Xu, H.K.: Strong convergence of an iterative method for nonexpansive and accretive operators. J. Math. Anal. Appl. **314**, 631–643 (2006)
54. Xu, H.K.: Viscosity approximation methods for nonexpansive mappings. J. Math. Anal. Appl. **298**, 279–291 (2004)
55. Xu, H.K.: Iterative algorithms for nonlinear operators. J. Lond. Math. Soc. **66**, 240–256 (2002)

Index

Symbols
1-local retract set, 116
$\mathbb{R}$-trees, 144
ρ-Cauchy sequence, 164
ρ-Lipschitzian mapping, 214
ρ-Lipschitzian semigroup, 214
ρ-a.e. Opial property, 190
  strong, 190
ρ-a.e. closed set, 164
ρ-a.e. compact set, 165
ρ-almost everywhere, 162
ρ-bounded set, 164
  strongly, 164
ρ-closed set, 164
ρ-compact set, 164
ρ-contractive semigroup, 214
ρ-convergent sequence, 164
ρ-distance, 165
ρ-hull, 162
ρ-nonexpansive mapping, 183
ρ-nonexpansive semigroup, 214
ρ-semicontinuous function, 185
τ-distance, 50
γ-condensing, 129
ε-constant mapping, 117
ε-solution, 94
r-parallel set, 116
w-distance, 49
1-local retract, 27
2-Helly property, 226
2-uniform convexity, 22

A
Absolute nonexpansive retract, 27
Absolute nonexpansive retract space, 115
Admissible cover, 110
Admissible hull, 110
Admissible map, 244
Admissible space, 107
AF point property, 277
Al Kashani, 7
Approximate ε-solution, 68
Approximate fixed point property, 8, 277
Approximate fixed point sequence, 9
Asymptotic center, 9
Asymptotic normal structure property, 11
Asymptotic pointwise ρ-contraction mapping, 183
Asymptotic pointwise ρ-nonexpansive mapping, 183
Asymptotic pointwise mapping, 183
Asymptotic radius, 9
Asymptotically ρ-nonexpansive mapping, 183
Asymptotically nonexpansive mapping, 151

B
Banach contraction principle, 2, 35
Banach contraction theorem, 75
Banach operator pair, 233
Best approximation, 276
Bilibear form, 263
Bounded away sequence, 173
Boyd-Wong theorem, 40
Brouwer fixed point theorem, 7, 239
Browder–Ky Fan fixed point theorem, 249

C
Cantor set, 6
Caristi mapping, 47
Caristi's fixed point theorem, 3, 47, 76
Caristi–Kirk's fixed point theorem, 76, 92

S. Almezel et al. (eds.), *Topics in Fixed Point Theory*, DOI 10.1007/978-3-319-01586-6,

CAT(0) spaces, 19
Chebyshev center, 11, 112
Chebyshev radius, 11, 23
Clarke's fixed point theorem, 78
Coefficient of uniform normal structure property, 12
Coincidence stable map, 251
Coincidence theorems, 251
Common fixed point, 232
Complete lattice, 227
Continuous form, 263
Contraction mapping, 2, 33, 183
Contractive mapping, 34
Convex generalized metric space, 226
Convex structure, 17
Convexity structure, 24, 230
CQ algorithm, 293

D
Dedekind complete, 234
Demiclosed mapping, 277
Demiclosed principle, 10, 277
Diametral sequence, 12
Directional contraction, 78

E
Ekeland's variational principle, 3
  converse, 74
  extended, 89, 92
  strong form, 68, 70
  weak form, 73
Equilibrium problem, 87
Extended Ekeland's variational principle, 89
Externally complete set, 233
Externally hyperconvex set, 233
Externally hyperconvex space, 107
Extremal function, 122

F
Finite intersection property, 141
Finitely closed set, 141
Fixed point, 2, 33, 228
Fixed point problem, 88
Fixed point property, 230
Fractal sets, 6

G
Generalized distance, 224
Generalized Ishikawa iteration process, 209
Generalized Mann iteration process, 204
Generalized metric spaces, 224
Geodesic, 144
Geodesic path, 144
Geodesic segment, 144
Geodesic space, 144
Geodesic triangle, 19

H
Hadamard space, 19
Hahn-Banach theorem, 25, 102
Halpern iteration, 291
Hausdorff measure of noncompactness, 128
Hausdorff metric, 6, 51, 83
Helly's theorem, 102
Helpern iterative method, 290
Hyperbolic metric space, 17
Hyperconvex, 226
Hyperconvex hull, 28, 122, 229, 230
Hyperconvex metric space, 25, 26
Hyperconvex space, 104, 226

I
Inequality of Bruhat and Tits, 19
Injective envelope, 230
Injective metric spaces, 27
Injective space, 114
Intersection theorems, 256
Isbell convex space, 154
Isbell's hyperconvex hull, 120
Ishikawa iteration, 288
Ishikawa iteration process, 288
Ishikawa iterative method, 288
Iterative sequence, 52

K
Kakutani maps, 243, 245
Kakutani–Ky Fan fixed point theorem, 250
Kakutani–Ky Fan–Himmelberg theorem, 250
Kannan mapping, 39
Karlovitz lemma, 14
Kirk fixed point theorem, 11, 230
KKM mapping, 142, 256
KKM principle, 142
Kuratowski embedding, 120
Kuratowski measure of noncompactness, 128
Ky Fan maps, 243, 244
Ky Fan's approximation principle, 137

L
LE property, 277
Limit compact, 132

Limit existence property, 277
Lin's fixed point theorem, 17
Lipschitzian, 2
Lipschitzian mapping, 2
Lipschtiz continuous mapping, 34
Lower semicontinuous function, 3, 47, 66, 262
  from above, 74
Luxemburg norm, 163

M
MacNeille completion, 230
Mann iteration process
  modified, 287
Mann iterative method, 285
Markov–Kakutani theorem, 268
Maurey fixed point theorem, 16
Mazur–Schauder theorem, 263
Menger convex space, 108
Menger convexity, 17
Metric projection, 118
Metric projection mapping, 276
Metric segement, 141
Metrically convex space, 104
Minimization problem, 87
Modular function space, 163
Modulus of (uniform) continuity, 106
Modulus of smoothness, 13
Multivalued mapping, 118
  ε-semicontinuous, 149
  lomost lower semicontinuous, 149
  uppersemicontinuous, 149

N
Nadler's fixed point theorem, 84
Nash equilibrium problem, 88
Nearly asymptotically nonexpansive mapping, 278
Nearly Lipschitz constant, 278
Nearly Lipschitzian mapping, 278
Nearly nonexpansive mapping, 278
Nearly uniform k-contraction mapping, 278
Nearly uniformly k-Lipschitzian mapping, 278
No retraction theorem, 240
Nondiametral point, 10
Nonexpansive mapping, 7, 34, 114, 119, 228
Nonexpansive retract set, 115
Nonlinear alternative for four functions, 269
Normal cone, 265
Normal structure property, 10, 230

O
Oettli–Théra theorem, 92
Opial condition, 276
Order preserving mapping, 229
Order-preserving retract, 229

P
P-uniform convexity, 22
Parallelogram inequality, 21
Partially ordered set, 225
Picard iterates, 7
Picard iterative method, 283
Picard iterative sequence, 283
Poincaré hyperbolic metric, 18
Pointwise contraction mapping, 183
Pompeiu–Hausdorff distance, 119
Proper function, 66
Proximinal nonexpansive retract set, 118
Proximinal set, 234
Proximinal space, 109

Q
Q-hyperconvexity, 154
Quasi-period, 207
Quasi-periodic, 207
Quasiconvex function, 262

R
Reflexive metric spaces, 21
Regular convex function modular, 162
Regular convex function pseudomodular, 162
Regular convex function semimodular, 162
Retract, 228, 276
Retraction, 239
Retraction mapping, 276
Retraction property, 228
River metric, 105

S
Saddle point problem, 87
Segment, 78
Selection, 118
Set-valued map, 243
  approachable, 248
  contraction, 83
  upper hemicontinuous, 265
  upper semicontinuous, 245
Sine – Soardi fixed point theorem, 231
Sion-von Neumann minimax theorem, 267
Spherically complete space, 4
Stampacchia theorem, 264
Star-shaped set, 258

Sunny mapping, 276
Sunny nonexpansive retract mapping, 276
Super-reflexive space, 13
Super-reflexive spaces, 14
Symmetric Banach operator pair, 233

T
Takahashi's minimization theorem, 80
  extended, 92
Tangent cone, 265
Tarski fixed point theorem, 232
Transfinite sequence, 131
Tree, 234

U
Ultimate compact operator, 132
Ultimate range, 132
Ultrafilter, 14
Ultrametric space, 4, 152
Ultrapower of Banach space, 14
Ultrapower space, 15
Uniform continuity of a modular, 188
Uniform convex metric space, 20
Uniform normal structure property, 12
Uniformly convex, 12
Uniformly convex metric spaces, 19
Uniformly convex space, 171
Uniformly Lipschitzian mapping, 22
Uniformly non-square, 12
Uniformly smooth, 13
Uniquely geodesic, 144
Upper semicontinuous function, 66, 262

V
Variational inequality problem, 88

W
Weak sharp minima, 85
Well defined Ishikawa process, 210
Well defined Mann process, 205

MIX
Papier aus verantwortungsvollen Quellen
Paper from responsible sources
FSC® C105338

FSC
www.fsc.org

If you have any concerns about our products,
you can contact us on
ProductSafety@springernature.com

In case Publisher is established outside the EU,
the EU authorized representative is:
Springer Nature Customer Service Center GmbH
Europaplatz 3, 69115 Heidelberg, Germany

Printed by Libri Plureos GmbH
in Hamburg, Germany